AF443676

FUNDAMENTALS OF ELECTROCHEMICAL DEPOSITION

THE ELECTROCHEMICAL SOCIETY SERIES

The Electrochemical Society
65 South Main Street
Pennington, NJ 08534-2839
http://www.electrochem.org

A complete list of the titles in this series appears at the end of this volume.

FUNDAMENTALS OF ELECTROCHEMICAL DEPOSITION

SECOND EDITION

MILAN PAUNOVIC
Formerly with IBM Research Division
Yorktown Heights, NY

MORDECHAY SCHLESINGER
University of Windsor
Windsor, Ontario, Canada

A JOHN WILEY & SONS, INC., PUBLICATION

Library of Congress Cataloging-in-Publication Data:

Paunovic, Milan.
    Fundamentals of electrochemical deposition / Milan Paunovic, Mordechay
  Schlesinger. — 2nd ed.
      p. cm.
  Includes bibliographical references and index.
  ISBN-13: 978-0-471-71221-3
  ISBN-10: 0-471-71221-3
    1. Electroplating. I. Schlesinger, Mordechay. II. Title.

  TS6 70.P29 2006
  671.7′32—dc22

                                                  2005058421

Printed in the United States of America

10 9 8 7 6 5 4 3 2 1

CONTENTS

PREFACE TO THE SECOND EDITION

The first edition of this book was published some eight years ago. During this period of time, a number of new significant developments have taken place in the field of electrochemical deposition. It is this progress that makes the second edition desirable and indeed, necessary.

Specifically, in this edition we have made *updates* to most chapters. We have also added *three new chapters* dealing with important applications of electrochemical deposition in the areas of semiconductor technology (Chapter 19), magnetism and microelectronics (Chapter 20), and medicine (Chapter 21). In Chapter 19 we describe development of electrodeposition of copper interconnects on chips as one major advance in microelectronics. In Chapter 20, applications relevant to magnetic information/data storage are highlighted. In Chapter 21 we discuss applications relating to issues of electrochemical deposition of medical devices, highlighting surface electrochemistry. In this context it is noted that in the last decade or so, a number of first-tier universities have been introducing courses dealing with electrochemical deposition and its many practical applications.

Last, but not least, we have added *problems* for the benefit of those who intend to use this book as a text in an appropriate university course at the graduate or senior undergraduate level.

We are indebted to a number of people, probably too many to be mentioned, with one exception. M.S. thanks his wife, Sarah, for always being there, and M.P. thanks his grandchildren, Nicole, Alexander, Daniel, and Isabel, for constant encouragement.

MILAN PAUNOVIC
Port Washington, New York

MORDECHAY SCHLESINGER
Windsor, Ontario, Canada

PREFACE TO THE FIRST EDITION

Electrochemical deposition has, over recent decades, evolved from an art to an exact science. This development is seen as responsible for the ever-increasing number and widening types of applications of this branch of practical science and engineering.

Some of the technological areas in which means and methods of electrochemical deposition constitute an essential component are all aspects of electronics—macro and micro, optics, optoelectronics, and sensors of most types, to name only a few. In addition, a number of key industries, such as the automobile industry, adopt the methods even when other methods, such as evaporation, sputtering, chemical vapor deposition (CVD), and the like, are an option. That is so for reasons of economy and/or convenience.

By way of illustration, it should be noted that modern electrodeposition equips the practitioner with the ability to predesign the properties of surfaces, and in the case of electroforming, those of the whole part. Furthermore, the ability to deposit multilayers of thicknesses in the nanometer region, via electrochemical methods, represents yet a new avenue of producing new materials.

This book, whose title and subject matter are fundamental to the science of electrochemical deposition, is intended for readers who are newcomers to the field as well as to those practitioners who wish to broaden and sharpen their skills in using this technology.

It may be considered a fortunate coincidence that this book is published at the time of the introduction of copper interconnection technology in the microelectronics industry. In 1998 the major electronic manufacturers of integrated circuits (ICs) are switching from aluminum conductors produced by physical methods (evaporation) to copper conductors manufactured by electrochemical methods (electrodeposition). This revolutionary change from physical to electrochemical techniques in the production of microconductors on silicon is bound to generate an increased interest and an urgent need for familiarity with the fundamentals of electrochemical deposition. This book should be of great help in this crucial time.

The book is divided into 18 chapters, presented in a logical and practical order as follows. After a brief introduction (Chapter 1) comes the discussion of ionic solutions (Chapter 2), followed by the subjects of metal surfaces (Chapter 3) and metal solution interphases (Chapter 4). Electrode potential, deposition kinetics, and thin-film nucleation are the themes of the next three chapters (5–7). Next come electroless and displacement-type depositions (Chapter 8 and 9), followed by the chapters dealing with the effects of additives and the science and technology of alloy deposition

(Chapters 10 and 11). Current distribution during deposition and both in situ and ex situ deposit characterization are the focus of Chapters 12–14. Electronic design (mathematical modeling) is the subject of Chapter 15, followed by the issues of structure, properties of deposits, multilayers, and interdiffusion (Chapters 16–18).

Each chapter is self-contained and independent of the other chapters; thus the chapters do not have to be read in consecutive order or as a continuum, and readers who are familiar with the material in certain chapters may skip those chapters and still derive maximum benefit from the chapters they read.

As the title page indicates, the two of us, the authors, come from different and at the same time complementing environments. One of us (M.P.) has been with industry and the other (M.S.), with academia for most of our respective carriers. This gave rise to the unique style and flavor of the book hereby presented to the reader.

To sum up, this book may and should be viewed as either a textbook for advanced science and engineering students, a reference book for the practitioners of deposition, or as a resource book for the science-minded individuals who desire to familiarize themselves with a modern, exciting, and ever-evolving field of practical knowledge.

It is a pleasure to gratefully acknowledge the professional help and support of the staff at John Wiley & Sons.

Finally, we would like to express our heartfelt gratitude to many individuals in The Electrochemical Society and in particular to members of the Electrodeposition Division as well as to our respective families for support and encouragement.

MILAN PAUNOVIC
Yorktown Heights, New York

MORDECHAY SCHLESINGER
Windsor, Ontario, Canada

1

Overview

1.1. INTRODUCTION

Electrochemical deposition of metals and alloys involves the reduction of metal ions from aqueous, organic, and fused-salt electrolytes. In this book we treat deposition from aqueous solutions only. The reduction of metal ions M^{z+} in aqueous solution is represented by

$$M^{z+}_{solution} + ze \rightarrow M_{lattice} \qquad (1.1)$$

This can be accomplished by means of two different processes: (1) an electrodeposition process in which z electrons (e) are provided by an external power supply, and (2) an electroless (autocatalytic) deposition process in which a reducing agent in the solution is the electron source (no external power supply is involved). These two processes, electrodeposition and electroless deposition, constitute the electrochemical deposition. In this book we treat both of these processes. In either case our interest is in a metal electrode in contact with an aqueous ionic solution. Deposition reaction presented by Eq. (1.1) is a reaction of charged particles at the interface between a solid metal electrode and a liquid solution. The two types of charged particles, a metal ion and an electron, can cross the interface.

Four types of fundamental subjects are involved in the process represented by Eq. (1.1): (1) metal–solution interface as the locus of the deposition process, (2) kinetics and mechanism of the deposition process, (3) nucleation and growth processes of the metal lattice ($M_{lattice}$), and (4) structure and properties of the deposits. The material in this book is arranged according to these four fundamental issues. We start by considering in the first three chapters the basic components of an electrochemical cell for deposition. Chapter 2 treats water and ionic solutions; Chapter 3, metal and metal surfaces; and Chapter 4, the metal–solution interface. In Chapter 5 we discuss the potential difference across an interface, and in Chapter 6,

Fundamentals of Electrochemical Deposition, Second Edition.
By Milan Paunovic and Mordechay Schlesinger
Copyright © 2006 John Wiley & Sons, Inc.

the kinetics and mechanisms of electrodeposition. Nucleation and growth of thin films and formation of the bulk phase are treated in Chapter 7. Electroless deposition and deposition by displacement are the subject of Chapters 8 and 9, respectively. In Chapter 10 we discuss the effects of additives in the deposition and nucleation and growth processes. Simultaneous deposition of two or more metals, alloy deposition, is discussed in Chapter 11. The manner in which current and metal are distributed on the substrate is the subject of Chapter 12. Characterization of metal surfaces before and/or after deposition and during the deposition process is treated in Chapters 13 and 14. Chapter 15 treats modeling of the deposition process. Structure and properties of deposits are treated in Chapters 16 to 18.

It is seen from the above that the present book contains a number of different types of material, and it is likely that on first reading, some readers, will want to use some chapters, whereas others may want to use different ones. For this reason the chapters and their various sections have been made independent of each other as far as possible. Certain chapters can be omitted without causing difficulties in reading succeeding chapters. For example, Chapters 3 (on metals and metal surfaces), 7 (on nucleation and growth models), 14 (on in situ characterization of deposition processes), and 15 (mathematical modeling in electrochemistry) can be omitted on first reading. Thus, the book can be used in a variety of ways to serve the needs of different readers.

1.2. RELATION OF ELECTROCHEMICAL DEPOSITION TO OTHER SCIENCES

The relation of electrochemical deposition to other sciences may be appreciated by considering the above-mentioned four types of fundamental problems associated with Eq. (1.1).

1. *The metal–solution interface as the locus of the deposition processes.* This interface has two components: a metal and an aqueous ionic solution. To understand this interface, it is necessary to have a basic knowledge of the structure and electronic properties of metals, the molecular structure of water, and the structure and properties of ionic solutions. The structure and electronic properties of metals are the subject matter of solid-state physics. The structure and properties of water and ionic solutions are (mainly) subjects related to chemical physics (and physical chemistry). Thus, to study and understand the structure of the metal–solution interface, it is necessary to have some knowledge of solid-state physics as well as of chemical physics. Relevant presentations of these subjects are given in Chapters 2 and 3.

2. *Kinetics and mechanism of the deposition process.* The rate of the deposition reaction ν [Eq. (1.1)] is defined as the number of moles of M^{z+} depositing per second and per unit area of the electrode surface:

$$\nu = k[M^{z+}] \tag{1.2}$$

where k is the rate constant of the reduction reaction and $[M^{z+}]$ represents the activity of M^{z+}. The rate constant k of electrochemical processes is interpreted on the basis of the statistical mechanics and is given by the expression

$$k = \frac{k_B T}{h}\left(-\frac{\Delta G_e^{\ddagger}}{RT}\right) \tag{1.3}$$

where k_B is the Boltzmann constant, T is the absolute temperature, h is the Planck constant, $\Delta G_e^{\ddagger}$ is the electrochemical activation energy, and R is the gas constant. The electrochemical activation energy is a function of the electrode potential E:

$$\Delta G_e^{\ddagger} = f(E) \tag{1.4}$$

This account of electrochemical kinetics shows why understanding and development of electrochemical deposition depends on statistical mechanics, which itself was developed by both physicists and chemists. The interpretation of $\Delta G_e^{\ddagger}$ is connected also to quantum mechanics.

3. *Nucleation and growth processes of the metal lattice.* Understanding of the nucleation and growth of surface nuclei, formation of monolayers and multilayers, and growth of coherent bulk deposit is based on knowledge of condensed-matter physics and physical chemistry of surfaces.

4. *Structure and properties of deposits.* These can be understood and interpreted on the basis of a variety of surface and bulk analytic techniques and methods that reveal electrical, magnetic, and physical properties of metals and alloys.

The authors of this book believe that this review of the relationship between the subject of electrochemical deposition and other sciences justifies one important general conclusion: that *electrochemical deposition is a fascinating field.* We hope that the readers will agree with this and work diligently on understanding, development, and/or applications of the fundamentals of electrochemical deposition.

1.3. BRIEF HISTORY OF ELECTROCHEMICAL DEPOSITION

An overview of a scientific subject must include at least two parts: retrospect (history) and the present status. The present status (in a condensed form) is presented in Chapters 2 to 21. In this section of the overview we outline (sketch) from our subjective point of view the history of electrochemical deposition science. In Section 1.2 we show the relationship of electrochemical deposition to other sciences. In this section we show how the development of electrodeposition science was dependent on the development of physical sciences, especially physics and chemistry in general. It is interesting to note that the electron was discovered in 1897 by J. J. Thomson, and the Rutherford–Bohr model of the atom was formulated in 1911.

The history of the subject matter of this book may be divided into three periods:

- *1905–1935.* The linear relationship between the overpotential η and log i (logarithm of the current density i; $i = I/A$, where I is the current and A is the surface area of the electrode) was established experimentally in 1905 by Tafel (1):

$$\eta = a + b \log i \qquad (1.5)$$

where a and b are constants. *Overpotential* η is defined as the difference between the potential of the electrode through which an external current I is flowing, $E(I)$, and the equilibrium potential of the electrode (potential in the absence of external current) E:

$$\eta = E(I) - E \qquad (1.6)$$

Erdey-Gruz and Volmer (2) derived the current–potential relationship in 1930 using the Arrhenius equation (1889) for the reaction rate constant and introduced the transfer coefficient. They also formulated the nucleation model of electrochemical crystal growth.

- *1935–1965.* Work on the development of the modern theory of the electrochemical activation energy (overpotential) started about 30 years after Tafel's formulation of Eq. (1.5). Eyring (3) and Wynne-Jones and Eyring (4) formulated the absolute rate theory on the basis of statistical mechanics [Eq. (1.3)]. Frank (6) and Burton et al. (7) realized that the real crystal surfaces (substrates for deposition) have imperfections and a variety of growth sites. This consideration introduced a major change in the theoretical interpretation of the deposition process and resulted in a series of new models. Lorenz (8) introduced the consideration of rate-determining surface diffusion of adions. Conway and Bockris (9) calculated probabilities of charge transfer to different sites on the metal surface.
- *1965–1997.* Damjanovic et al. (13) treated the optical determination of mechanisms of lateral and vertical step propagation. Dickson et al. (12) studied the nucleation and growth of electrodeposited gold on surfaces of silver by means of electron microscopy.

A series of nucleation and growth models was developed by, for example, Bewick et al. (11), Armstrong and Harrison (16), and Scharifker and Hills (17). Amblart et al. (18) have shown that nickel epitaxial growth starts with the formation of three-dimensional epitaxial crystallites. An electrochemical model for the process of electroless metal depositions (mixed-potential theory) was suggested by Paunovic (14) and Saito (14b).

Marton and Schlesinger (15) studied the nucleation and growth of electrolessly deposited thin nickel (Ni–P) films. These studies were later extended and complemented by the studies performed by Cortijo and Schlesinger (19,20) on radial distribution functions (RDFs). RDF curves were derived from electron diffraction data obtained from similar types of films as well as electrolessly deposited copper ones. Those studies, taken together, have elucidated the process of crystallization in the electroless deposition of thin metal films. Rynders and Alkire (26) studied propagation of copper microsteps on platinum surfaces using in situ atomic force microscopy.

It is possible that we have missed some important contributions in this sketch of history. However, we hope that readers will appreciate that this type of presentation is usually influenced by the personal interests of the authors.

1.4. NEW TECHNOLOGIES AND NEW INTEREST IN ELECTROCHEMICAL DEPOSITION

There has been a recent upsurge of interest in electrochemical deposition, due mainly to three new technologies: (1) metal deposition for the fabrication of integrated circuits, (2) deposition for magnetic recording devices (heads, disks), and (3) deposition of multilayer structures.

Electrochemical deposition for integrated circuits can be achieved through either electroless or electrodeposition. The feasibility of using selective electroless metal disposition for integrated circuit (IC) fabrication has been demonstrated by Ting et al. (21) and Shacham-Diamand (27).

Electrodeposition of Cu for IC fabrication has been used successfully since 1997 for the production of interconnection lines down to $0.20\,\mu$m width. Electrochemical metal deposition methods represent a very attractive alternative to the conventional IC fabrication processes (33). Development of electrochemical deposition technology for IC fabrication also represents an excellent opportunity for the electrochemists' community. This opportunity stems from the fact that new electrochemical deposition processes, producing deposits of different structure and properties, are needed to meet requirements of new, sub-micrometer-range computer technologies.

Another area with a large research activity is also related to computer technology. It is electrodeposition of magnetic alloys for thin-film recording heads and magnetic storage media. Here new magnetic materials are needed that have properties superior to those of electrodeposited NiFe (Permalloy). These activities are reviewed by Andricacos and Romankiw (25) and Romankiw (32).

The third example of new technology with increasing interest is electrodeposition of multilayers. For example, Schlesinger et al. (29) have shown that this technology can be applied to produce systems with nanometer-scale structural and compositional variations. Giant magnetoresistance (GMR) in electrodeposited Ni/Cu and Co/Cu multilayers was reported by Schlesinger et al. (28). Those constructs have a number of immediate applications in the areas of sensors as well as nanometer-scale electronic circuitry. For a more complete reference list as well as applications to date, see the review article by Schwartzacher and Lashmore (30).

We hope that readers will enjoy and learn from this book as much as we have enjoyed writing it.

REFERENCES AND FURTHER READING

1. J. Tafel, *Z. Phys. Chem.* **50**, 641 (1905).

2. T. Erdey-Gruz and M. Volmer, *Z. Phys. Chem.* **A150**, 203 (1930).

3. H. Eyring, *J. Chem. Phys.* **3**, 107 (1935).

4. W. F. K. Wynne-Jones and H. Eyring, *J. Chem. Phys.* **3**, 492 (1935).

5. W. Kauzmann, *Quantum Chemistry*, Academic Press, New York, 1957, pp. 1–10.

6. F. C. Frank, *Discuss. Faraday Soc.* **5**, 48 (1949).

7. W. K. Burton, N. Cabrera, and F. C. Frank, *Philos. Trans. R. Soc. London* **A243**, 299 (1951).

8. W. J. Lorenz, *Z. Naturforsch.* **9a**, 716 (1954).

9. B. Conway and J. O'M. Bockris, *Proc. R. Soc. London* **A248**, 394 (1958).

10. B. Conway and J. O'M. Bockris, *Electrochim. Acta* **3**, 340 (1961).

11. A. Bewick, M. Fleischmann, and H. R. Thirsk, *Trans. Faraday Soc.* **58**, 2200 (1962).

12. E. W. Dickson, M. H. Jacobs, and D. W. Pashley, *Philos. Mag.* **11**, 575 (1965).

13. A. Damjanovic, M. Paunovic, and J. O'M. Bockris, *J. Electroanal. Chem.* **9**, 93 (1965).

14. (a) M. Paunovic, *Plating* **55**, 1161 (1968); (b) M. Saito, *J. Met. Finish. Soc. Jpn.* **17**, 14 (1966).

15. J. P. Marton and M. Schlesinger, *J. Electrochem. Soc.* **115**, 16 (1968).

16. R. D. Armstrong and J. A. Harrison, *J. Electrochem. Soc.* **116**, 328 (1969).

17. B. Scharifker and G. Hills, *Electrochim. Acta* **28**, 879 (1983).

18. J. A. Amblart, M. Froment, G. Maurin, N. Spyrellis, and E. T. Trevisan-Souteyarand, *Electrochim. Acta* **28**, 909 (1983).

19. R. O. Cortijo and M. Schlesinger, *J. Electrochem. Soc.* **131**, 2800 (1984).

20. R. O. Cortijo and M. Schlesinger, *Solid State Commun.* **49**, 283 (1984).

21. C. H. Ting, M. Paunovic, and G. Chiu, *The Electrochemical Society Extended Abstracts*, Vol. 86-1, Abstract 239, p. 343, Philadelphia, May 10–15, 1987.

22. C. H. Ting and M. Paunovic, *J. Electrochem. Soc.* **136**, 456 (1989).

23. C. H. Ting, M. Paunovic, P. L. Pai, and G. Chiu, *J. Electrochem. Soc.* **136**, 462 (1989).

24. C. H. Ting and M. Paunovic, U.S. patent 5, 169, 680 (Dec. 8, 1992).

25. P. C. Andricacos and L. T. Romankiw, in *Advances in Electrochemical Science and Engineering*, Vol. 3, VCH, New York, 1994.

26. R. M. Rynders and R. C. Alkire, *J. Electrochem. Soc.* **141**, 166 (1994).

27. Y. Shacham-Diamand, in *Electrochemically Deposited Thin Films II*, M. Paunovic, ed., *Proceedings*, Vol. 94–31, Electrochemical Society, Pennington, NJ, 1995, p. 293.

28. M. Schlesinger, K. D. Bird, and D. D. Snyder, in *Electrochemically Deposited Thin Films II*, M. Paunovic, ed., *Proceedings*, Vol. 94–31, Electrochemical Society, Pennington, NJ, 1995, p. 97.

29. M. Schlesinger, D. S. Lashmore, and J. L. Schwartzendruber, *Scr. Met. Mater.* **33**, 1643 (1995).

30. W. Schwartzacher and D. S. Lashmore, *IEEE Trans. Magn.* **32**, 3133 (1996).

31. P. Singer, *Semicond. Int.* **20**, 67 (1997).

32. L. T. Romankiw, *J. Magn. Soc. Jpn.* **21** (Suppl. S2), 429 (1997).

33. C.-K. Hu and J. M. E. Harper, *Mater. Chem. Phys.* **52**, 5 (1998).

2

Water and Ionic Solutions

2.1. INTRODUCTION

Before discussing metal–solution interphase, we shall discuss the relevant properties of the individual components of an interphase. These individual components are at the same time also basic components of an electrodeposition cell (excluding the power supply). The basic components of an electrodeposition cell are, as shown in Figure 2.1, two metal electrodes (M_1 and M_2), water containing dissolved ions, and two metal–solution interfaces: M_1–solution and M_2–solution.

Successful use of this cell for electrodeposition in the production of electrodeposits of desired properties depends on understanding each component: specifically, components of the metal–solution interface. The metal–solution interface is the locus of the electrodeposition process and thus the most important component of an electrodeposition cell.

In this chapter we discuss water and ionic solutions, in Chapter 3, structure of metals and metal surfaces and in Chapter 4, the formation and structure of the metal–solution interface. Discussion is limited to those topics that are directly relevant to the electrodeposition processes and the properties of electrodeposits.

2.2. SINGLE WATER MOLECULE

Molecular Dimensions. The nuclei of oxygen and two hydrogens in the water molecule form an isosceles triangle (Fig. 2.2). The O—H bond length is 0.95718 Å, and the H—O—H angle is 104.523°.

Molecular Orbital Theory Model. Oxygen and hydrogen atoms in H_2O are held together by a covalent bond. According to the quantum molecular orbital theory of covalent bonding between atoms, electrons in molecules occupy molecular orbitals that are described, using quantum mechanical language, by a linear combination of

Fundamentals of Electrochemical Deposition, Second Edition.
By Milan Paunovic and Mordechay Schlesinger
Copyright © 2006 John Wiley & Sons, Inc.

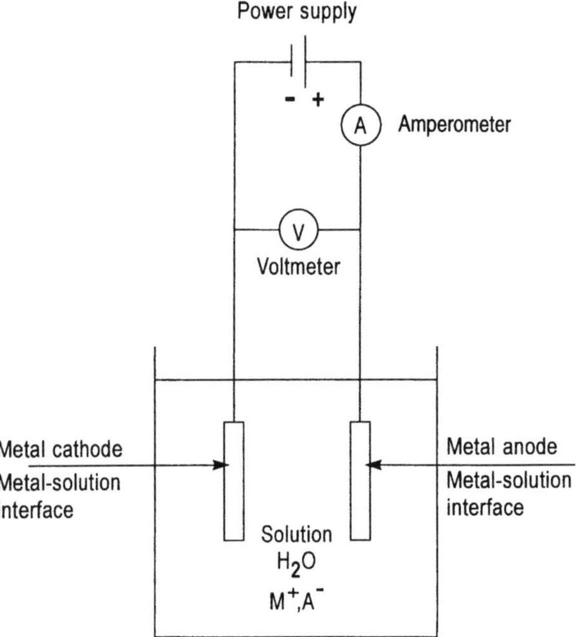

Figure 2.1. Electrolytic cell for electrodeposition of metal, M, from an aqueous solution of metal salt, MA.

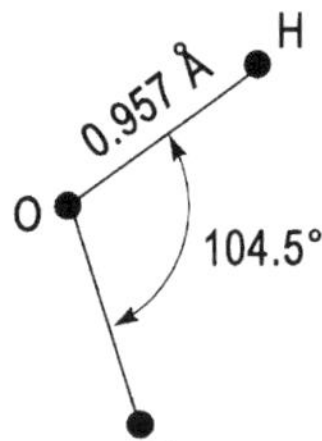

Figure 2.2. Dimensions of a single water molecule. The O—H distance and H—O—H angle are indicated.

atomic orbitals. Thus, for the H_2O molecule one considers the ground-state configurations of the individual H and O atoms:

$$H\ (1s)$$
$$O\ (1s^2 2s^2 2p_x^1 2p_y^1 2p_z^2)$$

Molecular orbital theory of the covalent bond shows a direct relationship between the extent of the overlap of two atomic orbitals and the bond strength. The larger the overlap, the stronger the bond. Maximum overlapping would produce the strongest bond and the most stable system. Maximum overlap of the H and O atomic orbitals

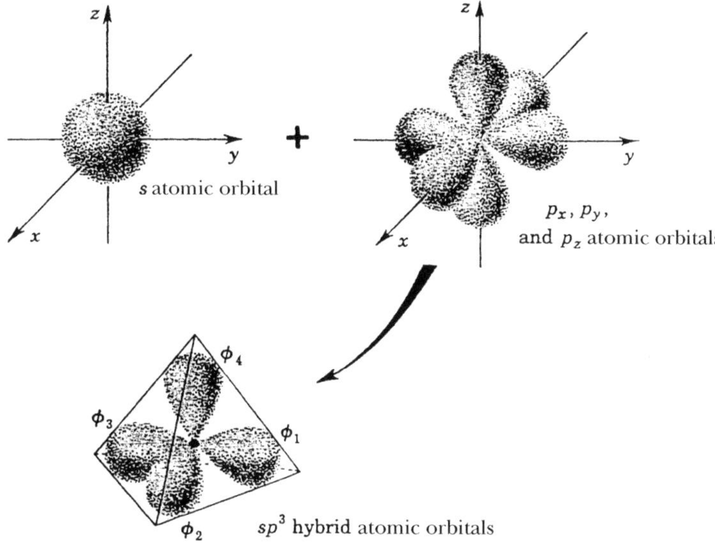

Figure 2.3. sp^3 hybrid orbitals.

can be obtained if one $2s$ and the three $2p$ atomic orbitals of oxygen are rearranged to form four equivalent sp^3 hybrid orbitals. Hybrid orbitals are described using linear combinations of atomic orbitals on the same atom. The four sp^3 orbitals are directed in space toward the corners of a regular tetrahedron, as illustrated in Figure 2.3.

The two bonding orbitals in the H_2O molecule are formed by overlapping of two sp^3 atomic orbitals of the one oxygen atom and the $1s$ orbitals of the two separate hydrogen atoms. Two lone electron pairs of oxygen atom occupy the remaining two sp^3 orbitals. The sp^3 model of the electronic structure of H_2O is shown in Figure 2.4. This model predicts an H—O—H bond angle of 109°28′.

Permanent Dipole Moment. In a pure, single covalent bond between two atoms, the bonding electrons are shared equally between the atoms; they belong equally to both

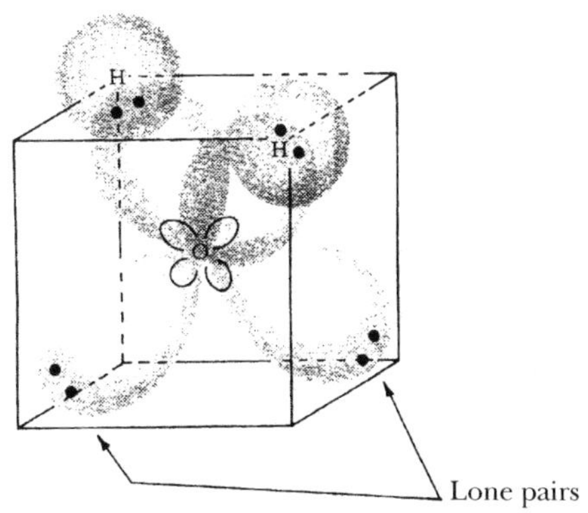

Figure 2.4. sp^3 model of the electronic structure of an H_2O molecule.

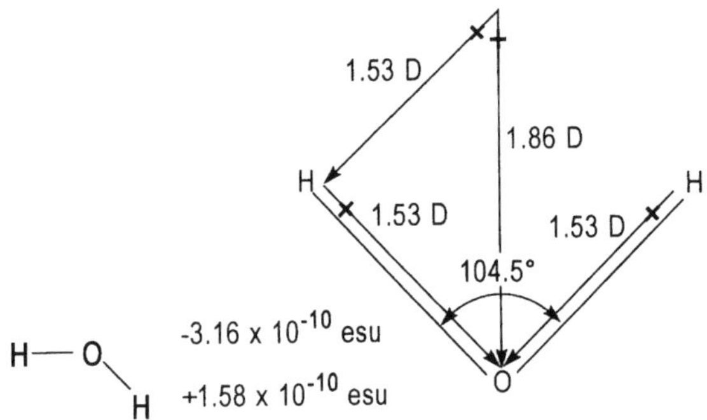

Figure 2.5. Water dipole. The observed dipole moment of 1.86 D for water is the vector sum of two O—H bond moments of 1.53 D.

nuclei. This equal sharing of the electron pair in the bonding molecular orbital is present in homonuclear molecules such as H_2 and O_2. However, in heteronuclear molecules this is not so, and in this case there is an unequal electron charge distribution in the bond. Unequal sharing is caused by differences in electron affinities of the two atoms. The atom with greater electron affinity attracts electrons to itself in a chemical bond. In the case of the O—H bond, the oxygen atom has greater affinity for electrons, and the result is separation of charge in the bond. The bonding electron pair spends more time near the oxygen than near the hydrogen. Thus, the O—H bond is polarized; that is, the hydrogen carries a small partial positive charge, δ^+, and oxygen carries a small partial negative charge, δ^-. This gives rise to a bond dipole moment. The separation of the charge in the H_2O molecule in the ground state is shown in Figure 2.5.

The product of the charge δe and the distance d between the charges, $\mu = \delta ed$, is called the *dipole moment*. The unit for the dipole moment is the debye (D), which is defined as μ for charge of the magnitude of the electronic charge e [4.803×10^{-10} esu (electrostatic unit)] and the separation d of 1 Å between the charges.

2.3. HYDROGEN BOND BETWEEN H_2O MOLECULES

According to the simple electrostatic model, the *hydrogen bond* between H_2O molecules consists of electrostatic interaction between the O—H bond dipole of one water molecule and the unshared lone pair of electrons on the oxygen of another water molecule (Fig. 2.6). The resulting bond is a *hydrogen bond*.

Each water molecule can form four hydrogen bonds since it contains two O—H bonds and two unshared electron pairs (Fig. 2.7). Thus, two bonds are formed by means of the molecule's own H atoms and two by means of two lone-pair electrons (Fig. 2.7). These four hydrogen bonds are directed in the four tetrahedral directions in space (Fig. 2.4).

The hydrogen bond energy between H and O is approximately 5 kcal/mol. Hydrogen bond is much weaker than the covalent bond since the typical covalent bond energy is about 100 kcal/mol.

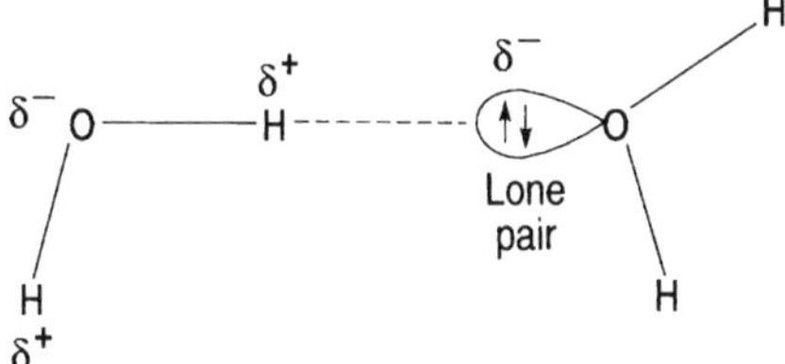

Figure 2.6. Hydrogen bond between two H_2O molecules.

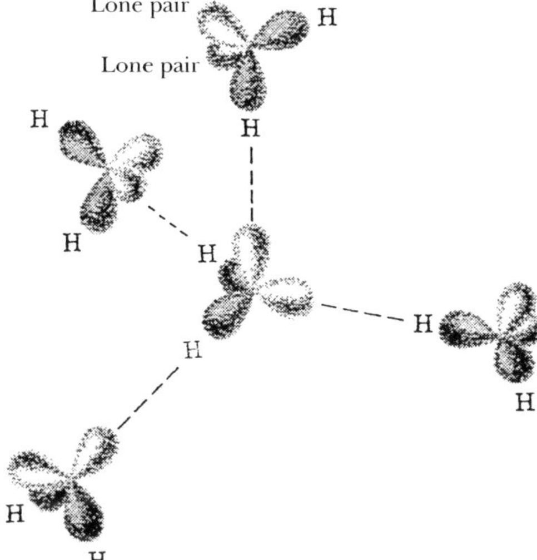

Figure 2.7. Four hydrogen bonds of an H_2O molecule.

2.4. MODELS OF LIQUID WATER

In light of the discussion of hydrogen bonds between water molecules, one can expect some degree of association of molecules in liquid water. Many models for the structure of water have been proposed, but we discuss only one.

Cluster Model. The cluster model is a version of a two-structure model. According to this model, liquid water consists of structured and unstructured regions. The structured regions are clusters of hydrogen-bonded water molecules. Large clusters made of 50 to 100 water molecules are probable. The unstructured regions are regions of independent single water molecules. A cluster exists until fluctuations in local energy break it up. But statistically, another cluster is formed elsewhere in the water through breaking and forming hydrogen bonds. Since the lifetime of a cluster is about 10^{-10} s, the model may be called a *flickering cluster model*. It is shown schematically in Figure 2.8.

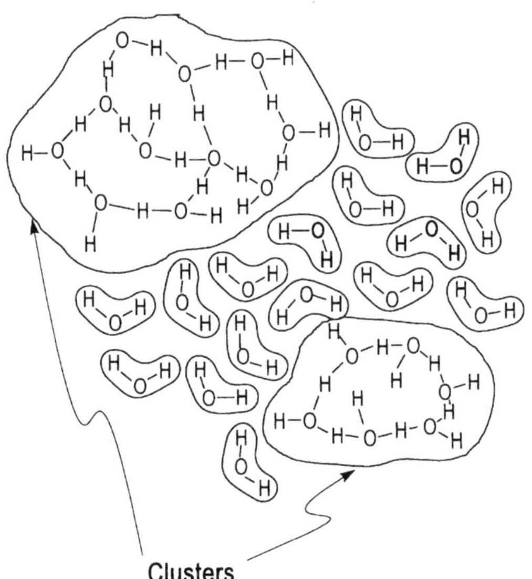

Figure 2.8. Flickering cluster model of liquid water. (From Ref. 1, with permission from *J. Chem. Phys.*)

2.5. IONIC DISSOCIATION OF WATER

Pure neutral water dissociates to a small extent, forming H^+ and OH^- ions:

$$H_2O \rightleftharpoons H^+ + OH^- \tag{2.1}$$

Like every other ion, H^+ and OH^- ions are hydrated in the aqueous solution. The concentration of H^+ ions in pure water is 1.0×10^{-14} mol/L (at 25°C). OH^- ions are naturally at the same concentration.

2.6. DIELECTRIC CONSTANT

In this section we discuss ion–ion interaction in solution. For this discussion we need to introduce the dielectric constant of water and basic models for interpretation of the dielectric constant.

According to Coulomb's law, the potential energy of electrostatic interaction U between two point charges, q_1 and q_2 in vacuum, is given by

$$U = \pm \frac{q_1 q_2}{r} \tag{2.2}$$

with $U = 0$ at $r \to \infty$ (in esu), where r is the distance between charges. If the same two charges are in a medium other than vacuum, the potential energy of the electrostatic interaction is

$$U = \pm \frac{q_1 q_2}{\epsilon r} \tag{2.3}$$

with $\epsilon > 1$, where ϵ is the dielectric constant of the medium. Thus, the surrounding medium reduces the potential energy of interaction between charges q_1 and q_2. If the medium is water, the coulombic interaction is reduced strongly, almost two orders of magnitude, since the dielectric constant of water is 78.5. The high dielectric constant of water can be interpreted in terms of the molecular properties and structure of water. The property that is related to the dielectric constant is the molecular dipole moment μ, $\epsilon = f(\mu)$. However, the dipole moment μ that determines the value of ϵ is not the dipole moment of an isolated water molecule but rather, the dipole moment of a group of water molecules (a dipole cluster), μ_{group}:

$$\mu_{\text{group}} = \mu(1 + g \cos \gamma) \tag{2.4}$$

where g is the number of water molecules in the cluster and $\cos \gamma$ is the average of the cosines of the angles between the dipole moment of the central water molecule and the group of water molecules bonded to the central water molecule. Thus, a decrease in g results in a decrease in μ_{group} and thus a decrease in the dielectric constant. This is, for instance, the case when ions enter liquid water (i.e., g decreases). Introduction of ions into water results in structure breaking (i.e., cluster breakup) and formation of independent water dipoles. The net result of this is a decrease in the g value and a subsequent decrease in the dielectric constant.

2.7. FORMATION OF IONS IN AQUEOUS SOLUTION

One method of introducing ions into solution is by the dissolution of an ionic crystal (e.g., NaCl). Ionic crystals are composed of separate positive and negative ions (Fig. 2.9). The overall dissolution process of an ionic crystal MA (M, M^{z+}; A, A^{z-}) can be represented by the reaction

$$M^{z+}A^{z-} + (m + a)H_2O \rightarrow M^{z+}(mH_2O) + A^{z-}(aH_2O) \tag{2.5}$$

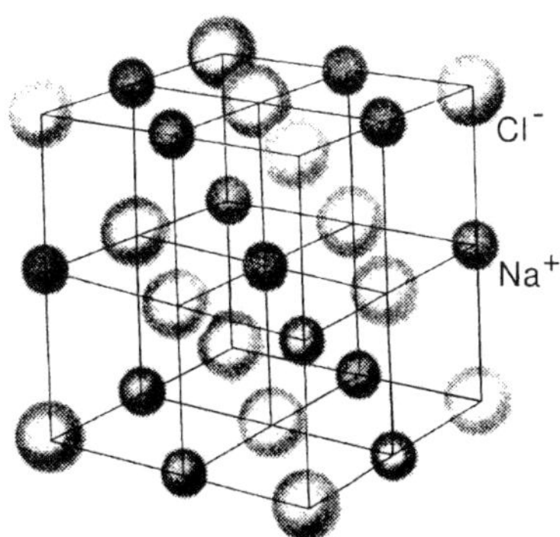

Figure 2.9. NaCl ionic crystal.

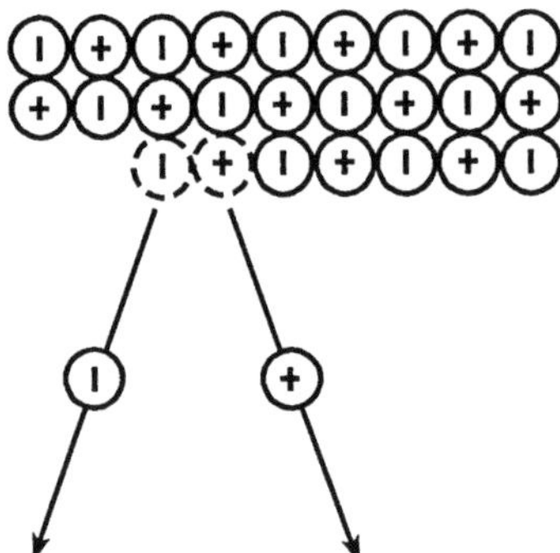

Figure 2.10. Formation of M^+ and A^- ions in aqueous solution.

This overall process can be considered as composed of two parts: (1) separation of ions from the lattice (breaking ion–ion bonds in the lattice), and (2) interaction of the ions with water molecules (hydration). Both processes involve ion–water interaction (Fig. 2.10). During crystal dissolution, the two processes are occurring simultaneously. Thus, we can write for the heat of solvation of a salt

$$\Delta H_{solv} = \Delta H_{subl} + \Delta H_{dis}$$

where ΔH_{solv} is the heat of solvation, ΔH_{subl} is the heat of sublimation, and ΔH_{dis} the heat of dissolution.

Another method of producing ions in solution is the dissolution of a potential electrolyte in water:

$$HA \rightarrow H^+ + A^- \tag{2.6}$$

In the pure state, a potential electrolyte such as oxalic acid (HOOCCOOH) consists of uncharged molecules. A true electrolyte such as NaCl in the pure state consists of two separate ions, Na^+ and Cl^-. The proton H^+ is a bare nucleus; it has no electrons. It is chemically unstable as an isolated entity because of its affinity for electrons. As a result, the proton reacts with the free electron pair of oxygen in the H_2O molecule,

$$H^+ + H_2O \rightarrow H_3O^+ \tag{2.7}$$

to form a hydronium ion. H_3O^+ is a simplified representation since the hydronium ion is hydrated in water. A better representation of dissolution of an acid is a proton transfer reaction,

$$HA + H_2O \rightarrow H_3O^+ + A^- \tag{2.8}$$

2.8. ION–WATER INTERACTION

Ion–Dipole Model. In this model ion–dipole forces are the principal forces in the ion–water interaction. The result of these forces is orientation of water molecules in

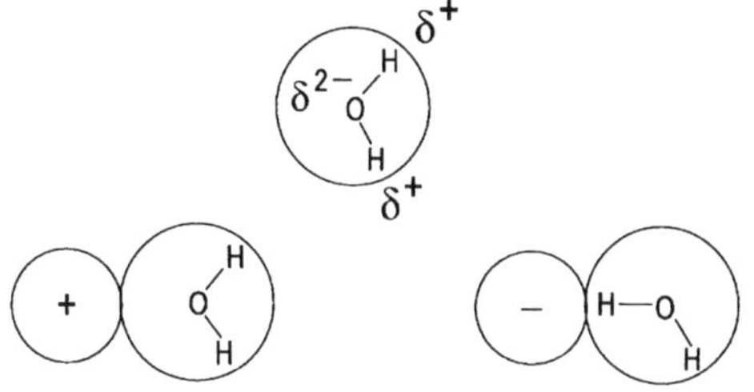

Figure 2.11. Orientation of water molecules with respect to positive and negative ions.

the immediate vicinity of an ion (Fig. 2.11). One end of the water dipole is attached electrostatically to the oppositely charged ion. The result of this orienting force is that a certain number of water molecules in the immediate vicinity of the ion are preferentially oriented, forming a primary hydration shell of oriented water molecules. These water molecules do not move independently in the solution. Rather, the ion and its primary water sheath form a single entity whose parts move together in the thermal motion and under the influence of an applied electric field. Next to the primary water of hydration is the shell of secondary water of hydration. In this region water molecules are under the influence of the orienting forces of the ion and the hydrogen-bond forces of bulk water molecules. Thus, water molecules in the shell of secondary water of hydration are partially oriented (Fig. 2.12). Beyond the secondary water of hydration is the bulk water. Agreement is good between experimentally determined heats of hydration and those calculated theoretically on the basis of the ion–dipole model (13,15).

Ion–Quadrupole Model. Improvement in agreement between experimental and theoretical values of heats of hydration is achieved by using the ion–quadrupole model. According to this model, the water molecule is represented as a *quadrupole*, an assembly of four charges: two positive charges each of value δ^+ due to the H atoms and two

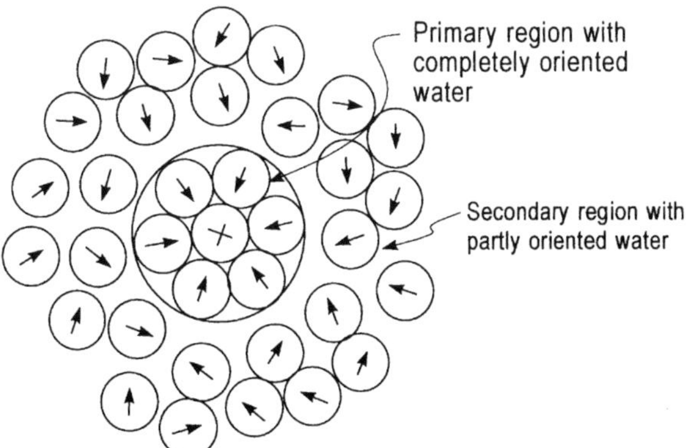

Figure 2.12. Ion–water interaction.

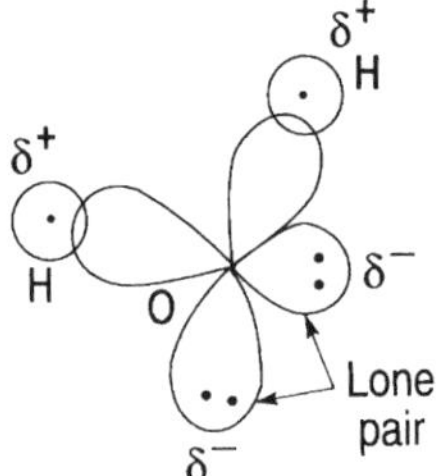

Figure 2.13. Quadrupole model of water molecule.

negative charges each of value δ^- due to the lone electron pair on the oxygen atom (Fig. 2.13).

2.9. ION–ION INTERACTION AND DISTRIBUTION OF IONS IN SOLUTION

Ionic Atmosphere. Before considering the distribution of ions in an ionic solution, it is instructive to consider the arrangement (distribution) of ions in an ionic crystal. For example, in a sodium chloride crystal, each ion is surrounded by six nearest neighbors of opposite charge. Each positive Na^+ ion is surrounded by six negative Cl^- ions, and each negative Cl^- ion is surrounded by six positive Na^+ ions (Figs. 2.9 and 2.14).

Thus, there is an apparent excess of ions of opposite charge around any ion. To some extent, a similar arrangement of ions is found in a dilute solution. Distribution of

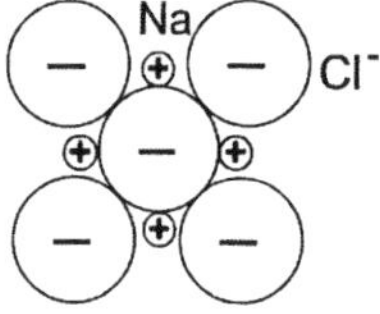

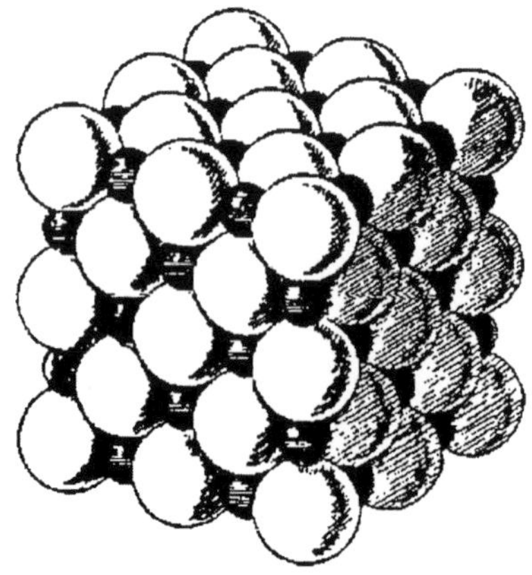

Figure 2.14. Cross-sectional view of a NaCl ionic crystal: one Na^+ above and one Na^+ below the central Cl^- ion, for a three-dimensional presentation.

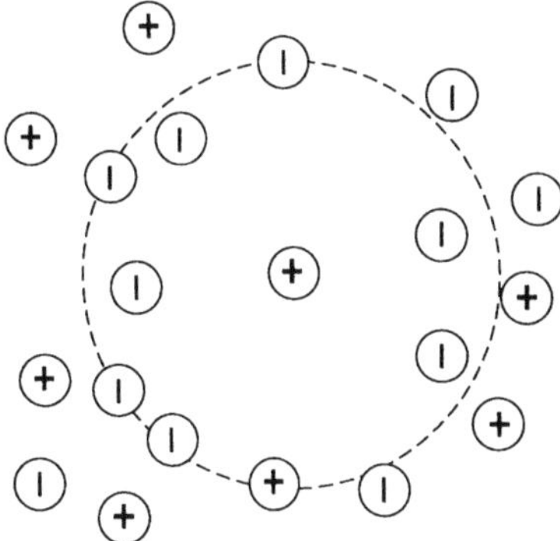

Figure 2.15. Ion–ion interaction.

cations and anions in a solution is such that on average, in time there is a statistical excess of ions of opposite charge around any ion. Each positive ion is surrounded by an atmosphere of negative charge, and each negative ion is surrounded by an atmosphere of positive charge. The local concentration of cations and anions can be evaluated by a Bolzmann type of statistic. Clearly, overall the solution is neutral. Thus, cations and anions are not distributed uniformly in an ionic solution. This is a result of the forces of interaction between ions (ion–ion interaction).

Ions of opposite charge are distributed in a spherical fashion around the central ion. This sphere around the central ion is called the *ionic atmosphere* (Fig. 2.15). This arrangement is dynamic; that is, there is a continuous interchange between ions contained in the ionic atmosphere and ions in the solution.

Debye–Huckel Theory. As shown above the cations and anions in an aqueous solution are not uniformly distributed due to forces of interaction between them (ion–ion interaction). There is a statistical excess (over bulk concentration) of opposite charges around a given ion. Thus, ions in solution are surrounded by an ionic atmosphere of an opposite charge. The total charge in this ionic atmosphere is of opposite sign and equal to the charge of the particular ion.

The discussion above is a description of problem that requires answers to the following: (1) the determination of the distribution of ions around a reference ion, and (2) the determination of the thickness (radius) of the ionic atmosphere. Obviously this is a complex problem. To solve this problem Debye and Huckel used a rather general approach: they suggested an oversimplified model in order to obtain an approximate solutions. The Debye–Huckel model has two basic assumptions. The first is continuous dielectric assumption. In this assumption water (or the solvent) is a continuous dielectric and is not considered to be composed of molecular species. The second, is a continuous charge distribution in the ionic atmosphere. Put differently, charges of the ions in the ionic surrounding atmosphere are smoothened out (continuously distributed).

Debye and Huckel applied the Boltzmann statistical distribution law and the Poisson equation for electrostatics in the model above (1, 6, 10). In the calculations using the model above they considered one particular ion (the reference ion, or central ion) with its ionic atmosphere. If there is no ion–ion interaction then

$$n_i(r) = n_i(b) \tag{2.9}$$

where $n_i(r)$ is the number of ions of type i per unit volume at the distance r from the reference ion and $n_i(b)$ is the number of ions of type i per unit volume in the bulk of the solution. In the presence of ion–ion interaction the value of $n_i(r)$ is determined by the Boltzmann distribution. According to the Boltzmann type distribution law

$$n_i(r) = n_i(b) \; \exp[-ze \; \psi(r)/k_B T] \tag{2.10}$$

where $\psi(r)$ is the electrical potential at a distance r from the reference ion and k_B is the Boltzmann constant.

Since each ion has charge $z_i e$, where z_i is the valence of the ion i and e electron charge

$$\rho(r) = n_1 z_1 e + n_2 z_2 e + \cdots + n_i z_i e \tag{2.11}$$

or

$$\rho(r) = \sum n_i z_i e \tag{2.12}$$

where $\rho(r)$ is the charge density at the distance r.

Introducing the $n_i(r)$ value as given in Eq. (2.10) into Eq. (2.12) one obtains the Boltzmann distribution law in terms of charge densities

$$\rho(r) = \sum n_i(b) \; z_i e \exp[-z_i e \; \psi(r)/k_B T] \tag{2.13}$$

In a further simplification, namely the expansion of the exponential function in Eq. (2.13) into a Taylor series up to the linear term only (neglecting other terms which may be shown to be negligible) yields

$$\exp[-z_i e \; \psi(r)/k_B T] \approx 1 - z_i e \; \psi(r)/k_B T \tag{2.14}$$

Inserting Eq. (2.14) into Eq. (2.13) we obtain

$$\rho(r) = \sum n_i(b) \; z_i e[1 - z_i e \psi(r)/k_B T] \tag{2.15}$$

and

$$\rho(r) = \sum n_i(b) \; z_i e - \sum n_i(b) \; z_i^2 e^2 \; \psi(r)/k_B T \tag{2.16}$$

The first term on the right-hand side of Eq. (2.16) is the charge on the uniform electrolytic solution as a whole and is equal to zero

$$\sum n_i(b) \; z_i e = 0 \tag{2.17}$$

This term is zero since the solution as whole must be electrically neutral (the principle of electrical neutrality). Thus, Eq. (2.16) is reduced to

$$\rho(r) = -\sum n_i^2(b)\, z_i^2 e^2\, \psi(r)/k_B T \tag{2.18}$$

Equation (2.18) represents a linearized Boltzmann distribution. It contains two unknown variables, $\rho(r)$ and $\psi(r)$. It is possible to reduce the problem of two unknown variables to a problem with one unknown variable by introducing a second equation expressing the relationship between the variables $\rho(r)$ and $\psi(r)$. This second equation is known as the Poisson equation. The Poisson equation for spherically symmetrical charge distribution is given as

$$\rho(r) = -\epsilon/4\pi[(1/r^2)d/dr(r^2 d\psi(r)/dr)] \tag{2.19}$$

where ϵ is the dielectric constant of water (an acceptable approximation for a dilute solution).

Combining Eqs. (2.18) and (2.19) yields the linearized Poisson–Boltzmann equation

$$(1/r^2)d/dr(r^2 d\psi(r)/dr) = \left[(4\pi/\epsilon k_B T)\sum n_i(b)\, z_i^2 e^2\right]\psi(r) \tag{2.20}$$

The constant multiplying $\psi(r)$ on the right-hand side of Eq. (2.20) may be represented by a single new constant named κ^2

$$\kappa^2 = (4\pi/\epsilon k_B T)\sum n_i(b)\, z_i^2 e^2 \tag{2.21}$$

In terms of this new constant the linearized Poisson–Boltzmann equation is

$$(1/r^2)d/dr(r^2 d\psi(r))/dr = \kappa^2 \psi(r) \tag{2.22}$$

Equation (2.22) may be written in the following form (see Problem 2.4)

$$[d^2(r\psi(r)/d^2)] = \kappa^2[r\psi(r)] \tag{2.23}$$

The solution of this differential equation is (6,10)

$$\psi(r) = [z_i e/\epsilon r]\exp[-\kappa r] \tag{2.24}$$

$\psi(r)$ in Eq. (2.24) represents the mean value of the potential at a point r produced by (1) the ionic atmosphere, and (2) the reference (central) ion. Thus, according to the principle of superposition of potentials

$$\psi(r) = \psi_{\text{ion}} + \psi_{\text{atm}} \tag{2.25}$$

The potential at distance r due to central ion is

$$\psi_{\text{ion}} = z_i e/\epsilon r \tag{2.26}$$

Using Eq. (2.24) for $\psi(r)$ in Eq. (2.25) and Eq. (2.26) for ψ_{ion} we obtain

$$\psi_{atm} = [z_i e/\epsilon r]\exp(-\kappa r) - z_i e/\epsilon r \tag{2.27}$$

$$\psi_{atm} = [z_i e/\epsilon r]\exp[(-\kappa r) - 1] \tag{2.28}$$

Expanding the exponential function into a series and neglecting the higher terms we obtain

$$\exp(-\kappa r) = 1 - \kappa r \tag{2.29}$$

Eq. (2.28) becomes

$$\psi_{atm} = -(z_i e \ \kappa/\epsilon) = -z_i e/\epsilon(1/\kappa) \tag{2.30}$$

Since the potential V is produced by charge Q at a distance, r is given by $V = Q/\epsilon r$.

Equation (2.30) represents the potential produced by a charge $z_i e$ of ionic atmosphere at a distance $1/\kappa$. The quantity $1/\kappa$ has the dimensions of length and is appropriately called the thickness (or radius) of the ionic atmosphere in a given solution. Also, κ^{-1} is called the Debye–Huckel length and is assigned symbol r_D. From Eq. (2.21)

$$\kappa^{-1} = r_D = \left[(\epsilon k_B T/4\pi)1/\sum n_i(b) \ z_i^2 e^2\right]^{1/2} \tag{2.31}$$

The above equation shows that the ionic atmosphere becomes thinner with increasing concentration of ions.

For a uni-univalent (1-1) electrolyte at 25°C ($T = 298°$) with water as solvent, considering 78.6 as the dielectric constant value, we obtain from Eq. (2.31)

$$r_D = 4.32/\left[\sum c_i z_i^2\right]^{1/2} \quad (\text{in Å}) \tag{2.32}$$

where c is the concentration of the salt in moles per liter.

Distance of the Closest Approach. The weak attractive force in an aqueous solution is unable to bring ions as close together as in the crystal lattice. The distance of closest approach for the same pair of ions in an aqueous solution is greater than that in the corresponding crystal lattice. According to Coulomb's law, the potential energy of the electrostatic attractive interaction between ions of opposite charge and the repulsive interaction between ions of the same charge in vacuum is given by $\pm q^2/r$, where r is the distance between the charges [Eq. (2.2)]. The potential energy of the electrostatic interaction between ions in a solution is given by $\pm q^2/r\epsilon$, where ϵ is the dielectric constant of the solution [Eq. (2.3)]. Thus, in an aqueous solution the interionic forces are weakened as a result of the high dielectric constant of water (i.e., solution). The result of this is an average increase in distance between ions in the solution. For example, the lattice spacing for NaCl is 2.81 Å, and the distance of the closest approach for the Na^+ ion and the Cl^- ion in water solution is 4.0 Å.

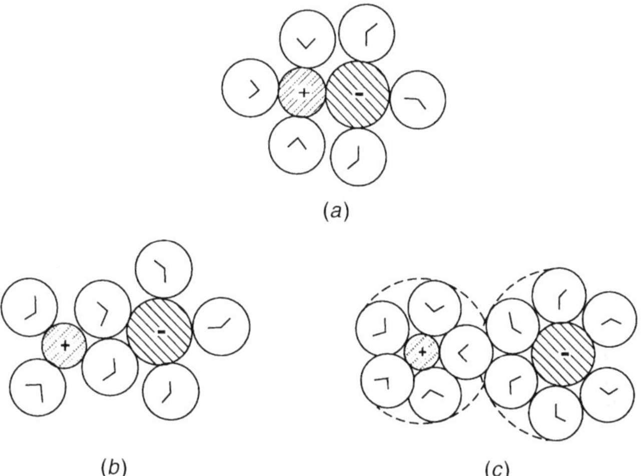

Figure 2.16. Ion pairs: (*a*) ion contact type; (*b*) shared hydration shells; (*c*) hydration shell contact type. (From Ref. 15, with permission from Academic Press.)

Ion Pairs. As a consequence of the thermal translation motion of ions in solution, ions of opposite charge may come sufficiently close so that the coulombic attractive force can be strong enough to overcome the random thermal agitation that eventually tends to scatter them apart. In this case the original ions lose their independence. Thus, in an ionic solution, one may expect to find a few ions of opposite charge in close proximity, forming ion pairs (Fig. 2.16).

An ion pair is electrically neutral, and when an external electric field is applied, it does not contribute to the electric current (conductivity). Ion pairs have a short lifetime however, since there is a continuous interchange between ions in the solution, due to random thermal agitation.

2.10. EFFECT OF IONS ON STRUCTURE AND DIELECTRIC CONSTANT OF WATER

In this section we link our discussion in two previous sections dealing with structure of water and with ion–water interaction. In the discussion on ion–water interaction it was shown that ions in water arrange their immediate neighboring water dipoles into a local structure of the primary water of hydration. Between this local structure and the bulk water is the nonstructured secondary water of hydration. Thus, the presence of ions in water will change the number of water molecules in both the structured and unstructured regions. Any decrease in the number of water molecules in a cluster will result in a corresponding decrease in the value of g and thus a decrease in the dielectric constant of water [Eq. (2.4)].

Thus, one may expect two changes in the dielectric constant of water due to the presence of ions: (1) lower dielectric constant in the primary and secondary hydration

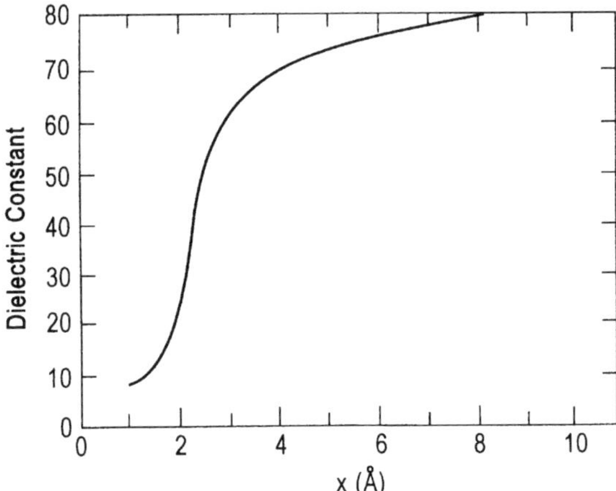

Figure 2.17. Variation of the dielectric constant around an ion in solution. (From Ref. 6, with permission from Plenum Press.)

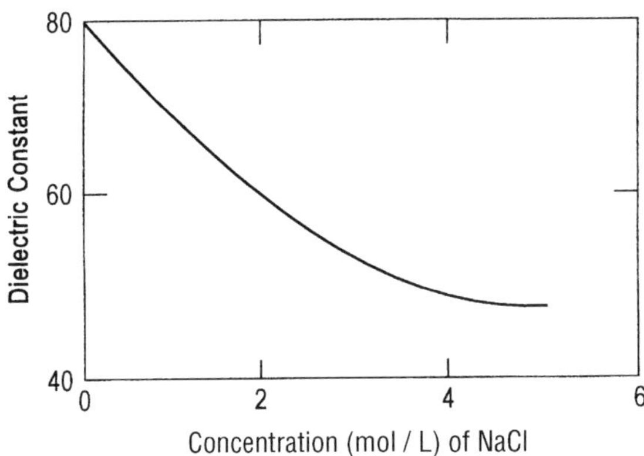

Figure 2.18. Dependence of the dielectric constant of an NaCl solution on ionic concentration. (From Ref. 6, with permission from Plenum Press.)

sheets, and (2) lower dielectric constant of the solution than in pure water. Variation of the dielectric constant around an ion is shown in Figure 2.17. Dependence of the dielectric constant of an NaCl solution on ionic concentration is shown in Figure 2.18.

REFERENCES AND FURTHER READING

1. G. Nemethy and H. A. Scheraga, *J. Chem. Phys.* **36**, 3382 (1962).

2. D. Eisenberg and W. Kauzmann, *The Structure and Properties of Water*, Oxford University Press, New York, 1969.

3. J. B. Hasted, in *Water: A Comprehensive Treatise*, Vol. 1, F. Franks, ed., Plenum Press, New York, 1972.

4. A. Ben-Naim, *Water and Aqueous Solutions*, Plenum Press, New York, 1974.

5. R. P. Feynman, R. B. Leighton, and M. Sands, *The Feynman Lectures on Physics*, Vol. II, Addison-Wesley, Reading, MA, 1964.

6. J. O'M. Bockris and A. K. N. Reddy, *Modern Electrochemistry*, 2nd ed., Vol. 1, Plenum Press, New York, 1998.

7. P. Debye and E. Hückel, Z. *Phys.* **24**, 185 (1923).

8. J. D. Bernal and R. H. Fowler, *J. Chem. Phys.* **1**, 515 (1933).

9. R. W. Gurney, *Ionic Processes in Solution*, Dover, New York, 1953.

10. B. E. Conway and J. O'M. Bockris, in *Modern Aspects of Electrochemistry*, J. O'M. Bockris, ed., Butterworth, London, 1954.

11. R. A. Robinson and R. H. Stokes, *Electrolyte Solutions*, Butterworth, London, 1959.

12. T. L. Hill, *Statistical Thermodynamics*, Addison-Wesley, Reading, MA, 1962.

13. B. E. Conway, in *Comprehensive Treatise of Electrochemistry*, Vol. 5, B. E. Conway, J. O'M. Bockris, and E. Yeager, eds., Plenum Press, New York, 1982.

14. J. E. Desnoyers and C. Jolicoeur, in *Comprehensive Treatise of Electrochemistry*, Vol. 5, B. E. Conway, J. O'M. Bockris, and E. Yeager, eds., Plenum Press, New York, 1982.

15. B. E. Conway, in *Physical Chemistry: An Advanced Treatise*, Vol. 9A, H. Eyring, ed., Academic Press, New York, 1970.

16. H. L. Friedman, *J. Electrochem. Soc.* **124**, 421C (1977).

17. J. O'M. Bockris and S. U. M. Khan, *Surface Electrochemistry*, Plenum Press, New York, 1993.

PROBLEMS

2.1. Calculate the heat of hydration of NaCl and KCl from the following experimental data. The heat of sublimation for NaCl is 185.0 and that for KCl is 169.2 kcal/mol; the heat of dissolution for NaCl is 1.3 and that for KCl is 4.4 kcal/mol.

2.2. Derive Eq. (2.23) from Eq. (2.22). [*Hint:* Introduce a new variable, $y = r\,\psi\,(r)$.]

2.3. Derive Eq. (2.32) from Eq. (2.31) using the following values for the required constants:

$k_B = 1.38 \times 10^{-16}$ erg/deg, $e = 4.803 \times 10^{-10}$ esu, $\epsilon = 78.6$, and $N = 6.023 \times 10^{23}\,mol^{-1}$. Note that n_i in Eq. (2.32) denotes the number of i-type ions per liter and c is the concentration in mol/L.

2.4. Calculate the radius of the ionic atmosphere r_D for uni-univalent (1:1) electrolytes at 25°C with water as a solvent using Eq. (2.32). The concentrations of the salt are 0.0001, 0.01, and 0.1 mol/L.

2.5. Derive an equation for the radius of ionic atmosphere r_D for bi-bivalent (2:2) electrolytes at 25°C with water as a solvent. Use the derived equation to calculate r_D values for the following concentrations of salt: 0.0001, 0.001, and 0.1 mol/L. Compare the results obtained with the results of Problem 2.4.

3

Metals and Metal Surfaces

3.1. INTRODUCTION

In this section we treat the bulk and surface properties of metals relevant to the problems of electrochemical deposition. First, we discuss briefly the bulk and electronic structure of metals and then analyze the surface properties. Surface properties of the greatest interest in electrodeposition are atomic and electronic structure, surface diffusion, and interaction with the metal surface (adsorption) of atoms and molecules in solution.

3.2. BULK STRUCTURE OF METALS

A metal can be considered as a fixed lattice of positive ions permeated by a gas of free electrons. Positive ions are the atomic cores; the electrons are the valence electrons. For example, copper has a configuration (electronic structure) $1s^2 2s^2 2p^6 3s^2 3p^6 3d^{10} 4s^1$ (superscripts designate the number of electrons in the orbit) with one valence electron ($4s$). The atomic core of Cu^+ is the configuration given above, less the one valence electron $4s^1$. The free electrons form an electron gas in the metal and move nearly freely through the volume of the metal. Each metal atom contributes its valence electrons to the electron gas in the metal. Interactions between the free electrons and the metal ions make a large contribution to the metallic bond.

Various lattices are described in Chapter 16. Since there are about 10^{22} atoms in $1\,cm^3$ of a metal, one can expect that some atoms are not exactly in their right place. Thus, one can expect that a real lattice will contain defects (imperfections).

Point Defects. The simplest defect is a *vacancy*, where an atom is missing from its site in the lattice (Fig. 3.1). Another point defect is an *interstitial*, where an extra atom (e.g., impurity), without a proper lattice site, is forced into the lattice (Fig. 3.2). The insertion of an extra atom results in a distortion of the lattice around the interstitial

Fundamentals of Electrochemical Deposition, Second Edition.
By Milan Paunovic and Mordechay Schlesinger
Copyright © 2006 John Wiley & Sons, Inc.

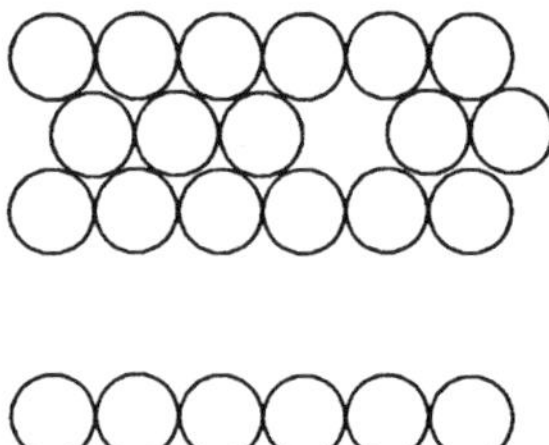

Figure 3.1. Lattice vacancy.

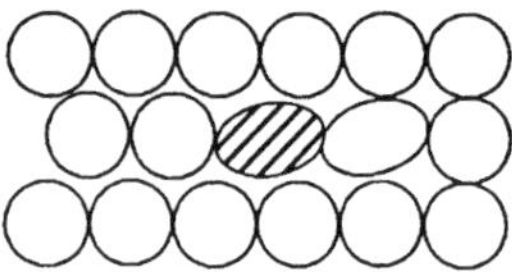

Figure 3.2. Interstitial atom in the crystal.

atom. These facts should be interpreted in terms of statistical mechanics. In general, a statistical system of large number of particles at finite temperature cannot have perfect order.

Grain Boundaries. Most metals are not single crystals but agglomerates of small crystallites packed together. These crystallites are randomly oriented with respect to one another. The boundaries between these crystallites are lattice defects and are called *grain boundaries* (Fig. 3.3). They are transitional regions between grains, about 2 to 10 atoms thick, where the atoms change from one orientation to another. In general, the structure of a grain boundary is complex. It is simple only in the case of low-angle boundaries, where the orientations of neighboring grains are very similar.

Dislocations. *Screw dislocations* are the most important defects when crystal growth is considered, since they produce steps on the crystal surface. These steps are crystal growth sites. Another type of dislocation of interest for metal deposition is the *edge dislocation*. Screw and edge dislocations are shown in Figure 3.4.

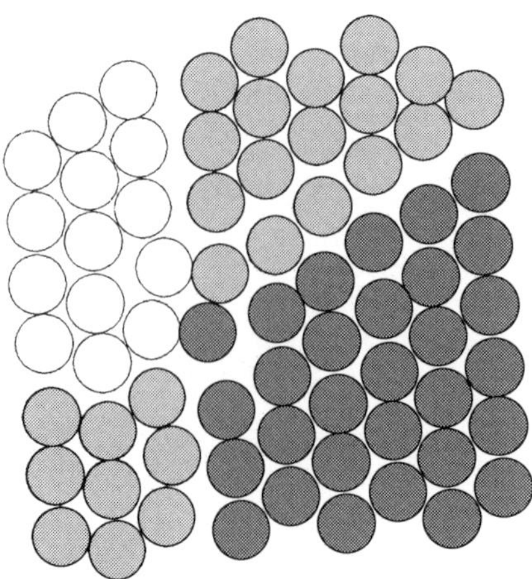

Figure 3.3. Grain boundaries.

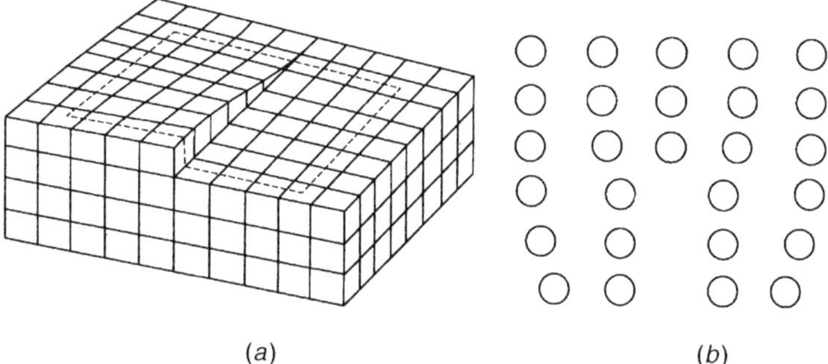

Figure 3.4. (*a*) Screw dislocation; (*b*) structure of an edge dislocation.

3.3. ELECTRONIC STRUCTURE OF METALS

The free-electron theory of metals was developed in three main stages: (1) classical free-electron theory, (2) quantum free-electron theory, and (3) band theory.

Classical Free-Electron Theory. Classical free-electron theory assumes the valence electrons to be virtually free everywhere in the metal. The periodic lattice field of the positively charged ions is evened out into a uniform potential inside the metal. The major assumptions of this model are that (1) an electron can pass from one atom to another, and (2) in the absence of an electric field, electrons move randomly in all directions and their movements obey the laws of classical mechanics and the kinetic theory of gases. In an electric field, electrons drift toward the positive direction of the field, producing an electric current in the metal. The two main successes of classical free-electron theory are that (1) it provides an explanation of the high electronic and thermal conductivities of metals in terms of the ease with which the free electrons could move, and (2) it provides an explanation of the Wiedemann–Franz law, which states that at a given temperature T, the ratio of the electrical (σ) to the thermal (κ) conductivities should be the same for all metals, in near agreement with experiment:

$$\frac{k}{\sigma T} = \text{constant} \tag{3.1}$$

The theory fails to explain the molar specific heat of metals since the free electrons do not absorb heat as a gas obeying the classical kinetic gas laws. This problem was solved when Sommerfeld (1) applied quantum mechanics to the electron system.

Quantum Free-Electron Theory: Constant-Potential Model. The simple quantum free-electron theory (1) is based on the *electron-in-a-box model*, where the box is the size of the crystal. This model assumes that (1) the positively charged ions and all other electrons (nonvalence electrons) are smeared out to give a constant background potential (a potential box having a constant interior potential), and (2) the electron cannot escape from the box; boundary conditions are such that the wavefunction ψ is zero at all faces of the box. The Schrödinger equation for a particle of mass m moving in one dimension with energy E is

$$-\frac{\hbar}{2m}\frac{d^2\psi}{dx^2} + V\psi = E\psi \tag{3.2}$$

where ψ is the wavefunction, V is the potential energy of the particle, and $\hbar$ ("aitch bar") is $h/2\pi$ (h *is* Planck's constant). Solution of the Schrödinger equation for this model is a wavefunction of the type

$$\psi = e^{ikx} \tag{3.3}$$

for each free electron. The result is quantization of the energy of the electrons. The quantization arises from the boundary conditions (ψ is zero at all faces of the box) that the wavefunction must satisfy to be acceptable. However, for an ordinary-size piece of metal, the allowed energy levels are too close together and the number of states permitted for the electrons are too numerous. The result of this is that the energy spectrum, the probability of occupation of states versus energy of states E, is a quasi-continuous curve (Fig. 3.5).

Electrons occupy states of lowest available energy (the *Pauli principle*). Two electrons are accommodated in each state (electron spin). Thus, at 0 K all permitted quantum states are filled, up to the limiting value, E_{max}. For metals, $E_{max} = E_F$ (Fermi energy, chemical potential). At temperatures above 0 K, some electrons possess thermal energy and move into higher quantum states so that the occupation of states in the region of E_{max} is smeared out (Fig. 3.5b). Figure 3.6 shows that the relationship between energy E and wavenumber k ($k = 2\pi/\lambda$; λ is the wavelength in the Schrödinger equation) is a parabola. The major success of the simple quantum free-electron theory is the explanation of the electronic specific heat. However, this theory must be improved further to explain the electrical conductivity for both metallic and nonmetallic crystals.

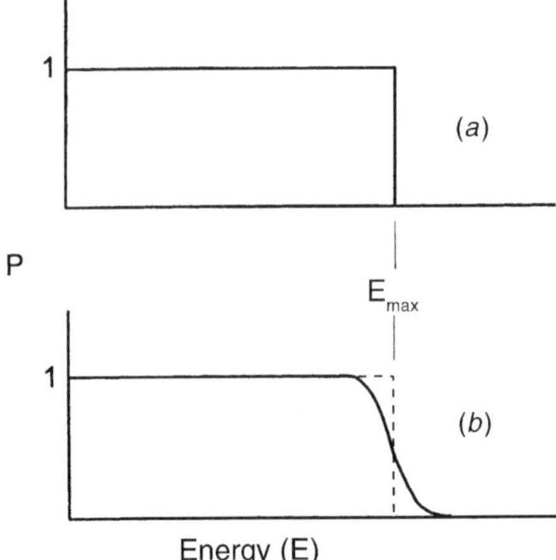

Figure 3.5. Probability of occupation of electron states versus energy of states E: (a) $T = 0$ K; (b) $T > 0$ K.

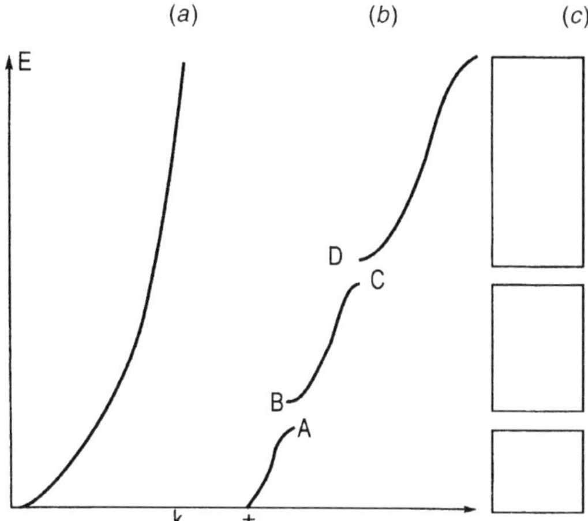

Figure 3.6. Plot of energy E versus wavenumber k (a) for a free electron moving in a constant background potential and (b) for an electron moving in a periodic field in an one-dimensional crystal. (c) The energy bands for (b).

Band Theory of Metals. Three approaches predict the electronic band structure of metals. The first approach (Kronig–Penney), the periodic potential method, starts with free electrons and then considers nearly bound electrons. The second (Ziman) takes into account Bragg reflection as a strong disturbance in the propagation of electrons. The third approach (Feynman) starts with completely bound electrons to atoms and then considers a linear combination of atomic orbitals (LCAOs).

1. *Kronig–Penney model.* The simple quantum free-electron theory assumes that electrons in metal move in a constant background potential V (a potential box having a constant internal potential). The result of this assumption is that the (E,k) relationship is a parabola (Fig. 3.6a). An extension of the theory is obtained if one assumes that electrons in a metal move in a periodic field resulting from the periodic structure of the crystal. Considering a one-dimensional crystal (atoms arranged on a straight line) for simplicity, the potential energy of an electron is as shown in Figure 3.7a. The highest potential is between the ions, and then the potential tends to minus infinity as the position of the ions approached. The essential features of this function are that (1) it has the same period as the lattice, and (2) the potential is lower in the vicinity of the lattice ion and higher between ions. Kronig and Penney replaced this relatively complicated function with a simpler one (Fig. 3.7b) having the same essential features as the function in Figure 3.7a. For this model the solution of the Schrödinger equation for a periodic V is

$$\psi = e^{ikx}u_k(x) \tag{3.4}$$

where u is a function depending on k, which is periodic in x with the period of V, that is, with the period of the lattice. Thus, the wavefunction is modulated by the periodic field of the lattice. The relationship between E and k is shown in Figure 3.6b. A comparison of Figure 3.6a with b shows that the most important difference between the

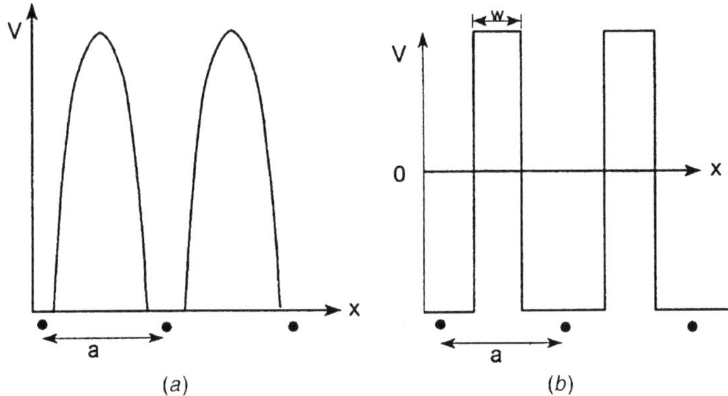

Figure 3.7. (*a*) Potential energy of an electron in a one-dimensional crystal; (*b*) Kronig–Penney model of the potential energy of an electron in a one-dimensional crystal (square-well periodic potential model).

simple quantum free-electron theory (motion of free electrons in a constant field) and the periodic potential theory is that the (*E,k*) relationship for the first is a simple parabolic function, and for the latter, the (*E,k*) relationship is a parabolic function with discontinuities, with bands of energy that are not possible. This model is called the *band model* of metals. Figure 3.6*b* shows that there are critical values of *k* at which the free-electron level is split into two distinct levels, such as *A–B* and *C–D*. Energy levels *A–B* and *C–D* are separated by a range of energies in which there is no allowed state for the electrons. At those critical values $\pm k_1$, $\pm k_2$, and so on, the parabola "flattens" off, producing discontinuities, energy gaps, between *A–B* and *C–D* and at higher energies not shown on the diagram. These gaps, not allowed (forbidden) ranges, divide the energy spectrum into bands of allowed and forbidden regions (Fig. 3.6*b*).

2. *Ziman model.* This model is based on a consideration of Bragg reflection for electron waves (Fig. 3.8). When the individual reflections add in phase, the Bragg relationship

$$\lambda = 2a \cos \theta, \quad n = 1, 2, \dots \tag{3.5}$$

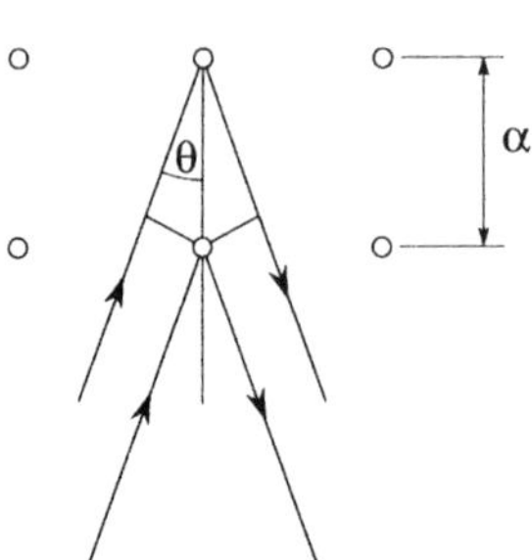

Figure 3.8. Geometry of reflection from atomic planes.

holds. In a one-dimensional crystal

$$n\lambda = 2a \tag{3.6}$$

and from the relationship between the wavelength and the wavenumber ($k = 2\pi/\lambda$), it follows that

$$k = \frac{n\pi}{a}, \quad n = 1, 2, 3, \ldots \tag{3.7}$$

Thus, the free-electron model is not valid when Eq. (3.7) applies since the wave is reflected. The (E,k) curve constructed on this basis is like that obtained from the Kronig–Penney model: bands of allowed and forbidden energy regions.

3. *Feynman model.* The Feynman approach, or LCAO (linear combination of atomic orbitals) method, assumes that a wavefunction of valence electrons ψ in a metal is a linear combination of atomic functions:

$$\psi = \sum_l c_l \phi_l \tag{3.8}$$

where l stands for the three location indices l_1, l_2, l_3; c is a constant; and ϕ is an atomic wavefunction localized at the lattice point l. Solutions based on this equation show that the energy levels for electrons in an infinite periodic crystal are arranged into energy bands. The formation of energy bands using the LCAO method can be illustrated qualitatively by considering a simple infinitely long line of monovalent atoms (a one-dimensional metal), each having a single valence electron in an s orbital (e.g., Cu, Ag, Au). Thus, each atom contributes one s orbital at a given energy (Fig. 3.9a). When a second atom is brought to the line of one-dimensional metal, its s orbital overlaps with the s orbital of the first and forms two LCAO molecular orbitals (MOs): a bonding orbital and an antibonding orbital ($\psi_A \pm \psi_B$) (Fig. 3.9b). Thus, the energy

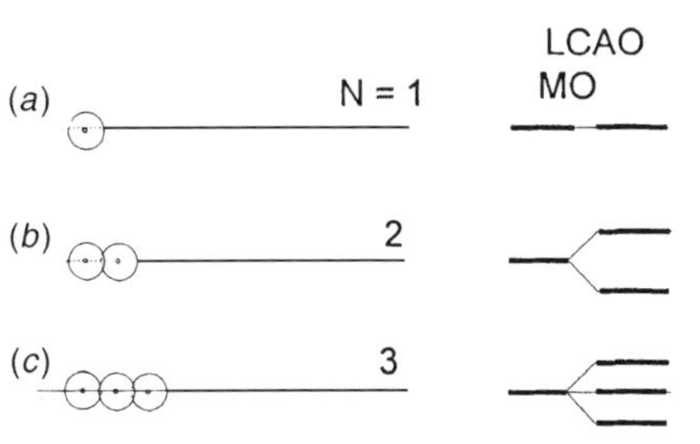

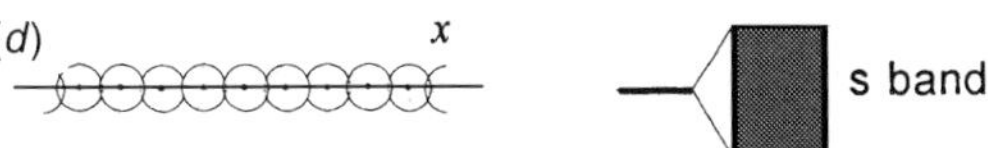

Figure 3.9. Formation of a band of N orbitals by successive addition of atoms to a line.

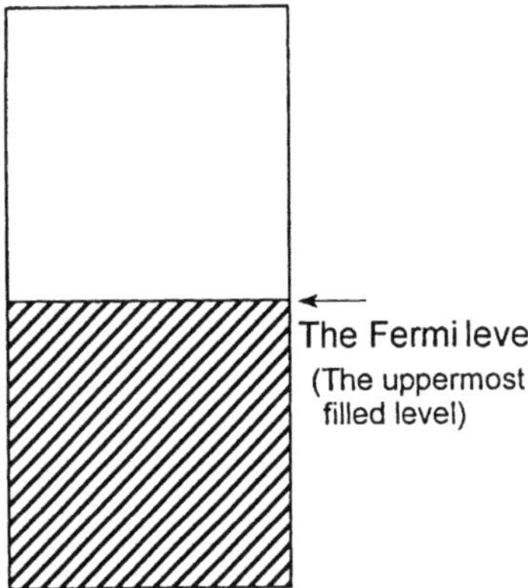

Figure 3.10. Fermi surface separating unfilled orbitals from filled orbitals (the shaded area) at absolute zero.

levels of the two interacting atoms are split; one is slightly higher, the other is slightly below the original level. The third atom overlaps its nearest neighbor, and from these, three LCAO MOs are formed (Fig. 3.9*c*). Thus, the addition of one atom to the one-dimensional metal adds one orbital and causes a slight spread of the range of energies covered by the MOs. Addition of N atoms to the line results in N different orbitals covering a band of finite width. When N is very large ($N \to \infty$), the difference between neighboring energy levels is infinitely small (Fig. 3.9*d*). According to the Pauli exclusion principle, each energy level can have at most two electrons (spin). When there are N electrons, only the lowest $\frac{1}{2} N$ MOs are occupied. The band is only partially occupied by electrons (Fig. 3.10).

Electrons near the Fermi level are mobile and give rise to electrical conductivity. The band formed from overlap of *p* atomic orbitals is called the *p band*. The overlap of *s* and *p* orbitals produces *s* and *p* bands. If separation between *s* and *p* atomic orbitals is large, the separation between s and *p* bands in metal is large. There is a bandgap in this case (Fig. 3.11). When the *s–p* separation is smaller, the bands overlap.

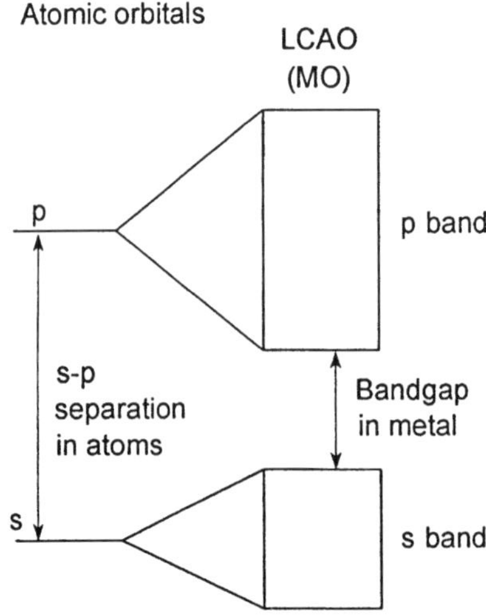

Figure 3.11. Overlap of *s* orbitals gives rise to an *s* band, and overlap of *p* orbitals gives rise to a *p* band. The *s* and *p* orbitals of the atoms can be so widely spaced that there is a bandgap. In many cases the separation is less and the bands overlap.

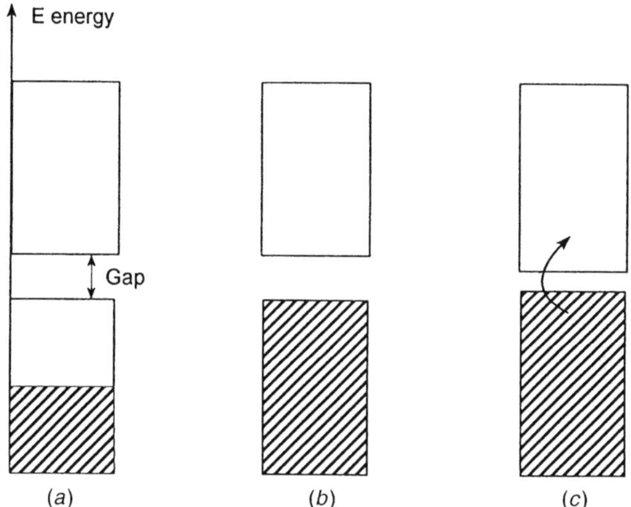

Figure 3.12. Schematic electron occupancy of allowed energy bands for a metal (*a*), an insulator (*b*), and a semiconductor (*c*). The shaded areas indicate the regions filled with electrons. The vertical extent of the boxes indicates the energy regions allowed.

Next, we consider the electronic structure of a metal formed from atoms each contributing two electrons. We have seen that overlap of *s* orbitals in N atoms produces N molecular orbitals and that each orbital can accommodate two electrons. The maximum number of electrons that can be placed in N orbitals is $2N$. When each atom contributes two electrons, there are $2N$ electrons to be placed in N molecular orbitals. Thus, when each atom contributes two electrons, the band is full and the material is an insulator (Fig. 3.12*b*). The major success of band theory rests on the explanation of the three types of electrical conductors (Fig. 3.12).

3.4. ATOMIC STRUCTURE OF SURFACES

The term *surface of a metal* usually means the top layer of atoms (ions). However, in this book the term *surface* means the top few (two or three) atomic layers of a metal. Surfaces can be divided into ideal and real. *Ideal surfaces* exhibit no lattice defects (vacancies, impurities, grain boundaries, dislocations, etc.). *Real surfaces* have all types of defects. For example, the density of metal surface atoms is about $10^{15}\,\mathrm{cm}^{-2}$ and the density of dislocations is on the order of magnitude $10^{8}\,\mathrm{cm}^{-2}$.

Ideal Surfaces. A model of an ideal atomically smooth (100) surface of a face-centered cubic (fcc) lattice is shown in Figure 3.13. If the surface differs only slightly in orientation from one that is atomically smooth, it will consist of flat portions called terraces and atomic steps or *ledges*. Such a surface is called *vicinal*. The steps on a vicinal surface can be completely straight (Fig. 3.13*a*) or they may have kinks (Fig. 3.13*b*).

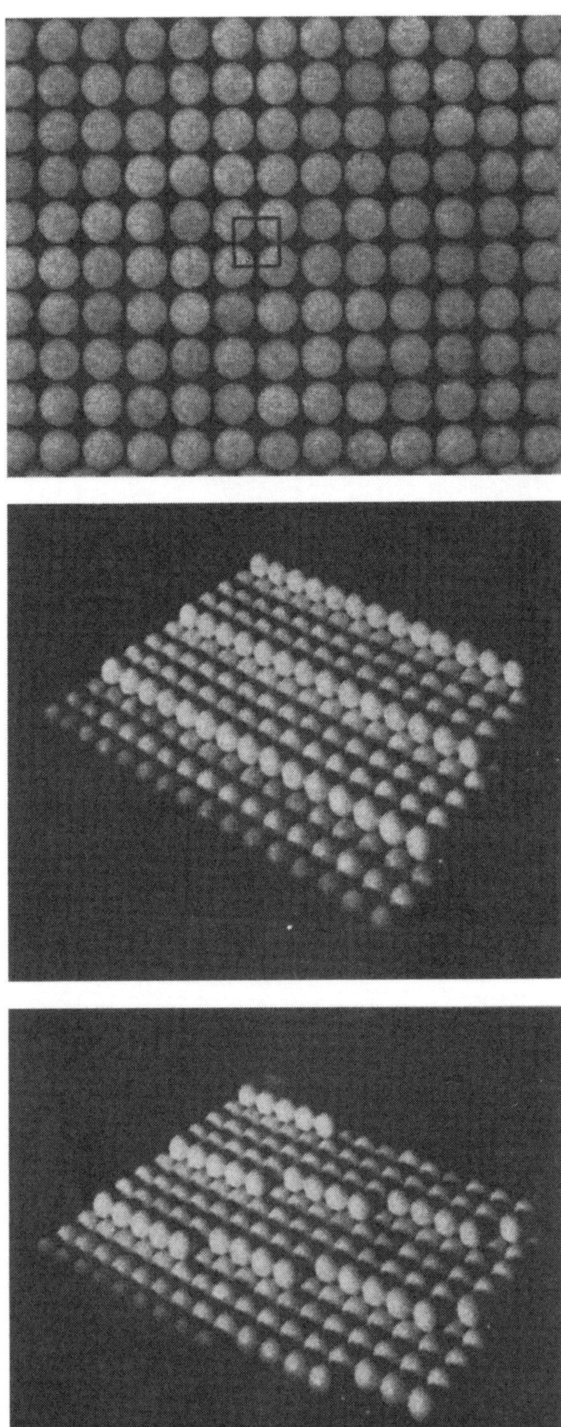

(a)

(b)

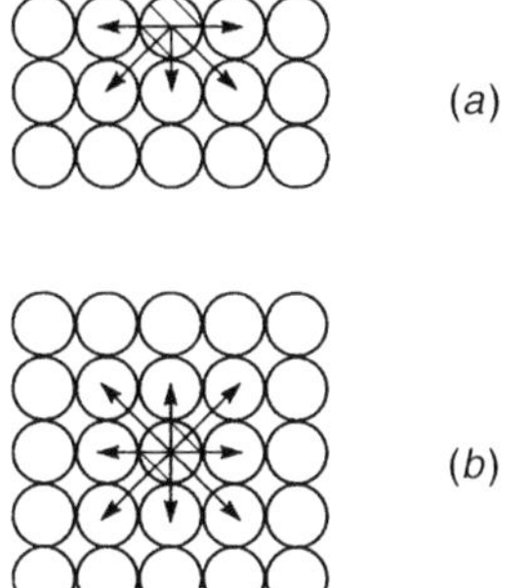

Figure 3.14. Schematic presentation of binding forces on an atom in the surface (*a*) and in the bulk (*b*).

Real Surfaces. The atomic arrangement at the surface can deviate from the bulk arrangement. The simplest deviation (disturbance) can be caused by the absence of the bonding forces of nearest neighbors on one side of the surface atoms. An atom on the surface is joined by metallic bonding forces to other atoms in the same plane and the plane below it. Thus, an atom in the surface has fewer nearest neighbors than does an atom in the bulk. Since surface atoms have matter on one side and not on the other, the electron distribution around surface atoms is unsymmetric with respect to the positive ions. The result of this imbalance in the binding force is a net resultant force on each surface atom acting toward the bulk (Fig. 3.14). A consequence of this imbalance of forces (this disturbance) is a new equilibrium position for atoms in the two or three last surface layers of metal. The simplest change toward a new equilibrium is the *surface relaxation of the lattice*, in which the distance (separation) between surface planes is increased over the corresponding distance in the bulk metal (Fig. 3.15b). However, the structure of the surface is unchanged. The more complex change in reaching a new equilibrium position for atoms in the surface is *surface reconstruction* (Fig. 3.15c), in which atoms (ions) move from positions they would have in the bulk into a new surface structure that differs from the bulk. The surface reconstruction can be caused by interaction of the surface with a solution, which may modify the bond strength of surface atoms with their neighbors.

The structure of real surfaces differs from the structure of ideal surfaces by the surface roughness. Whereas an ideal surface is atomically smooth, a real surface has defects, steps, kinks, vacancies, and clusters of adatoms (Fig. 3.16).

Surface Defects. Dislocations that exist in the bulk of crystal can extend onto the surface. Dislocation density N_d is defined as the number of dislocations that cut through a unit area. N_d in metals is usually on the order of $10^8 \, \text{cm}^{-2}$. This dislocation density can be reduced by annealing. In a well-annealed crystal, N_d is in the range 10^4 to $10^6 \, \text{cm}^{-2}$. Screw dislocation free surfaces (about 0.01 mm^2) can be produced by

Figure 3.13. *Top*: Model of an ideal (100) surface of a face-centered crystal (fcc) lattice. *Center and bottom*: Model of a vicinal surface of an fcc cut at 12° to the (100) plane: (*a*) with straight monatomic steps and (*b*) monatomic steps with kinks along the steps. (From Ref. 11, with permission from Pergamon Press.)

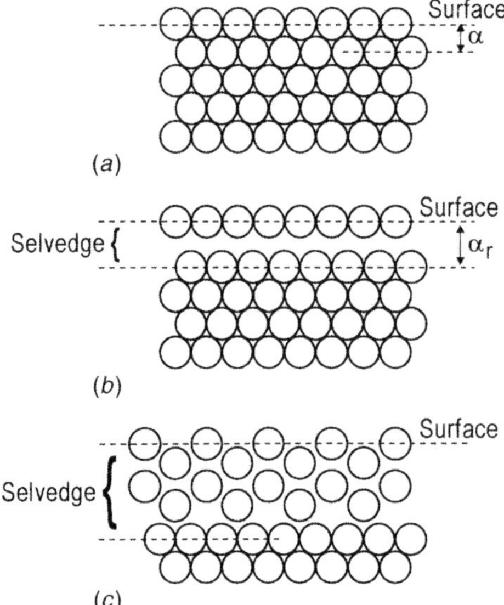

Figure 3.15. Rearrangement of atomic positions at a solid surface: (*a*) the bulk exposed plane; (*b*) relaxation of the surface plane outward; (*c*) reconstruction (hypothetical) of the outer four atomic planes. (From Ref. 12, with permission from Oxford University Press.)

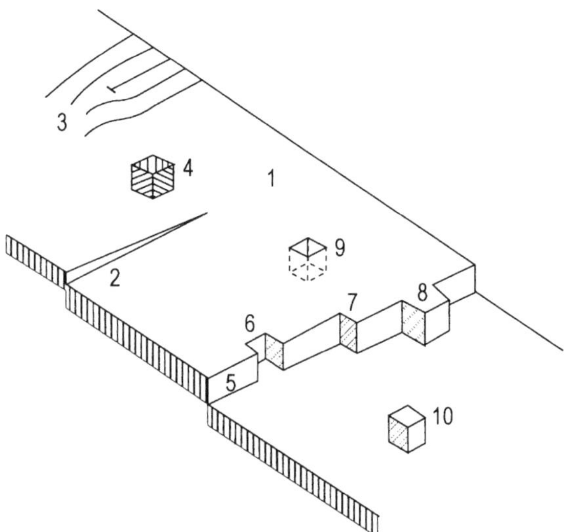

Figure 3.16. Some simple defects found on a low-index crystal face: 1, the perfect flat face, a terrace; 2, an emerging screw dislocation; 3, the intersection of an edge dislocation with the terrace; 4, an impurity adsorbed atom; 5, a monatomic step in the surface, a ledge; 6, a vacancy in the ledge; 7, a kink, a step in the ledge; 8 an adatom of the same type as the bulk atoms; 9, a vacancy in the terrace; 10, an adatom on the terrace. (From Ref. 12, with permission from Oxford University Press.)

electrodeposition (19). Dislocation–surface intersections (Fig. 3.4) play a significant role in surface processes such as adsorption, nucleation, and crystal growth.

Steps. In a real crystal where dislocations are present, there are two types of steps: the step that begins and ends on the boundary of the surface (Fig. 3.13*a*), and the step that starts on the surface and terminates on a boundary (Fig. 3.4). If a step starts on a surface, this is a place where a screw dislocation meets the surface. At 0 K, steps tend to be straight, but as the temperature is raised ($T > 0$ K), step roughness develops and the structure of the step includes a number of kinks, adsorbed atoms (adatoms or adions), and vacancies (Fig. 3.16). Steps can be of monatomic height or, as in the case of a real crystal surface, polyatomic height.

3.5. ELECTRONIC STRUCTURE OF SURFACES

We have shown in Section 3.3 that solutions of the Schrödinger equation for bulk electrons in an infinite periodic crystal represent the energy spectrum divided into bands of allowed and forbidden regions (bandgaps). Introduction of the metal surface into the problem of the electronic structure of metals changes the boundary conditions for the Schrödinger equation and results in a new solution of this equation. These new solutions represent surface states that are localized at the surface and can have energies within the bandgap of the band structure. If there is one surface state per surface atom, there are about 10^{15} states per square centimeter (the density of surface atoms is about 10^{15} atoms/cm^2). These surface states are distributed into surface bands. Another result of introduction of a surface is a change in distribution of valence electrons so that there is an outward spread of charge, which results in a positive charge inside the surface and a negative charge outside the surface. The distortion creates a surface double layer (Fig. 3.17).

3.6. ATOMIC PROCESSES AT SURFACES

The most important processes at surfaces are adsorption and surface diffusion. Here we mention only that adsorbed molecules can be bound to the metal by covalent bonding involving metal *d* orbitals. One example is given in Figure 3.18. Electron

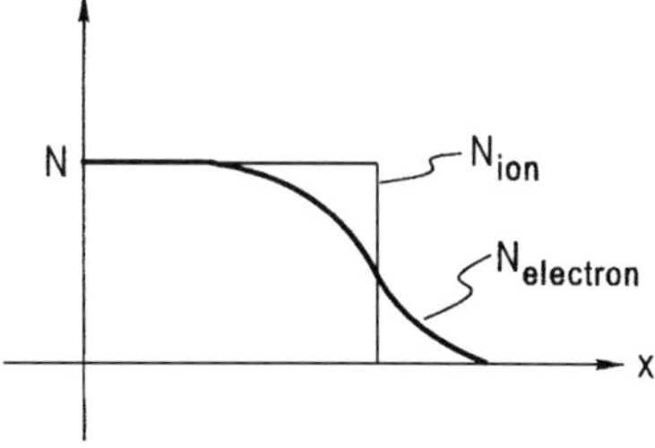

Figure 3.17. Jelium model of a metal surface. The ion density N_{ion} terminates abruptly at the surface, but the electron density $N_{electron}$ extends beyond it. The net charge density $N_{ion} - N_{electron}$ gives a dipole layer.

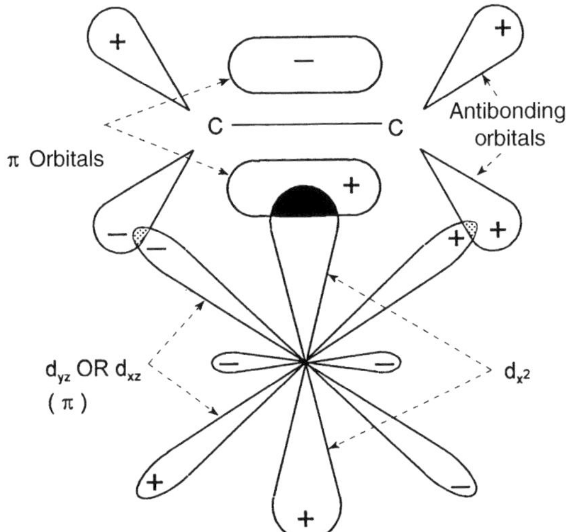

Figure 3.18. π-adsorbed ethylene on the (100) face of nickel. (From Ref. 8, with permission from Academic Press.)

tails ("spillover" electrons, Fig. 3.17) can also be involved in bonding of adsorbed molecules (atoms). Adsorption is discussed further in Chapter 10.

REFERENCES AND FURTHER READING

1. A. Sommerfeld, *Z. Phys.* **47**, 1 (1928).
2. J. Bardeen, *Phys. Rev.* **49**, 653 (1936).
3. J. Frenkel, *J. Phys.* **9**, 392 (1945).
4. L. Pauling, *The Nature of the Chemical Bond*, Cornell University Press, Ithaca, NY, 1948.
5. N. F. Mott and H. Jones, *Theory of the Properties of Metals and Alloys*, Dover, New York, 1958.
6. T. L. Hill, *An Introduction to Statistical Thermodynamics*, Addison-Wesley, Reading, MA, 1960.
7. C. Kittel, *Quantum Theory of Solids*, 2nd rev. ed., Wiley, New York, 1987.
8. A. Clark, *The Theory of Adsorption and Catalysis*, Academic Press, New York, 1970.
9. N. D. Lang and W. Kohn, *Phys. Rev.* **B1**, 4555 (1970).
10. N. D. Lang and W. Kohn, *Phys. Rev.* **B3**, 1215 (1971).
11. J. M. Blakely, *Introduction to the Properties of Crystal Surfaces*, Pergamon Press, Elmsford, NY, 1973.
12. M. Prutton, *Surface Physics*, Oxford University Press, Oxford, 1975.
13. J. Goodisman, *Electrochemistry: Theoretical Foundations*, Wiley, New York, 1987.
14. L. Solymar and D. Walsh, *Lectures on the Electrical Properties of Materials*, Oxford University Press, Oxford, 1988.
15. H. M. Rosenberg, *The Solid State*, Oxford University Press, Oxford, 1988.

16. W. A. Harrison, *Electronic Structure and the Properties of Solids*, Dover, New York, 1989.

17. J. E. Inglesfield, in *Interaction of Atoms and Molecules with Solid Surfaces*, V. Bortolani, N. H. March, and M. P. Tosi, eds., Plenum Press, New York, 1990.

18. C. Kittel, *Introduction to Solid State Physics*, 7th ed., Wiley, New York, 1996.

19. E. Budevski, G. Staikov, and W. J. Lorenz, *Electrochemical Phase Formation*, VCH, Weinheim, Germany, 1996, p. 203.

PROBLEMS

3.1. Copper and gold, among a number of metals, crystallize into a face-centered cubic lattice (fcc). The unit cell edge is of length $a = 3.62 \times 10^{-8}$ cm for Cu and 4.08×10^{-8} cm for Au. Calculate (a) the radius r of each of these atoms in the unit cell, and (b) the density ρ of Cu and Au in g/cm^3.

[*Hints*: (a) The face diagonal in the fcc lattice is equal to $4r$. (b) $\rho =$ by definition, (weight of atoms in unit cell)/(volume of unit cell), or

$$\rho = \frac{\sum A/N}{V_c}$$

where A is the atomic weight (weight of 1 mol of the atoms), N is Avogadro's number, and V_c is the volume of the unit cell (cm^3).]

3.2. As stated in Problem 3.1, copper crystallizes into a face-centered cubic lattice with a unit cell edge length $a = 3.62$ Å. Calculate the number of Cu atoms per cm^2 exposed on each surface of the (100), (110), and (111) planes.

3.3. The *packing fraction* (PF) is defined as the ratio (volume of atoms in the unit cell)/(volume of the unit cell).

(a) Calculate the general value of the PF for a face-centered cubic (fcc) crystal.

(b) Calculate PF for Cu, Ni, and Au using the following data:

	Unit Cell Side Length, a (Å)	Radius of Atom (Ion), r (Å)
Cu	3.62	1.28
Ni	3.52	1.24
Au	4.08	1.44

(c) Compare the PF calculated for individual metals with the general PF result for the fcc crystal structure. Comment on the validity of the general formula.

3.4. The number of vacancies per cubic centimeter is a very strong function of temperature and is given by

$$n_v = N \exp\left(\frac{E_v}{RT}\right)$$

where N is the number of atoms per cm^3, E_v is the energy of formation of 1 mol of vacancies, and R and T have their usual meaning. In deriving this equation it was taken into account that N is far larger than n_v and thus for the total number of sites in a crystal, $N + n_v \approx N$. E_v is in the range 20,000 to 40,000 cal/mol. Thus, in your calculations assume that $E_v = 25,000$ cal/mol and that the unit cell of Cu has an edge length of 3.62 Å.

(a) Calculate the number of vacancies per cubic centimeter (n_v) in the face-centered cubic lattice of Cu at 25, 100, 700, and 1000°C.

(b) Calculate the N/n_v ratio and explain its meaning.

4

Metal–Solution Interphase

4.1. INTRODUCTION

In Chapters 2 and 3 we have described basic structural properties of the components of an interphase. In Chapter 2 we have shown that water molecules form clusters and that ions in a water solution are hydrated. Each ion in an ionic solution is surrounded predominantly by ions of opposite charge. In Chapter 3 we have shown that a metal is composed of positive ions distributed on crystal lattice points and surrounded by a free-electron "gas" which extends outside the ionic lattice to form a surface dipole layer.

In Section 4.2 we describe what happens when these two phases come in contact and the nature of the structure of the interphase between the metal and the ionic solution (Fig. 4.1). Figure 4.1b defines an interphase as a region between two phases that has a different composition from that of bulk phases (here, bulk metal and bulk solution).

We show that the electric field in the metal–solution interphase is very high (e.g., 10^6 or 10^7 V/cm). The importance of understanding the structure of the metal–solution interphase stems from the fact that the electrodeposition processes occur in this very thin region, where there is a very high electric field. Thus, the basic characteristics of the electrodeposition processes are that they proceed in a region of high electric field and that this field can be controlled by an external power source. In Chapter 6 we show how the rate of deposition varies with the potential and structure of the double layer.

4.2. FORMATION OF METAL–SOLUTION INTERPHASE

Charging of Interphase. Let us consider a case where a metal M is immersed in the aqueous solution of its salt, MA. Both phases, metal and the ionic solution, contain M^+ ions, as discussed earlier. At the metal–solution interface (physical boundary) there will be an exchange of metal ions M^+ between the two phases (Fig. 4.2).

Fundamentals of Electrochemical Deposition, Second Edition.
By Milan Paunovic and Mordechay Schlesinger
Copyright © 2006 John Wiley & Sons, Inc.

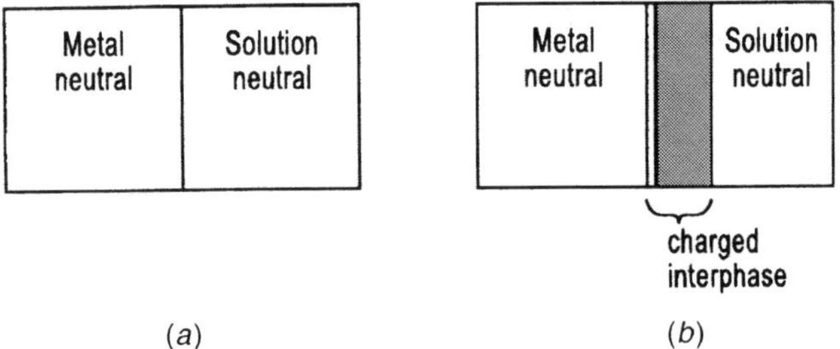

Figure 4.1. Two phases in contact: (*a*) at $t = 0$, moment of contact; (*b*) at equilibrium.

Some M^+ ions from the crystal lattice enter the solution, and some ions from the solution enter the crystal lattice. Let us assume that conditions are such that more M^+ ions leave than enter the crystal lattice. In this case there is an excess of electrons on the metal and the metal acquires negative charge, q_M^- (charge on the metal per unit area). In response to the charging of the metal side of the interface, there is a rearrangement of charges on the solution side of the interface. The negative charge on the metal attracts positively charged M^+ ions from the solution and repels negatively charged A^- ions. The result of this is an excess of positive M^+ ions in the solution in the vicinity of the metal interface. If the number per square centimeter of ionic species i in the bulk of solution is n_i^b and the number per square centimeter of these species in the interphase is n_i, the excess charge of ionic species in the interphase is

$$\Delta n_i = n_i - n_i^b$$

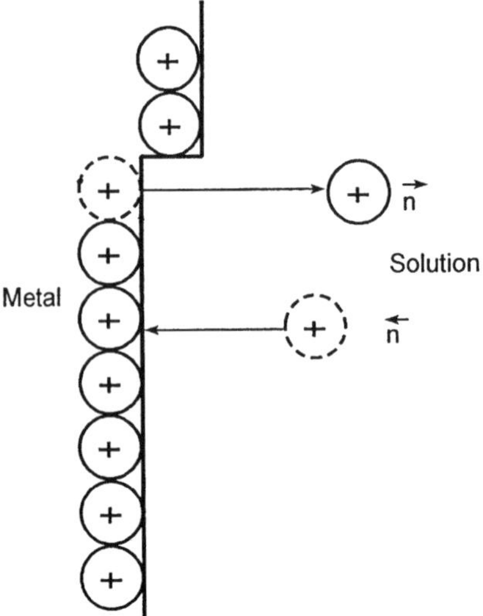

Figure 4.2. Formation of metal–solution interphase; equilibrium state: $\vec{n} = \overleftarrow{n}$.

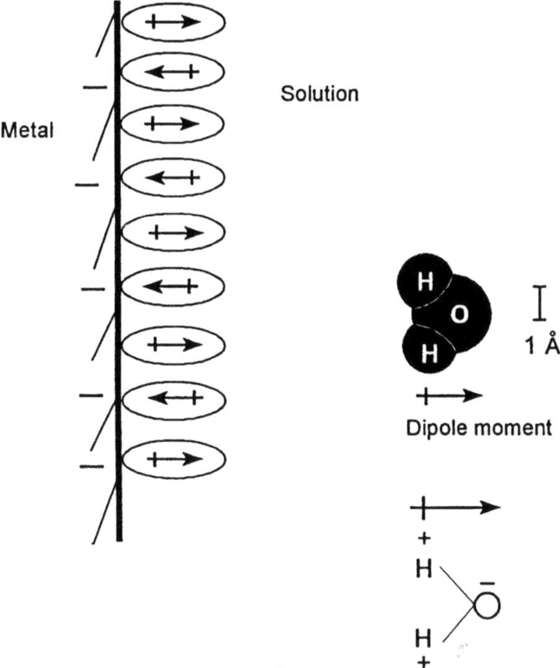

Figure 4.3. Structure of water in the interphase. At a negatively charged electrode, there is an excess of water dipoles with their positive hydrogen ends oriented toward the metal.

Thus, in this case the solution side of the interphase acquires opposite and equal charge, q_s^+ (the charge per unit area on the solution side of the interphase). At equilibrium the interphase region is neutral:

$$q_M = -q_s$$

The next question concerns how these excess charges are distributed on the metal and solution sides of the interphase. We discuss these topics in the next four sections. Four models of charge distribution in the solution side of the interphase are discussed: the Helmholtz, Gouy–Chapman, Stern, and Grahame models.

Water Structure at the Interphase. The presence of excess charge on a metal produces at least two effects: ion redistribution and reorientation of water dipoles in the solution. Thus, in the vicinity of the charged metal the structure of water is changed because of the presence of the electric field in the interphase (Fig. 4.3).

4.3. HELMHOLTZ COMPACT DOUBLE-LAYER MODEL

The simplest model of the structure of the metal–solution interphase is the Helmholtz compact double-layer model (1879). According to this model, all the excess charge on the solution side of the interphase, q_s, is lined up in the same plane at a fixed distance away from the electrode, the Helmholtz plane (HP; Fig. 4.4). This fixed distance x_{HP} is determined by the hydration sphere of the ions. It is defined as the plane of the centers of the hydrated ions. All excess charge on the metal, q_M, is located at the metal surface.

Thus, according to this model, the interphase consists of two equal and opposite layers of charges, one on the metal (q_M) and the other in solution (q_s). This pair of charged layers, called the *double layer*, is equivalent to a parallel-plate capacitor (Fig. 4.5). The variation of potential in the double layer with distance from the electrode is linear (Fig. 4.4). A parallel-plate condenser has capacitance per unit area given by the equation

$$c = \frac{\epsilon}{4\pi d} \tag{4.1}$$

where ϵ is the dielectric constant of the material (the dielectric) between the plates and d is the distance between the plates. For constant values of ϵ and d, the Helmholtz

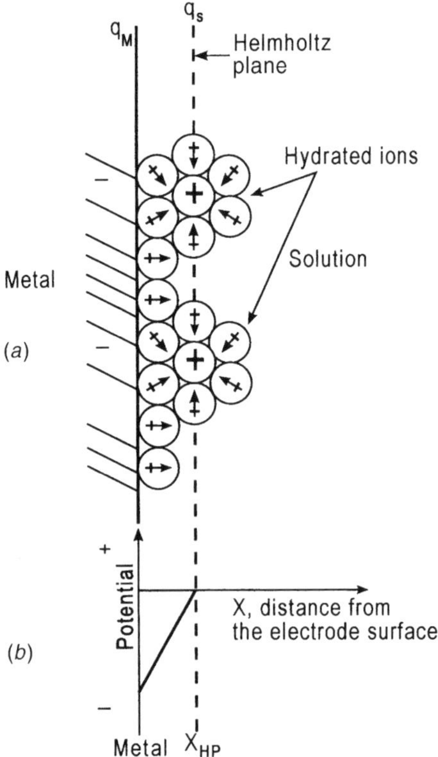

Figure 4.4. (*a*) Helmholtz model of a double layer: q_M, excess charge density on metal; q_s, excess charge density in solution, on HP; (*b*) linear variation of potential in the double layer with distance from the electrode.

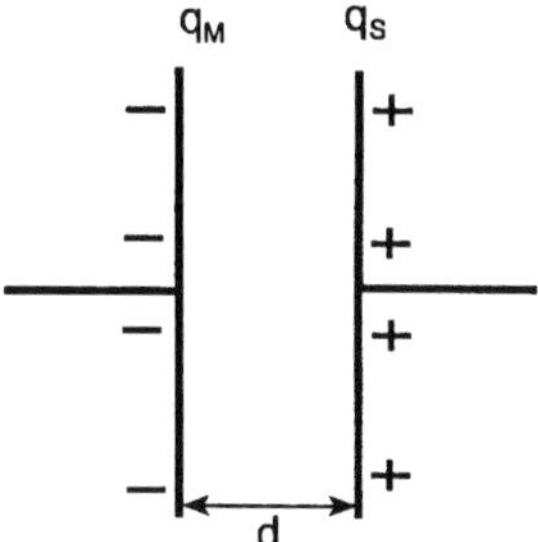

Figure 4.5. Electrical equivalent of the Helmholtz double layer: a parallel-plate capacitor.

model predicts a potential-independent capacitance. This is in contradiction with experiment. Experiments show that the double-layer capacitance is a function of potential. Thus, the interphase does not behave as a simple double layer. A new (improved) model is necessary.

4.4. GOUY–CHAPMAN DIFFUSE-CHARGE MODEL

In the Helmholtz model the excess charges in the solution were restricted to a single plane close to the metal. Gouy (2) and Chapman (3) independently proposed a new model that removed this restriction and allowed for the statistical potential-dependent distribution of ions in the solution side of the double layer. They assumed that this distribution obeys Boltzmann's distribution law and that ions can be modeled as point charges. The Gouy–Chapman model is illustrated in Figure 4.6. For a double layer in which the metal has a positive charge and the electrolyte solution consists of two types of ions of equal and opposite charge ($+z$ and $-z$), the number of positive ions per unit volume, at a distance x from the electrode, is given as

$$n^+(x) = n^+(b)\exp\left(-\frac{ze\psi(x)}{k_\mathrm{B}T}\right) \tag{4.2}$$

and the number of negative ions per unit volume is given by

$$n^-(x) = n^-(b)\exp\left(\frac{ze\psi(x)}{k_\mathrm{B}T}\right) \tag{4.3}$$

where $n(b)$ is the number of corresponding ions per unit volume in the bulk of the solution, $\psi(x)$ is the local potential at the distance x, and other symbols have their usual meaning: e, the charge of the electron; k_B, the Boltzmann constant; and T, temperature (Kelvin).

Equation (4.2) shows to what extent positive ions are repelled from the surface, producing a deficit of ($+$) ions, and Eq. (4.3) shows to what extent negative ions are

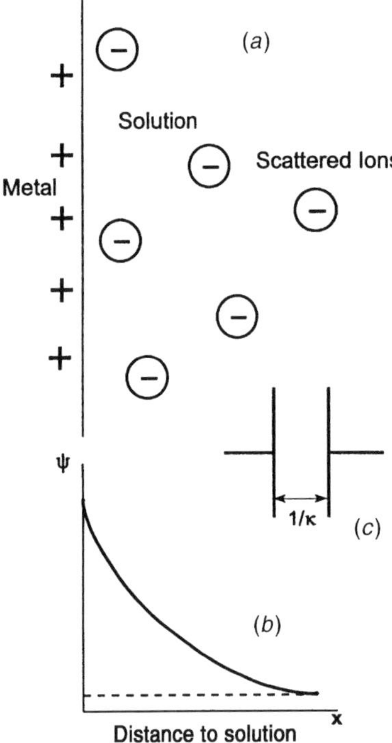

Figure 4.6. Gouy–Chapman model: (*a*) model; (*b*) variation of the potential with distance from the electrode; (*c*) equivalent capacitor.

attracted to the surface, producing an excess of $(-)$ ions. The net charge per unit volume at the point x, $q(x)$ in the double-layer region, is given by

$$q(x) = ze[n^-(x) - n^+(x)] \tag{4.4}$$

For the case shown in Figure 4.6, $n^-(x) > n^+(x)$, and there is a net negative charge on the solution side of the interphase. But as a whole, the interphase is neutral since $q_M = -q_s$. In the bulk, far away from the surface,

$$n^+(b) = n^-(b) \tag{4.5}$$

and there is no net charge in the solution.

This breakdown of electroneutrality in the solution, in the vicinity of the electrode, is a fundamental characteristic of the double-layer region. The next question is how far this double-layer region (interphase) extends out from the electrode into the solution. This question can be answered on the basis of analysis of the potential variation (distribution) in the double layer.

Using Eqs. (4.2)–(4.4) and the one-dimensional Poisson equation

$$\frac{\partial^2 \psi_x}{\partial x^2} = -\frac{4\pi \rho_x}{\epsilon} \tag{4.6}$$

for the relationship between the potential ψ and the charge distribution ρ_x, Gouy and Chapman derived Eq. (4.7), which describes the variation of potential in the double layer in the direction perpendicular to the electrode (x, Fig. 4.6):

$$\psi(x) = \psi(0)\exp(-\kappa x) \tag{4.7}$$

where $\psi(x)$ is the potential at a point x, $\psi(0)$ the potential at $x = 0$, and κ is given by

$$\kappa = \left[\frac{4\pi e^2}{\epsilon k_B T}\, n(b)\right]^{1/2} \tag{4.8}$$

for a uni-univalent electrolyte and bulk concentration $n(b)$.

The basic characteristics of Eq. (4.7) are: (1) the potential $\psi(x)$ decays exponentially from the electrode into the solution (Fig. 4.6); and (2) an increase in the solution concentration $n(b)$ results in a faster potential decay, since κ increases with an increase in $n(b)$ (Fig. 4.7). For example, when $x = 1/\kappa$, the potential $\psi(x) = \psi(0)\exp(-1) = 0.37$, or 37% of $\psi(0)$; when $x = 3(1/\kappa)$, $\psi(x) = 0.050\,\psi(0)$, or 5% of $\psi(0)$.

These results of the Gouy–Chapman theory are illustrated in Figure 4.7, where it can be seen that a large potential change, strong field, is located across a distance $1/\kappa$, which will be designated d_{dl}. The distance d_{dl} is taken as the thickness of the double

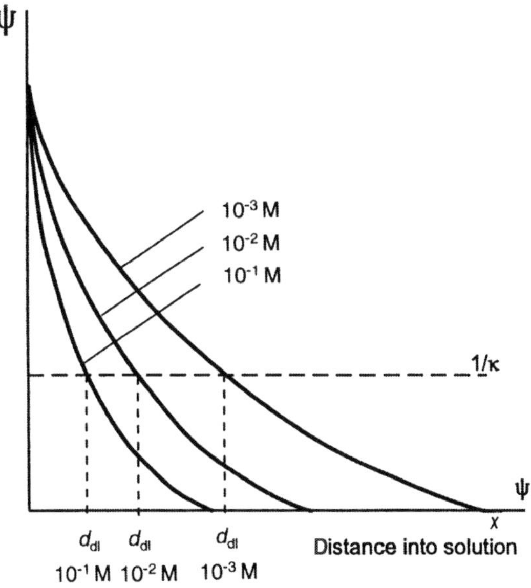

Figure 4.7. Variation of the potential with distance (in the diffuse double layer) for different concentrations and thicknesses of the double layer, d_{dl}.

layer in the Gouy–Chapman model. The thickness of the double layer, at 298 K and for $\epsilon = 78$, on the basis of Eq. (4.8) is given by

$$d_{\mathrm{dl}} = \frac{1}{3 \times 10^7 |z|(c_s)^{1/2}} \, \mathrm{cm} \tag{4.9}$$

where c_s is the bulk concentration in mol/L. For example, for a 1 M solution of 1 : 1 electrolyte, the thickness of the diffuse double layer is approximately 3 Å, which is approximately equal to the distance of closest approach of hydrated ions at the electrode. Thus, in concentrated solutions the diffuse double layer does not exist. On the other hand, if the solution concentration is 10^{-2} M, the thickness of the diffuse double layer is approximately 30 Å.

We can calculate the electric field (V/cm) in a double layer, since we have a value for the thickness of the double layer. If the thickness of the double layer is 10 Å and the potential difference is 0.1 V, the electric field in the double layer is $(0.1 \, \mathrm{V}/10 \times 10^{-8}) = 1 \times 10^7 \, \mathrm{V/cm}$.

The Gouy–Chapman theory was tested experimentally on the basis of the double-layer capacity measurements. This theory predicts a parabolic capacitance–potential relationship and a square-root dependence on concentration at constant ϵ and T:

$$C = A\sqrt{n}(b) \cosh \frac{e\psi_{\mathrm{M}}}{k_{\mathrm{B}}T} \tag{4.10}$$

where $\cosh x = \frac{1}{2}[\exp(x) + \exp(-x)]$ and A is a constant.

Variation of the double-layer capacity with applied potential according to the Gouy–Chapman theory is shown in Figure 4.8. Equation (4.10) includes the approximation $\psi_{\mathrm{M}} = \psi(x = 0)$, which is in harmony with the basic assumption of this

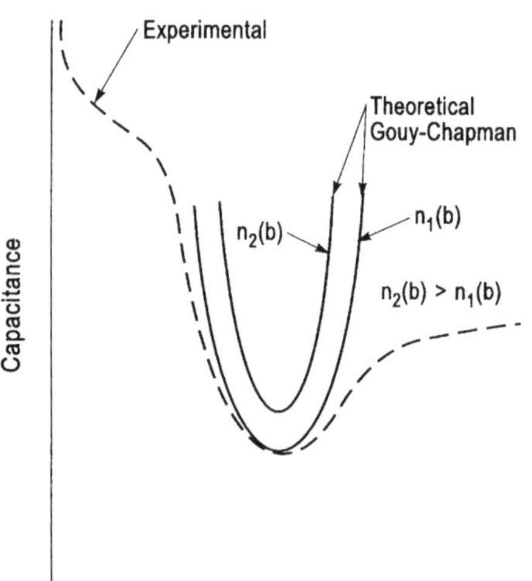

Figure 4.8. Variation of the double-layer capacity with applied potential according to the Gouy–Chapman theory.

model, which considers ions as point charges. Figure 4.8 indeed shows a parabolic relationship, but only in a narrow potential range.

The potential dependence of C is the basic improvement of the model, in comparison with the Helmholtz model, which predicts potential-independent capacity. However, a comparison of experimental data and values calculated on the basis of Eq. (4.10) shows that the function $C = f(\psi)$ behaves according to the Gouy–Chapman model in very dilute solutions and at potentials near the minimum (Fig. 4.8). In concentrated solutions, on the other hand, and at potentials farther away from the minimum, the theory is in disagreement with experimental results. Once again, a new theory is called for.

4.5. STERN MODEL

The Stern (4) model is a combination of the Helmholtz fixed (compact) layer and the Gouy–Chapman diffuse layer model. According to the Stern model, some ions of excess charges are fixed, restricted to a single plane close to the metal, the Helmholtz excess charge q_H, and others are statistically distributed into the solution, the Gouy–Chapman excess charge q_{GC}. Thus, in this model the double layer is divided into two regions: the compact and the diffuse double layers:

$$q_s = q_H + q_{GC} \tag{4.11}$$

The compact double layer extends from the electrode to the plane of the fixed charges at a distance $x = x_H$ from the electrode. The diffuse double layer extends from the distance x_H to the bulk of the solution. This is shown schematically in Figure 4.9.

The plane at the distance x_H from the electrode is called the *Helmholtz plane* (HP) or the *plane of closest approach*. The dividing HP is taken to be the locus of centers of fixed hydrated ions (Fig. 4.9). According to Stern, ions cannot come closer to the electrode than the plane of closest approach (HP). This postulate eliminates the point-charge approximation of the Gouy–Chapman theory. The separation of the interphase into two regions is equivalent to separation of the total double-layer capacitance as due to two contributions: C_H, the Helmholtz capacity, and the C_{GC}, the Gouy–Chapman capacity. Thus, the interphase according to the Stern model is equivalent to two capacitors in series (Fig. 4.9). The total capacitance of the interphase C is related to the Helmholtz capacitance C_H and the Gouy–Chapman capacitance C_{GH} by the expression for the total capacity of two capacitors in series:

$$\frac{1}{C} = \frac{1}{C_H} + \frac{1}{C_{GC}} \tag{4.12}$$

C_H may be evaluated using Eq. (4.1) and C_{GC} using Eq. (4.10). For low solution concentration, C_{GC} is very small and $1/C_{GC} \gg 1/G_H$; then $1/C \simeq 1/C_{GC}$, or $C \simeq C_{GC}$. Thus, for low solution concentration the double-layer capacitance behaves as the

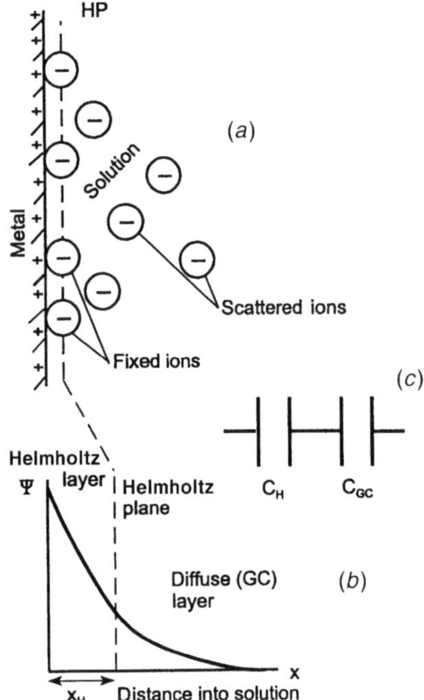

Figure 4.9. Stern model: (*a*) the model; (*b*) variation of the potential with distance from the electrode; (*c*) equivalent capacitor.

Gouy–Chapman capacitance. At high solution concentration, C_{GC} is large and $1/C_{GC} \ll 1/C_H$, resulting in $1/C \approx 1/C_H$, or $C \approx C_H$.

A comparison between the Stern theory and experiment was reported by Grahame (5), who found very close agreement between experimental and calculated double-layer capacities when one and the same solution is considered: for example, Figure 4.10. However, when solutions of different electrolytes are compared, the theory fails. Thus, once more, a new model is needed.

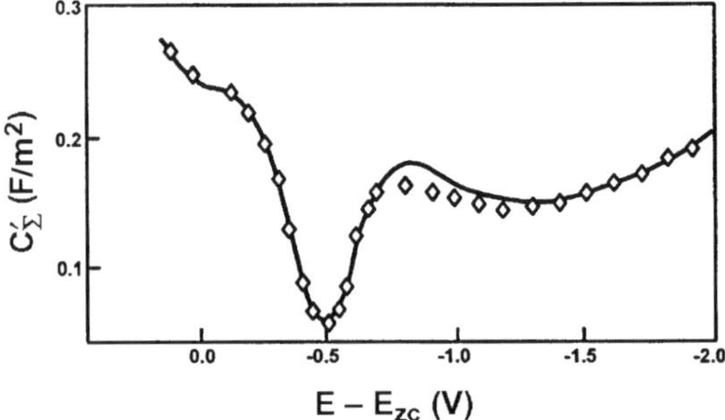

Figure 4.10. Comparison of theoretical and experimental capacitance for the $1 \times 10^{-3}\,\text{mol/L}$ of NaF. (From J. Albery, *Electrode Kinetics*, Oxford University Press, Oxford, 1975, with permission from Oxford University Press.)

4.6. GRAHAME TRIPLE-LAYER MODEL

In the Gouy–Chapman and Stern theories—equations showing the variation of potential with distance and the dependence of capacitance on the potential—ions in the interphase are characterized by one parameter only, the valence, z. Thus, according to these theories, all univalent (1 : 1) electrolytes should behave the same way. However, this is not what was observed experimentally. Solutions of different 1 : 1 electrolytes (e.g., NaCl, NaBr, NaI, KI) show species-specific behavior. To interpret this specific behavior, Grahame (5) proposed a new model of the interphase: the *triple-layer model*. The basic idea in the interpretation of the ion-specific behavior is that when attracted into the interphase, anions may become dehydrated and thus get closer to the electrode. Each anion undergoes this to a different extent. This difference in the degree of dehydration and the difference in the size of ions results in the specific behavior of the anions. Ions that are partially or fully dehydrated are in contact with the electrode. This contact adsorption of ions allows short-range forces (e.g., electric image forces) to act between the metal electrode and the ions, in addition to the conventional electrostatic coulombic forces. Models of the metal–solution interphase described in previous sections have been derived on the basis of the assumption that forces operating between the electrode and the ions in the interphase are the electrostatic coulombic forces.

Thus, Grahame modified Stern's model by introducing the *inner plane of closest approach* [inner Helmholtz plane (IHP)], which is located at a distance x_1 from the electrode (Fig. 4.11). The IHP is the plane of centers of partially or fully dehydrated

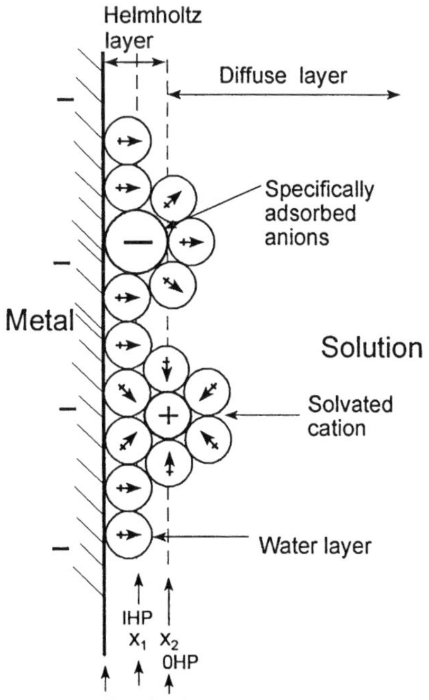

Figure 4.11. Triple-layer model (Grahame): IHP, inner Helmholtz plane; OHP, outer Helmholtz plane ($\rightarrow$, water dipole; +, positive end of the dipole).

specifically adsorbed ions. The closest approach of fully hydrated ions is at a distance x_2, called the *outer plane of closest approach* [outer Helmholtz plane (OHP)] (Fig. 4.11). The fully hydrated ions cannot approach the electrode closer than the OHP. The OHP is the plane of centers of hydrated ions. Grahame's model differs from previous models because it involves two distinct planes of closest approach, whereas only one such plane was postulated by Stern and Helmholtz. A comparison of theoretical and experimental data shows that this model represents an improvement.

The effect of the orientation of water dipoles on the electrode on the properties of the interphase was studied by Macdonald (6) and Mott and Watts-Tobin (7). Bockris et al. (8), in a modification of the Grahame model, considered adsorption of completely hydrated ions at the electrode with the water dipole layer present.

4.7. DETERMINATION OF THE DOUBLE LAYER'S CAPACITANCE

The simplest and most commonly used method for determination of a double layer's capacitance is the galvanostatic (constant-current) transient technique.

Galvanostatic Transient Technique. In the galvanostatic method a constant-current pulse is applied to the cell at equilibrium state and the resulting variation of the potential with time is recorded. The total galvanostatic current i_g is accounted for (1) by the double-layer charging, i_{dl}, and (2) by the electrode reaction (charge transfer), i_{ct}:

$$i_g = i_{dl} + i_{ct} \qquad (4.13)$$

The total current i_g is kept constant, but its two components, i_{dl} and i_{ct}, do change in time. At the beginning of the pulse (microseconds range) the total current i_g is utilized mainly for the double-layer charging and Eq. (4.13) reduces to

$$i_g \approx i_{dl} \qquad (4.14)$$

The double-layer charging current i_{dl} may be calculated from the definitions of capacitance,

$$C = \frac{dq}{dV} \qquad (4.15)$$

and that of current,

$$i = \frac{dq}{dt} \qquad (4.16)$$

where q is the charge on the capacitor.

In Section 4.3 it was shown that the electrical equivalent of the Helmholtz double layer is a parallel-plate capacitor (Fig. 4.5). In Section 4.5 (Fig. 4.9) it was shown that the Stern double-layer model is equivalent to two capacitors in series. From Eqs. (4.14) to (4.16) we have

$$C_{dl} = i_{dl}\,\frac{dt}{dV} = i_{dl}\left(\frac{dV}{dt}\right)^{-1} \tag{4.17}$$

The plot $V = f(t)$, in the microseconds range, is a straight line with slope dV/dt. Thus, the double-layer capacitance C_{dl} may be calculated by means of Eq. (4.17) using the slope (dV/dt) provided by the experimental data. One example of such calculations is presented in Problem 4.2. Galvanostatic transient technique is discussed in detail in Section 6.9.

REFERENCES AND FURTHER READING

1. H. L. von Helmholtz, *Wied. Ann.* **7**, 337 (1879).
2. G. Gouy, *J. Chim. Phys.* **9**, 457 (1910).
3. D. L. Chapman, *Philos. Mag.* **25**, 475 (1913).
4. O. Stern, *Z. Elektrochem.* **30**, 508 (1924).
5. D. C. Grahame, *Chem. Rev.* **41**, 441 (1947).
6. J. R. Macdonald, *J. Chem. Phys.* **22**, 1857 (1954).
7. N. F. Mott and R. J. Watts-Tobin, *Electrochim. Acta* **4**, 79 (1961).
8. J. O'M. Bockris, M. A. Devanathan, and K. Muller, *Proc. R. Soc. London* **274**, 55 (1963).
9. K. J. Vetter, *Electrochemical Kinetics: Theoretical and Experimental Aspects*, Academic Press, New York, 1967.
10. J. O'M. Bockris, A. K. N. Reddy, and M. E. Gamboa-Adelco, *Modern Electrochemistry*, Vol. 2, Plenum Press, New York, 2001.
11. J. O'M. Bockris and S. U. M. Khan, *Surface Electrochemistry*, Plenum Press, New York, 1993.
12. J. A. Greathouse and D. A. McQuarrie, *J. Phys. Chem.* **100**, 1847 (1996).
13. E. Mattson and J. O'M. Bockris, *Trans. Faraday Soc.* **55**, 1586 (1959).

PROBLEMS

4.1. The double-layer capacity of an electrode immersed in a $10^{-2}\,M$ aqueous solution is found to be $50\ \mu F/cm^2$. The potential difference at the interphase is 0.1 V. Assume that the dielectric constant ϵ in the double layer is 5.9.

    (a) Calculate the thickness of the double layer according to the Helmholtz model, d_{Hdl}.

    (b) Calculate the electric field between the layers in the double layer (V/cm) using the values calculated for d_{Hdl}.

(c) Calculate the areal charge density q (C/cm^2) on the electrode. Use the equivalency $1\,\mu F = 9 \times 10^5\,$esu.

4.2. The slope of the galvanostatic charging curve near the reversible potential $(d\eta/dt)_{\eta \to 0}$ for a copper electrode immersed in $1\,N$ CuSO$_4$ and $1\,N$ H$_2$SO$_4$ at 30°C was found to be $1.3 \times 10^3\,$V/s (13). The galvanostatic current density was $9.2 \times 10^{-2}\,$A/cm^{-2}. The slope of the galvanostatic transient was determined using the range 0 to 50 μs in the galvanostatic $V = f(t)$ curve.

(a) Draw the equivalent circuit for this galvanostatic experiment.

(b) Calculate the double-layer capacity C_{dl} using the experimental data given above.

4.3. (a) Calculate the thickness of the diffuse double layer, d_{Gdl}, for an electrode immersed in 0.01 M and 0.001 M aqueous solutions of a uni-univalent ($1:1$) electrolyte.

(b) Calculate the variation of the potential ψ with distance x in the diffuse double layer in a direction perpendicular to the electrode; calculate ψ for the following x-values: 10, 20, 30, 40, 50, and $100 \times 10^{-8}\,$cm and the following bulk concentrations $c(b)$: 0.01 and 0.001 M. Discuss your results in relation to the effect of concentration of the solutions on d_{Gdl} and $\psi(x)$. The temperature is 25°C, $\psi(0) = 25 \times 10^{-3}\,$V (25 mV), and assume that the average value of the dielectric constant ϵ of the solution in the diffuse double layer is 50.

4.4. Calculate the extent of charging of the capacitor C during the charging time t expressed in terms of the capacitive time constant τ ($\tau = RC$); take $t = \tau, 2\tau, 3\tau, 4\tau, 5\tau$, and 10τ. Express the results of charging in as a percentage with respect to the fully charged capacitor. Use the equation of charging a capacitor in an RC (resistance–capacitor) series circuit:

$$q_t = C\zeta \left[1 - \exp\left(-\frac{t}{RC} \right) \right]$$

where q is the charge on the plates of the capacitor, C is the capacitance, and ζ is the voltage imposed across the capacitor (the potential difference across the battery poles of the pole charging the capacitor). At $t = 0$ the term $\exp(-t/RC)$ is equal to 1 and $q_{t=0} = 0$. After a long time, $t \approx \infty$, the term $\exp(-t/RC)$ becomes zero and the equation represents the value for the full charge on the capacitor (i.e., $q = C\zeta$). The product RC has the dimensions of time and is called the *capacitive time constant* τ of the RC circuit, $RC = \tau$.

5

Equilibrium Electrode Potential

5.1. INTRODUCTION

Here we are interested in the potential difference across an interphase. Let us consider the interphase shown in Figure 5.1, where the potential of the solution is ϕ_S and that of the metal is ϕ_M. The potential difference across the interphase is $\Delta\phi(M,S) = \phi_M - \phi_S$. This potential difference cannot be measured directly since instruments that measure potential difference require two terminals and we have only one terminal: the metal M. Thus, to measure the potential difference of an interphase, one should connect it to another interphase and thus form an electrochemical cell. Potential difference across such an electrochemical cell can be measured. We discuss two types of electrode potentials: metal/metal-ion and redox potentials.

5.2. CELL VOLTAGE AND ELECTRODE POTENTIALS

Let us consider the general electrochemical cell shown in Figure 5.2. The potential difference across the electrochemical cell, denoted $\mathcal{E}$, is a measurable quantity called the *electromotive force* (EMF) of the cell. The potential difference $\mathcal{E}$ in Figure 5.2 is made up of four contributions since there are four phase boundaries in this cell: two metal–solution interphases and two metal–metal interfaces. The cell in Figure 5.2 can be represented schematically as Pt/M′/S/M/Pt.

Starting from the right-hand electrode in Figure 5.2 and proceeding clockwise, keeping the order of the symbols of substances the same as written in the schematic representation of the cell, one obtains

$$\mathcal{E} = \Delta\phi(\text{Pt},\text{M}) + \Delta\phi(\text{M},\text{S}) + \Delta\phi(\text{S},\text{M}') + \Delta\phi(\text{M}',\text{Pt})$$

Fundamentals of Electrochemical Deposition, Second Edition.
By Milan Paunovic and Mordechay Schlesinger
Copyright © 2006 John Wiley & Sons, Inc.

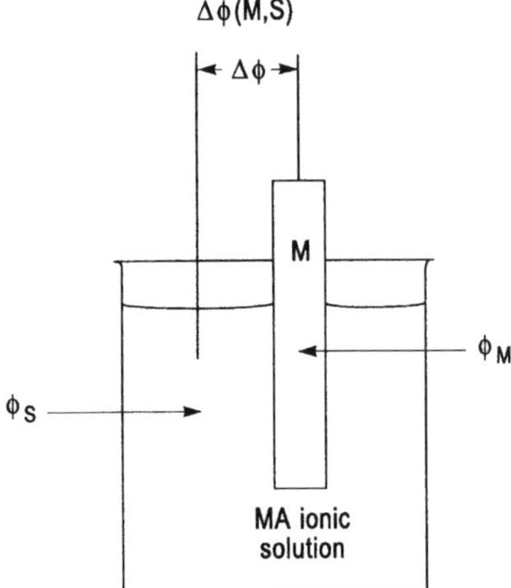

Figure 5.1. The potential difference across the interphase $\Delta\phi$ (M,S) is the difference $\phi_M - \phi_S$.

Since $\Delta\phi(S,M') = -\Delta\phi(M',S)$ and $\Delta\phi(M',Pt) = -\Delta\phi(Pt,M')$, the preceding equation can be written as the difference between two electrode potentials:

$$\mathcal{E} = \{\Delta\phi(Pt,M) + \Delta\phi(M,S)\} - \{\Delta\phi(M',S) + \Delta\phi(Pt,M')\}$$

or

$$\mathcal{E} = E_r - E_l$$

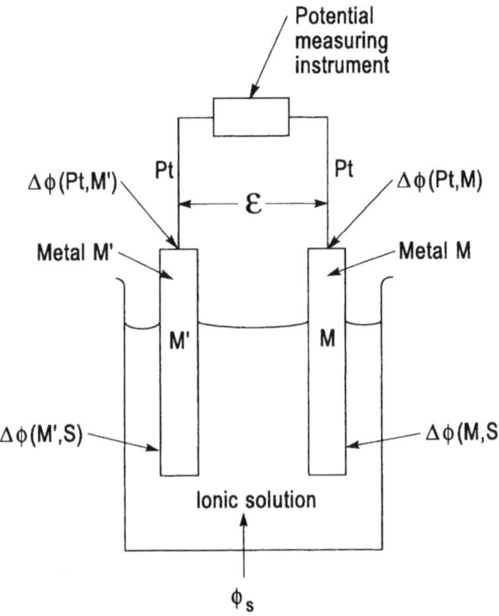

Figure 5.2. The potential difference across the electrochemical cell, $\mathcal{E}$, is the difference between the potential of the right-hand electrode E_r and the potential of the left-hand electrode E_l.

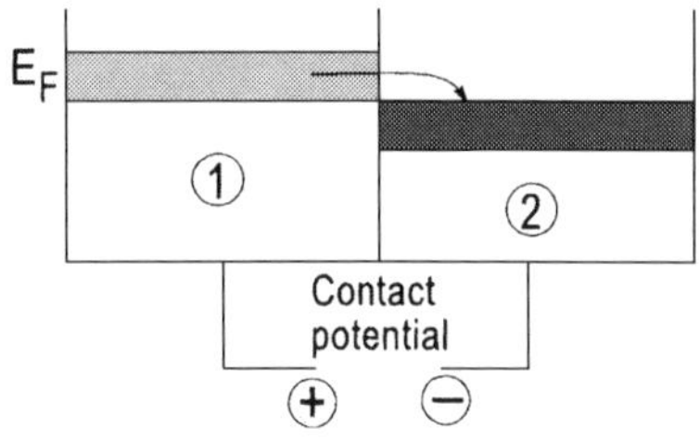

Figure 5.3. Contact potential difference between two dissimilar conductors; E_F, Fermi level.

where the right-hand electrode (Fig. 5.2) potential is

$$E_r = \Delta\phi(\text{Pt,M}) + \Delta\phi(\text{M,S})$$

and the left-hand electrode (Fig. 5.2) potential is

$$E_l = \Delta\phi(\text{Pt,M}') + \Delta\phi(\text{M}',\text{S})$$

If E_l is taken as a reference electrode and set arbitrarily to $E_l = 0$, then $\mathcal{E} = E_r$. Thus, in the relative scale of potential where $E_l = 0$, the electrode potential E_r is equal to the measured cell voltage $\mathcal{E}$, and we obtain

$$\mathcal{E} = E_r = \Delta\phi(\text{Pt,M}) + \Delta\phi(\text{M,S})$$

since $\mathcal{E} = E_r - E_l$ and $E_l = 0$.

The term $\Delta\phi(\text{Pt,M})$ appears in all measurements and thus does not influence the order of the measured electrode potentials. It is the potential difference that appears when two dissimilar conductors come into contact. Since the Fermi energies of two different metals are in general different, a flow of electrons occurs that tends to equalize the Fermi energies (i.e., their chemical potential). The Fermi level is either (1) the uppermost (the top) filled energy level in a partially occupied valence band of electrons in a solid, or (2) the boundary between the filled and the empty states in a band of electrons in a solid (Chapter 3). This electron flow charges up one conductor relative to the other and the contact potential difference results (Fig. 5.3).

5.3. CONCENTRATION DEPENDENCE OF EQUILIBRIUM CELL VOLTAGE: THE GENERAL NERNST EQUATION

The Nernst equation can be derived by considering a general cell reaction with A, B, ... reactants and M,N, ... products:

$$aA + bB + \cdots = mM + nN + \cdots \tag{5.1}$$

and applying to this reaction the following two equations for the free-energy change, ΔG.

Equation 1.   The free-energy change (ΔG) as a function of cell voltage ($\mathcal{E}$)

$$\Delta G = -zF\mathcal{E} \tag{5.2}$$

where z, F, and $\mathcal{E}$ are the number of electrons involved in the reaction, the Faraday number (96,500 C), and the cell voltage, respectively. Equation (5.2) represents work done by the cell in terms of $W = qV$, where W is the work done (ΔG), q is the charge (zF), and V is the voltage ($\mathcal{E}$).

Equation 2.   The free-energy change as a function of concentration (activity) of the reactants

$$\Delta G = \Delta G^0 + RT \ln Q \tag{5.3}$$

where

$$Q = \frac{a_M^m a_N^n, \ldots}{a_A^a a_B^b, \ldots} = \frac{\Pi\,(\text{products})}{\Pi\,(\text{reactants})} \tag{5.4}$$

R is the gas constant, T is the absolute temperature, ln is the natural logarithm, and Π represents the product of the concentrations (activities, a) raised to the power of their stoichiometric numbers.

For standard states, when the activities of the reactants and products equal 1, from Eqs. (5.2) and (5.3) ($\Delta G = \Delta G^0$, when $Q = 1$) the following equation holds:

$$\Delta G^0 = -zF\mathcal{E}^0 \tag{5.5}$$

where ΔG^0 is the standard free-energy change and $\mathcal{E}^0$ is the cell voltage for standard states of reactants and products.

Introducing Eqs. (5.2) and (5.5) into Eq. (5.3), one obtains

$$-zF\mathcal{E} = -zF\mathcal{E}^0 + RT \ln Q \tag{5.6}$$

Dividing by $-zF$ yields

$$\mathcal{E} = \mathcal{E}^0 - \frac{RT}{zF} \ln Q \tag{5.7}$$

or

$$\mathcal{E} = \mathcal{E}^0 + \frac{RT}{zF} \ln \frac{1}{Q} \tag{5.8}$$

Substituting the value of Q [Eq. (5.4)] into (5.8), one obtains

$$\mathcal{E} = \mathcal{E}^0 + \frac{RT}{zF} \ln \frac{\Pi[\text{reactants}]}{\Pi[\text{products}]} \tag{5.9}$$

Equation (5.9) is the general Nernst equation giving the concentration dependence of the equilibrium cell voltage. It will be used in Section 5.4 to derive the equilibrium electrode potential for metal/metal-ion and redox electrodes.

5.4. METAL/METAL-ION (M/M^{Z+}) POTENTIALS

A metal/metal-ion electrode consists of a metal immersed in a solution containing ions of the metal. The electrode potential of this electrode depends on the concentration (more exactly, the activity) of the metal ions M^{z+} in solution. An example is Cu immersed in a $CuSO_4$ solution, Cu/Cu^{2+}.

Mechanism of Formation of an Equilibrium Metal/Metal-Ion Potential. The basic mechanism of the formation of an M/M^{z+} equilibrium potential was discussed in Section 4.2 and shown in Figure 4.2 and is discussed further in Chapter 6.

Nernst Equation for the Concentration Dependence of Metal/Metal-Ion Potential. In the general case of a metal/metal-ion electrode, a metal M is in an equilibrium with its ions in the solution:

$$M^{z+} + ze \rightleftharpoons M \tag{5.10}$$

Reaction from left to right consumes electrons and is called *reduction*. Reaction from right to left liberates electrons and is called *oxidation*.

The potential of this electrode is defined (Section 5.2) as the voltage of the cell $Pt\,|\,H_2(1\text{ atm})\,|\,H^+(a = 1)\,|\,M^{z+}\,|\,M$, where the left-hand electrode, $E_l = 0$, is the normal hydrogen reference electrode (described in Section 5.6). In Chapter 6, we derive the Nernst equation on the basis of the electrochemical kinetics. Here we use a simplified approach and consider that Eq. (5.9) can be used to determine the potential E of the M/M^{z+} electrode as a function of the activity of the products and reactants in the equilibrium equation (5.10). Since in reaction (5.10) there are two reactants, M^{z+} and e, and only one product of reaction, M, Eq. (5.9) yields

$$E = E^0 + \frac{RT}{zF} \ln \frac{[M^{z+}][e]^z}{[M]} \tag{5.11}$$

where the brackets [ ], in general, signify activity of the species inside the brackets; when the concentration of solution is low, for example, 0.001 M or lower, activity can be replaced by the concentration in mol/L (the relationship between concentration and activity is discussed in Section 5.8). Since the activities of metal (M) and electrons (e) in the metal lattice both equal 1 by convention, Eq. (5.11) yields

$$E = E^0 + \frac{RT}{zF} \ln[\mathrm{M}^{z+}] \qquad (5.12)$$

Equation (5.12) is the Nernst equation for the variation of $\mathrm{M/M}^{z+}$ potential E with concentration M^{z+}. Converting the natural logarithm into a decimal logarithm gives us

$$E = E^0 + 2.303\,\frac{RT}{zF} \log[\mathrm{M}^{z+}] \qquad (5.13)$$

When the activity of M^{z+} in the solution is equal to 1,

$$E = E^0, \qquad [\mathrm{M}^{z+}] = 1 \qquad (5.14)$$

where E^0 is the relative standard electrode potential of the $\mathrm{M/M}^{z+}$ electrode. The quantity RT/F has the dimension of voltage and at 298 K (25°C) has the value $RT/F = 0.0257\,\mathrm{V}$ and the quantity $2.303(RT/F) = 0.0592\,\mathrm{V}$. With these values, Eq. (5.13) for 298 K (25°C) is

$$E = E^0 + \frac{0.0592}{z} \log[\mathrm{M}^{z+}] \qquad (5.15)$$

when $T = 298\,\mathrm{K}$ (25°C).

Example 5.1. Calculate the reversible electrode potential for Cu immersed in a $\mathrm{CuSO_4}$ solution having concentrations of 0.01, 0.001, and 0.0001 mol/L at 25°C, neglecting ion–ion interaction (using concentrations instead of activities). The standard electrode potential for $\mathrm{Cu/Cu}^{2+}$ electrode is 0.337 V.
   Using Eq. (5.15), for $z = 2$, $E^0 = 0.337\,\mathrm{V}$, and $[\mathrm{M}^{2+}] = 0.01$, one obtains

$$E = 0.337 + \frac{0.0592}{2} \log 0.01$$

Since $\log 0.01 = -2$,

$$E = 0.337 - 0.0592 = 0.278\,\mathrm{V}$$

Using the same procedure for $[\mathrm{Cu}^{2+}] = 0.001\,\mathrm{mol/L}$, one obtains $E = 0.248\,\mathrm{V}$, and for $[\mathrm{Cu}^{2+}] = 0.0001\,\mathrm{mol/L}$, one obtains $E = 0.219\,\mathrm{V}$. $\qquad\square$

This example illustrates a method for the determination of E^0 by plotting log a against E and extrapolating to $a = 1$.

5.5. RedOx POTENTIALS

RedOx electrode potentials are the result of an exchange of electrons between metal and electrolyte. In Section 5.4 we have shown that the metal/metal-ion electrode potentials are the result of an exchange of metal ions between metal and electrolyte. In the RedOx system the electrode must be made of an inert metal, usually platinum, for which there is no exchange of metal ions between metal and electrolyte. The electrode acts as a source or sink for electrons. The electrolyte in the RedOx system contains two substances: electron donors (electron-donating species) and electron acceptors (electron-accepting species). One example of a RedOx system is shown in Figure 5.4. In this case the electron donor is Fe^{2+}, the electron acceptor is Fe^{3+}, the electrode is Pt, and the electrode process is

$$Fe^{3+} + e \rightleftharpoons Fe^{2+}$$

This electrode reaction consists of changing the valence state of ions; in this example, Fe^{3+} is the oxidized form and Fe^{2+} is the reduced form.

In general, if the oxidized ions are designated Ox and the reduced ions Red, the general RedOx electrode reaction is

$$Ox + ze \rightleftharpoons Red \tag{5.16}$$

Reaction from left to the right consumes electrons and is called *reduction*. Reaction from right to the left liberates electrons and is called *oxidation*.

Complex RedOx reactions involve H^+ ions. For example,

$$MnO_4^- + 5e + 8H^+ \rightleftharpoons Mn^{2+} + 4H_2O \tag{5.17}$$

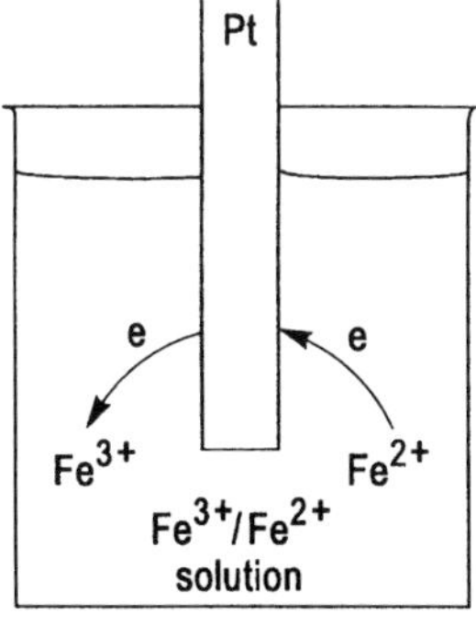

Figure 5.4. Fe^{3+}/Fe^{2+} RedOx system; RedOx reaction $Fe^{3+} + e \rightleftharpoons Fe^{2+}$.

Nernst Equation for Concentration Dependence of RedOx Potential.   Equation
(5.9) applied to the general RedOx electrode (5.16) yields

$$E = E^0 + \frac{RT}{zF} \ln \frac{[\text{Ox}]}{[\text{Red}]} \tag{5.18}$$

or in decimal logarithmic form,

$$E = E^0 + 2.303 \frac{RT}{zF} \log \frac{[\text{Ox}]}{[\text{Red}]} \tag{5.19}$$

When activities of the reactant (Ox) and the product (Red) are equal to 1, then

$$E = E^0, \qquad \text{when } [\text{Ox}] = 1, [\text{Red}] = 1 \tag{5.20}$$

where E^0 is the standard electrode potential of the RedOx electrode.

Example 5.2.   Consider a Red/Ox electrode consisting of a Pt electrode immersed
in a solution where Ox is 1×10^{-1} mol/L $KMnO_4$ and Red is 1×10^{-4} mol/L $MnSO_4$.
Calculate the Red/Ox potential of this electrode at 25°C for pH 2 and pH 4 assuming that
concentrations can be used instead of activities.

Red/Ox reaction of this electrode is given by Eq. (5.17). The standard electrode
potential of this electrode is 1.51 V. At 25°C, the potential of the electrode reaction
(5.17) is given by Eq. (5.19):

$$E = E^0 + \frac{0.0592}{5} \log \frac{[\text{MnO}_4^-][\text{H}^+]^8}{[\text{Mn}^{2+}]}$$
$$= E^0 + \frac{0.0592}{5} \log[\text{H}^+]^8 + \log \frac{[\text{MnO}_4^-]}{[\text{Mn}^{2+}]}$$

Since $\text{pH} = -\log[\text{H}^+]$ and $\log[\text{H}^+]^8 = -8\,\text{pH}$,

$$E = E^0 - \frac{8(0.0592)\text{pH}}{5} + \frac{0.0592}{5} \log \frac{[\text{MnO}_4^-]}{[\text{Mn}^{2+}]}$$

In this example, $[\text{MnO}_4^-] = 10^{-1}$ mol/L and $[\text{Mn}^{2+}] = 10^{-4}$ mol/L. Thus, $\log [\text{MnO}_4^-]/[\text{Mn}^{2+}] = \log(10^{-1}/10^{-4}) = \log 10^3 = 3$ and

$$E = E^0 - 0.0947\,\text{pH} + 0.0355$$
$$= 1.545 - 0.0947\,\text{pH}$$

For pH 2, $E = 1.545 - 0.189 = 1.356\,\text{V}$, and for pH $= 4$, $E = 1.545 - 0.379 = 1.166\,\text{V}$. Thus, in this case, the potential of the Red/Ox electrode depends on
the ratio [Ox/Red] and the pH. $\square$

5.6. MEASUREMENT OF EQUILIBRIUM ELECTRODE POTENTIALS

The equilibrium potential of an electrode (e.g., M/M^{z+}) is defined in Section 5.2 as the voltage of the cell, $Pt\,|\,H_2(1\ atm)\,|\,H^+(a = 1)\,\|\,M^{z+}\,|\,M$, where a is the activity. Three issues have to be resolved to measure this equilibrium electrode potential: (1) the selection of a reference electrode; (2) the coupling of the reference electrode with the electrode whose potential is being measured, in this case M/M^{z+}; and (3) the experimental method for the voltage measurement.

The basic requirement in selection of a reference electrode is that it should not change its potential during the measurement procedure. A reference electrode, connected to an electrode at equilibrium such as the M/M^{z+} electrode, make up an electrochemical cell with a characteristic cell voltage. If a voltage-measuring instrument (voltmeter) with an internal resistance R (the input impedance) is used to measure the potential difference between these two electrodes (Fig. 5.2), the instrument draws current according to Ohm's law, $I = V/R$. For example, if the cell voltage V is 0.5 V and the internal resistance of the voltmeter is $1000\,\Omega$, then $I = V/R = 0.5 \times 10^{-3}\,A$. Many electrodes will change their potential during measurement as a result of this high current. However, if the cell voltage V is 0.5 V and the internal resistance of the voltmeter is $1 \times 10^{10}\,\Omega$, then $I = V/R = 5 \times 10^{-11}\,A$. This very low current will not change the potential of a reference electrode. Thus, a voltage-measuring instrument should have high internal resistance and the reference electrode should be nonpolarizable. The ideally nonpolarizable electrode is always at the equilibrium potential. It resists changes in potential; the potential across an interface changes under extreme conditions only, a large change of input potential. We discuss the difference between a polarizable and nonpolarizable electrode in Chapter 6.

Next we discuss four types of reference electrodes: hydrogen, calomel, silver–silver chlorie, and mercury–mercurous sulfate electrodes.

Hydrogen Electrode. The hydrogen electrode is made of a platinum wire in contact with hydrogen gas and solution containing hydrogen ions (Fig. 5.5). Since hydrogen gas and hydrogen ions are present at the electrode–solution interface, this electrode can be represented as $Pt\,|\,H^+\,|\,H_2$, and the electrode reaction is

$$2H^+ + 2e \rightleftharpoons H_2$$

Platinum in the hydrogen electrode acts as a source or sink of electrons but does not take part in the reaction. It provides an electrical contact between H_2 and the solution containing H^+ ions and serves as a catalyst for the electrode reaction. Equilibrium between the hydrogen gas and H^+ ions in this reaction is established slowly when a bright Pt (or Pd) is used. Equilibrium in the hydrogen electrode reaction is established faster if the effective area of the Pt electrode is large. A large surface area is produced by electrolytic deposition of a finely divided layer of platinum ("platinum black"; platinized platinum).

From Eq. (5.9) the electrode potential of the hydrogen electrode is

$$E = E^0 + \frac{RT}{2F} \ln \frac{[H^+]^2}{p(H_2)} \tag{5.21}$$

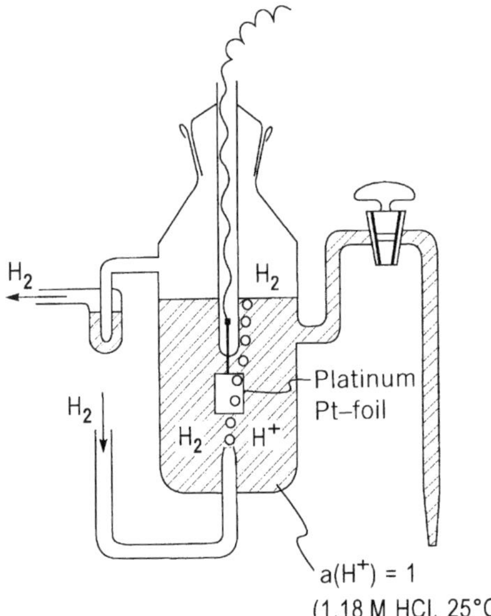

Figure 5.5. Hydrogen electrode.

where it is seen that the potential of the hydrogen electrode depends on the hydrogen-ion activity and the partial pressure of the hydrogen gas, $p(H_2)$. For a partial hydrogen pressure of 1 atm and a concentration of HCl (the source of H^+ ions) adjusted to give the activity of hydrogen ion equal to unity (1.18 M HCl at 25°C), we obtain $E = E^0 =$ the standard hydrogen electrode potential. By convention, the standard hydrogen electrode potential $E^0(H^+/H_2)$ is taken as zero at all temperatures.

Metal/Insoluble Salt/Ion Electrodes. Electrode potentials are usually reported relative to the normal hydrogen electrode [NHE; $a(H^+) = 1$, $p(H_2) = 1$], but they are actually measured with respect to a secondary reference electrode. Frequently used secondary reference electrodes are calomel, silver–silver chloride, and mercury–mercurous sulfate electrodes. These secondary reference electrodes consist of a metal M covered by a layer of its sparingly soluble salt MA immersed in a solution having the same anion (A^{z-}) as the sparingly soluble MA. The generalized reference electrode of this type may be represented as $M\,|\,MA\,|\,A^{z-}$ and may be considered to be composed of two interfaces: one between the metal electrode M and the metal ions M^{z+} in the salt MA:

$$M^{z+} + ze \rightleftharpoons M$$

and the other between A^{z-} anions in the solution and the A^{z-} anions in the salt MA:

$$MA \rightleftharpoons M^{z+} + A^{z-}$$

The overall electrode reaction is thus

$$MA + ze \rightleftharpoons M + A^{z-}$$

The equation for the electrode potential of this overall reaction is obtained from Eq. (5.9):

$$E = E^0 + \frac{RT}{zF} \ln \frac{[MA][e^z]}{[M][A^{z-}]} \tag{5.22}$$

Since the insoluble salt MA and the metal M are pure solids in their standard state ($a = 1$), Eq. (5.22) reduces to

$$E = E^0 + \frac{RT}{zF} \ln \frac{1}{[A^{z-}]} \tag{5.23}$$

or

$$E = E^0 - \frac{RT}{zF} \ln[A^{z-}] \tag{5.24}$$

Thus, the electrode potential of an electrode of the type $M\,|\,MA\,|\,A^{z-}$ depends on the activity of anion of the sparingly soluble compound of the electrode metal.

In an alternative presentation, the $M\,|\,MA\,|\,A^{z-}$ electrode can be considered to be of the type

$$M^{z+} + ze \rightleftharpoons M$$

and the potential of this electrode is given by the general equation for the metal/metal-ion electrode [Eq. (5.12)]. The activity of the metal ions M^{z+} is determined by the solubility product S of the salt MA and is given as

$$[M^{z+}] = \frac{S}{[A^{z-}]} \tag{5.25}$$

Substituting this value of $[M^{z+}]$ into Eq. (5.12) and noting that MA is a pure solid with $a = 1$, one has a final result that is the same as above, Eqs. (5.23) and (5.24).

Calomel Electrode. The calomel electrode consists of mercury covered with mercurous chloride (calomel) in contact with a solution of KCl:

$$Hg\,|\,Hg_2Cl_2\,|\,Cl^-$$

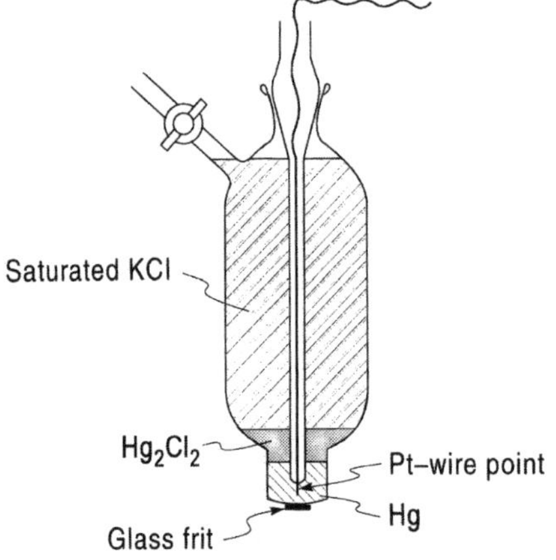

Figure 5.6. Saturated calomel electrode.

The overall electrode reaction in the calomel electrode is

$$Hg_2Cl_2 + 2e \rightleftharpoons Hg + 2Cl^-$$

and the potential is, from Eqs. (5.22) and (5.24),

$$E = E^0 - \frac{RT}{2F}\ln[Cl^-] \tag{5.26}$$

The most frequency used calomel electrode is the *saturated calomel electrode* (SCE), in which the concentration of KCl is at saturation (about 3.5 M) (Fig. 5.6). The potential of the SCE, at 25°C, is 0.242 V versus NHE. SCE has a large temperature coefficient, however, making it less frequently used in some applications:

$$E = 0.242 - 7.6 \times 10^{-4}(t - 25) \tag{5.27}$$

where t is temperature in Celsius. Equation (5.27) is given here with the linear terms only.

Silver–Silver Chloride Electrode. This reference electrode consists of a pure silver wire in a solution of KCl saturated with solid silver chloride. The electrode reaction is

$$AgCl + e \rightleftharpoons Ag + Cl^-$$

and the potential as a function of activity of Cl^- ions at 25°C is given by

$$E = 0.222 - 0.0592\log[Cl^-] \tag{5.28}$$

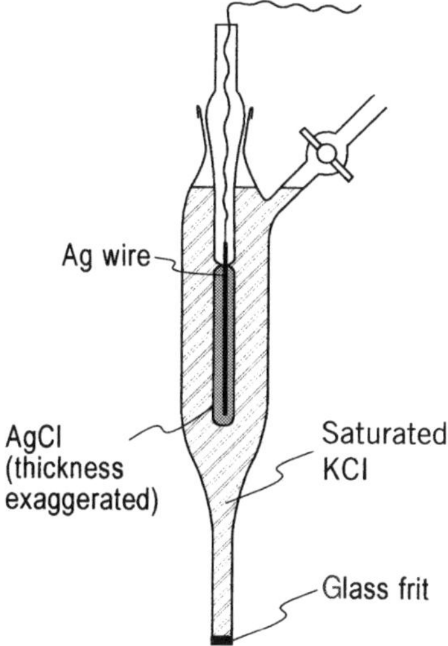

Figure 5.7. Silver–silver chloride electrode.

If saturated KCl solution is used (Fig. 5.7), the potential of this electrode at 25°C is 0.197 V versus NHE or −0.045 V versus SCE.

Mercury–Mercurous Sulfate Electrode. In this reference electrode the metal is mercury, the sparingly soluble compound is mercurous sulfate (Hg_2SO_4), and the source of SO_4^{2-} anions is sulfuric acid or potassium sulfate. The electrode is made in the same way as a calomel electrode, and it is represented as

$$Hg\,|\,Hg_2SO_4\,|\,SO_4^{2-}$$

and its potential is a function of the activity of SO_4^{2-} ions according to

$$E = 0.6156 - 0.0296\,\log[SO_4^{2-}]$$

When saturated potassium sulfate solution is used, the potential is 0.64 V versus NHE and 0.40 V versus SCE. The relationships between reference electrodes discussed above are shown in Figure 5.8.

5.7. STANDARD ELECTRODE POTENTIALS

We have seen in Section 5.2 that one can determine the relative electrode potential by measuring cell voltage. To form a series of relative electrode potentials, one has to select a reference electrode and standard conditions of components of an electrode/electrolyte interphase.

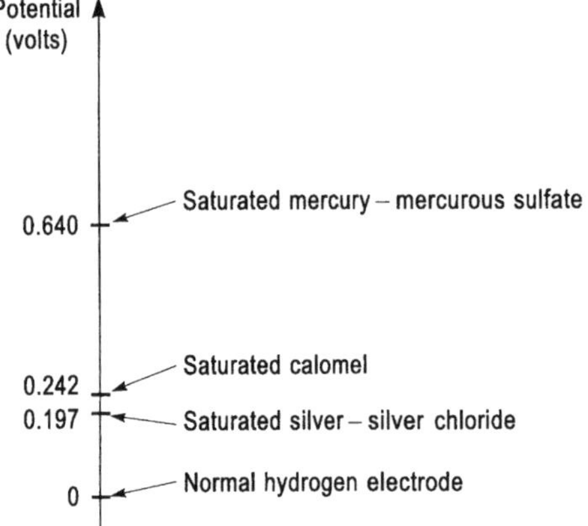

Figure 5.8. Electrode potentials of reference electrodes at 25°C.

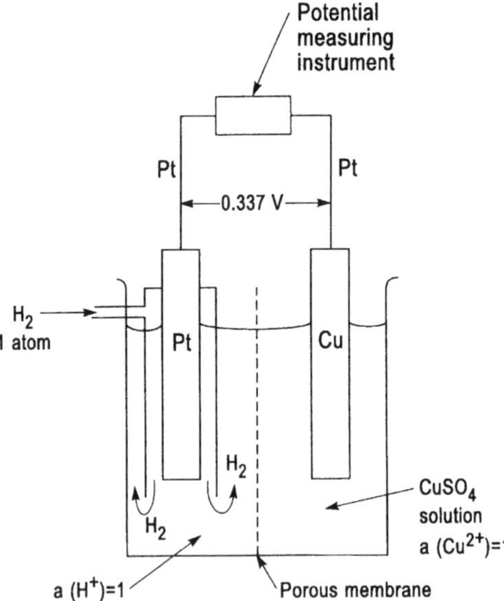

Figure 5.9. Relative standard electrode potential E^0 of a Cu/Cu^{2+} electrode.

The standard hydrogen electrode (Fig. 5.5) is chosen as the reference electrode when a series of relative electrode potentials is presented. By convention, the standard potential of this electrode is set to zero. Connecting this reference electrode with other electrodes into a cell, one can determine a series of relative values of electrode potentials (potential differences across interphases). For example, consider the cell shown in Figure 5.9. This cell can be represented schematically in the following way:

$$\text{Pt, H}_2(p = 1)\,|\,\text{H}^+(a = 1)\,|\,\text{Cu}^{2+}(a = 1)\,|\,\text{Cu}\,|\,\text{Pt}$$

TABLE 5.1. Standard Electrode Potentials

Metal/Metal-Ion Couple	Electrode Reaction	Standard Value (V)
Au/Au^+	$Au^+ + e \rightleftharpoons Au$	1.692
Au/Au^{3+}	$Au^{3+} + 3e \rightleftharpoons Au$	1.498
Pd/Pd^{2+}	$Pd^{2+} + 2 \rightleftharpoons Pd$	0.951
Cu/Cu^+	$Cu^+ + e \rightleftharpoons Cu$	0.521
Cu/Cu^{2+}	$Cu^{2+} + 2e \rightleftharpoons Cu$	0.3419
Fe/Fe^{3+}	$Fe^{3+} + 3e \rightleftharpoons Fe$	-0.037
Pb/Pb^{2+}	$Pb^{2+} + 2e \rightleftharpoons Pb$	-0.1262
Ni/Ni^{2+}	$Ni^{2+} + 2e \rightleftharpoons Ni$	-0.257
Co/Co^{2+}	$Co^{2+} + 2e \rightleftharpoons Co$	-0.28
Fe/Fe^{2+}	$Fe^{2+} + 2e \rightleftharpoons Fe$	-0.447
Zn/Zn^{2+}	$Zn^{2+} + 2e \rightleftharpoons Zn$	-0.7618
Al/Al^{3+}	$Al^{3+} + 3e \rightleftharpoons Al$	-1.662
Na/Na^+	$Na^+ + e \rightleftharpoons Na$	-2.71

Source: G. Milazzo and S. Caroli, *Tables of Standard Electrode Potentials*, Wiley, New York, 1978, with permission from Wiley.

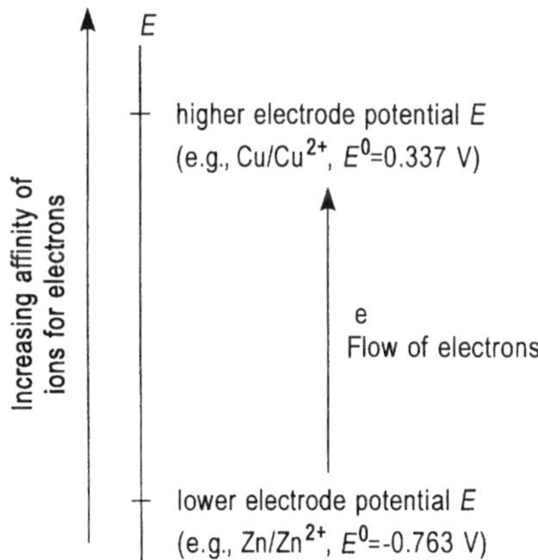

Figure 5.10. An electrode with lower electrode potential will reduce ions of an electrode with higher electrode potential.

where p is the pressure of H_2 and a is the activity. The measured value of the potential difference of this cell is $+0.337\,V$ at 25°C. This measured cell potential difference, $+0.337\,V$, is called the *relative standard electrode potential* of Cu and is denoted E^0 (the word *relative* is usually omitted). The standard electrode potential of other electrodes is obtained in a similar way, by forming a cell consisting of the standard hydrogen electrode (SHE) and the electrode under investigation. Standard electrode potentials at 25°C are listed in Table 5.1.

In general, an electrode with a lower electrode potential in Table 5.1 will reduce the ions of an electrode with a higher electrode potential (Fig. 5.10): or, a high positive standard electrode potential indicates a strong tendency toward reduction, whereas a low negative standard electrode potential indicates a strong tendency toward the

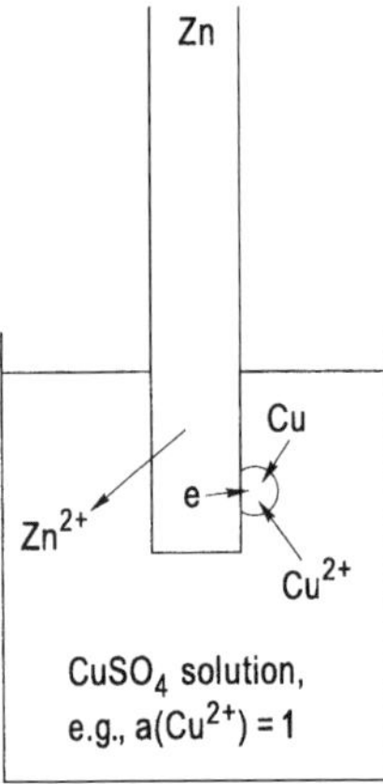

Figure 5.11. Displacement deposition of Cu on Zn.

oxidized state. For example, consider a strip of Zn placed in a solution of CuSO$_4$ (Fig. 5.11). Consider what reactions will occur in this system if the activity of Cu^{2+} ions is 1. This problem can be resolved by considering standard electrode potentials. The standard electrode potential for Cu/Cu^{2+} is $E^0 = 0.337$ V, and that for Zn/Zn^{2+} is $E^0 = -0.763$ V. Since Zn/Zn^{2+} has a lower electrode potential than that of the Cu/Cu^{2+} system, Zn will reduce Cu^{2+} ions in the solution. Thus, partial reactions in the system shown in Figure 5.11 are

$$Zn \rightarrow Zn^{2+} + 2e \quad \text{and} \quad Cu^{2+} + 2e \rightarrow Cu$$

and the overall reaction is

$$Zn + Cu^{2+} \rightarrow Zn^{2+} + Cu$$

Thus, a layer of metallic Cu is deposited on the zinc while Zn dissolves into solution; or, metallic Zn under these conditions reduces Cu^{2+} ions. This reaction is called a *displacement deposition* of Cu on Zn.

5.8. CONCENTRATION AND ACTIVITY

According to Nernst's equation, there should be a linear relationship between the equilibrium potential of the metal/metal-ion electrode (M/M^{z+}) and the logarithm of the concentration of M^{z+} ions [Eq. (5.13)]. This linear relationship was observed experimentally for a low concentration of the solute MA (e.g., 0.01 mol/L and lower). For higher concentrations, a deviation from linearity was observed (see, e.g., Fig. 5.12). The deviation from linearity is due to ion–ion interactions. In the example in Figure 5.12, the ion–ion interactions include interaction of the hydrated Ag$^+$ ions with one

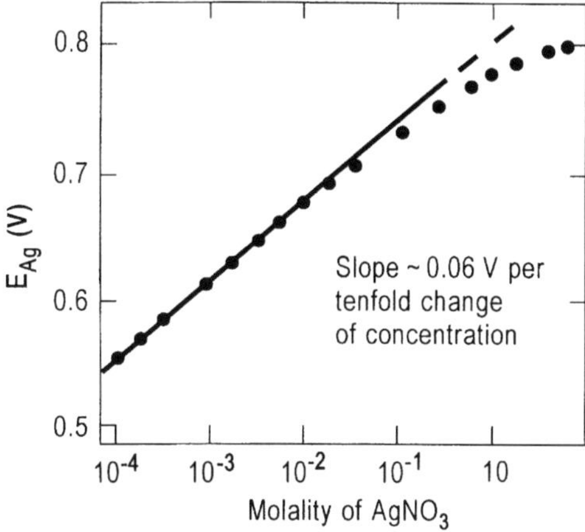

Figure 5.12. Electrode potential of Ag/Ag$^+$ electrode as a function of Ag$^+$ concentration. Molality is the concentration expressed as gram-molecules or gram-ions solute per liter (55 mol) of water. (From Ref. 5, with permission from Van Nostrand Reinhold.)

another and with NO$_3^-$ ions. The linear relationship between the equilibrium potential E and the log of concentration is obtained if the brackets in Eq. (5.13) signify the activity of species within the brackets. The activity of species i is defined by the equation

$$a_i = c_i \gamma_i \tag{5.29}$$

where c_i is the concentration of species i in mol/L and γ_i is the activity coefficient of species i. The activity coefficient γ is a dimensionless quantity that depends on the concentration of all ions present in the solution (ionic strength I, defined below). The activity coefficient of individual ionic species cannot be measured experimentally, but it can be calculated.

The Debye–Hückel model for ion–ion interaction yields the following equation for the activity coefficient of species i:

$$RT \ln \gamma_i = -\frac{N(z_i e)^2}{2\epsilon \kappa^{-1}} \tag{5.30}$$

where N is the Avogadro number, z_i is the electronic charge of ionic species i, e is the charge of the electron, ϵ is the dielectric constant of the medium (aqueous solution), and κ^{-1} is the thickness, or average radius, of the ionic cloud around a reference ion.

The experimentally measurable quantity is the mean ionic activity coefficient,

$$\gamma_\pm = \sqrt{\gamma_+ \gamma_-} \tag{5.31}$$

which is the geometric mean (the square root of the product) of activity coefficients of individual ionic species, γ_+ and γ_-.

Example 5.3. Calculate the reversible electrode potential for a Cu electrode immersed in a CuSO$_4$ aqueous solution with concentrations 1.0, 0.01, and 0.001 mol/L at 25°C. The mean activity coefficients of these solutions at 25°C are 0.043, 0.387, and 0.700, respectively. The standard electrode potential for the Cu/Cu^{2+} electrode is 0.337 V.

Using Eq. (5.15) for $z = 2$, $E^0 = 0.337$ V, 1.0 mol/L solution, and $\gamma_\pm = 0.043$, one obtains the following for $[M^{2+}] = [a(M^{2+})] = [c\gamma] = 1 \times 0.043 = 0.043$:

$$E = 0.337 + \frac{0.0592}{2} \log 0.043$$

$$= 0.337 - 0.0400 = 0.297 \text{ V}$$

If ion–ion interaction is neglected and the concentration is used instead of activity, then

$$E = 0.337 + \frac{0.0592}{2} \log 1, \quad E = 0.337 \text{ V}$$

Thus, the difference between E calculated without and with activity coefficient is 40 mV (0.337 − 0.297) for a 1 mol/L solution of $CuSO_4$ at 25°C.

For a 0.01 mol/L solution with $\gamma_\pm = 0.387$, $a = c\gamma_\pm = 0.01 \times 0.387 = 3.87 \times 10^{-3}$, and

$$E = 0.337 + \frac{0.0592}{2} \log(3.87 \times 10^{-3})$$

$$= 0.337 - 0.071 = 0.266 \text{ V}$$

In Example 5.1 we found that E for this solution is 0.278 V if ion–ion interaction is neglected. Thus, the difference between E as calculated without considering the activity coefficient compared to calculating with the activity coefficient is 12 mV (0.278 − 0.266).

For a 0.001 mol/L solution with $\gamma_\pm = 0.700$,

$$E = 0.337 + \frac{0.0592}{2} \log(0.001 \times 0.700)$$

$$= 0.244 \text{ V}$$

In Example 5.1, again, we found that E for this solution is 0.248 V when ion–ion interaction is neglected. The difference between E calculated with concentration and activity is 4 mV.   □

Thus, Examples 5.1 and 5.3 illustrate that the effect of the activity coefficient on the electrode potential value (for practical purposes) is not large and decreases with decrease in concentration.

Dubye–Hückel Theory of Activity Coefficient: Point-Charge Model. The Debye–Hückel theory of ion–ion interactions (Chapter 2) gives the following theoretical expression for the relationship between activity coefficient and ionic strength I, for water at 25°C and dielectric constant 78.54:

$$\log \gamma_{\pm} = -0.509 |z_+ z_-| \sqrt{I} \tag{5.32}$$

where I is the ionic strength of the medium. The ionic strength I is defined by the equation

$$I = \tfrac{1}{2}(c_i z_i^2 + c_2 z_2^2 + c_3 z_3^2 + \cdots) \tag{5.33a}$$

or

$$I = \tfrac{1}{2} \sum_i c_i z_i^2 \tag{5.33b}$$

where c_i, c_2, c_3, and c_i are the molar concentrations of various ions in the solution and z_1, z_2, and z_i are their respective charges. The summation is taken over all the various ions in a solution. Equation (5.33) quantifies the total concentration of ions, that is, the charge in an electrolytic solution.

Equation (5.32) is the *Debye–Hückel limiting law*. According to this law, the activity coefficient of an electrolyte is determined by the ionic strength I of the medium and the charge on the ions, z_+, z_-. The variation of activity coefficient $\gamma_{\pm}$ with the square root of concentration is shown in Figure 5.13. The figure shows that uni-univalent electrolytes (e.g., HCl, KCl, HNO_3) have similar activity coefficients and that Eq. (5.32) approximates $\gamma_{\pm}$ well at low concentrations. Equation (5.32) applied to uni-univalent electrolytes gives

$$\log \gamma_{\pm} = -0.509 \sqrt{I} \tag{5.34}$$

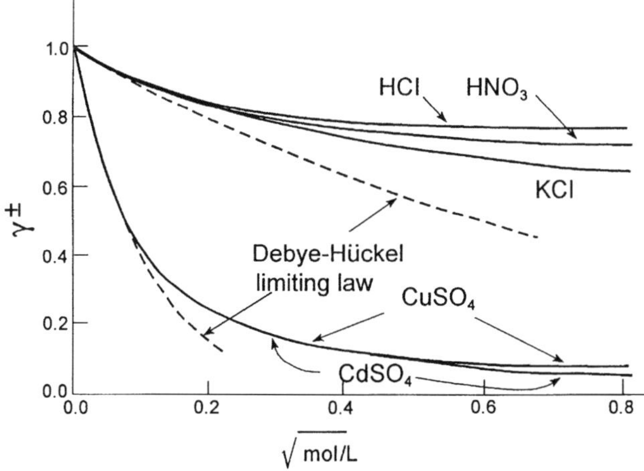

Figure 5.13. Variation of the activity coefficient $\gamma_{\pm}$ with $\sqrt{\text{mol/L}}$ (25°C). The solid curves are experimental values of $\gamma_{\pm}$. (From L. Pauling, *General Chemistry*, Dover, New York, 1970, with permission from Dover.)

and for a bi-bivalent electrolyte (e.g., $CuSO_4$, $CdSO_4$),

$$\log \gamma_\pm = -0.509 \times 4\sqrt{I} \tag{5.35}$$

Figure 5.13 also shows that bi-bivalent electrolytes deviate from ideal behavior, $\gamma_\pm = 1$, to a larger extent than do uni-univalent electrolytes. This is expected since interionic forces between bi-bivalent electrolytes are four times greater than those for uni-univalent electrolytes ($F = q_1 q_2 / Dr^2$). Theoretical curves, calculated from Eq. (5.32) and shown in Figure 5.13, agree well with experimental values at low concentrations. The Debye–Hückel limiting law was derived with the approximation of negligible ion size (ions were treated as point charges).

Example 5.4. Calculate the ionic strength of a 0.1 M solution of NaCl and a 0.1 M solution of Na_2SO_4.

For the NaCl solution, $c_i = c(K^+) = c(Cl^-) = 0.1$, and from Eq. (5.33),

$$I = \tfrac{1}{2}(c_1 z_1^2 + c_2 z_2^2) = \tfrac{1}{2}(0.1 \times 1^2 + 0.1 \times 1^2) = 0.1$$

For the Na_2SO_4 solution, we obtain $c(Na^+) = 0.2$, $c(SO_4^{2-}) = 0.1$, $z(Na^+) = 1$, $z(SO_4^{2-}) = 2$, and $I = \tfrac{1}{2}(0.2 \times 1^2 + 0.1 \times 2^2) = 0.3$. $\square$

Example 5.5. Calculate the ionic strength of a solution that is both 0.01 M in NaCl and 0.1 M in Na_2SO_4.

$$I = \tfrac{1}{2}(0.01 \times 1^2 + 0.01 \times 1^2 + 0.2 \times 1^2 + 0.1 \times 2^2) = 0.31$$

From Example 5.4 and 5.5 it follows that the ionic strength of a strong electrolytic solution consisting of singly charged ions is equal to the molar salt concentration and that the ionic strength of solutions consisting of multiply charged ions is greater than the molar concentration. $\square$

Debye–Hückel Theory: Finite-Ion-Size Model. If the approximation of the point charge is removed, the extended form of the Debye–Hückel law is obtained:

$$\log \gamma_\pm = -\frac{A}{1 + r / r_D} z_+ z_- \sqrt{I} \tag{5.36}$$

where r is the radius of the ion and r_D is the Debye length.

In further developments (9), the sizes of ions are considered as concentration-dependent parameters, and new expressions are derived for the activity coefficients.

Stokes–Robinson Modification of Debye–Hückel Theory: Effect of Ion–Solvent Interaction. Debye–Hückel theory explains the activity and activity coefficient data on the basis of ion–ion interaction for dilute solution. According to Eqs. (5.29) and

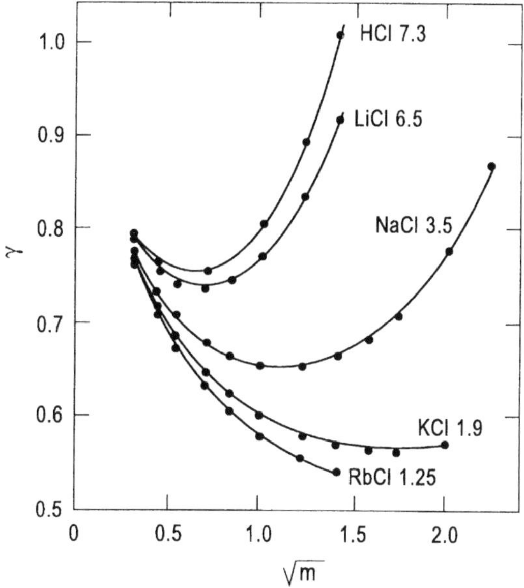

Figure 5.14. Comparison of experimental activity coefficients (circles) with theoretical values using the hydration correction (curves). (From Ref. 1, with permission from the American Chemical Society.)

(5.33), the activity coefficient is a decreasing function of concentration. However, experimentally observed $\gamma = f(c)$ functions usually first decrease, pass through a minimum, and then increase at high concentrations. To explain the increase of γ with concentration, Stokes and Robinson modified Debye–Hückel theory by introducing the effect of ion–solvent interaction. Thus, the modified theory is based on ion–ion and ion–solvent interactions. The modified theory is in good agreement with experimental results, up to an ionic strength of about 4, as shown in Figure 5.14.

REFERENCES AND FURTHER READING

1. R. H. Stokes and R. A. Robinson, *J. Am. Chem. Soc.* **70**, 1870 (1948).

2. K. J. Vetter, *Electrochemical Kinetics: Theoretical Aspects*, Academic Press, New York, 1967.

3. B. E. Conway, in *Physical Chemistry: An Advanced Treatise*, Vol. 9A, H. Eyring, ed., Academic Press, New York, 1970.

4. J. O'M. Bockris and A. K. N. Reddy, *Modern Electrochemistry*, 2nd ed., Vol. 1, Plenum Press, New York, 1998.

5. J. M. West, *Electrodeposition and Corrosion Processes*, Van Nostrand Reinhold, London, 1970.

6. J. Albery, *Electrode Kinetics*, Oxford University Press, Oxford, 1975.

7. S. Trasatti, in *Comprehensive Treatise of Electrochemistry*, Vol. 1, J. O'M. Bockris and B. Conway, eds., Plenum Press, New York, 1980.

8. J. Goodisman, *Electrochemistry: Theoretical Foundations*, Wiley, New York, 1987.

9. J.-P. Simonin, L. Blum, and P. Turq, *J. Phys. Chem.* **100**, 7704 (1996).

PROBLEMS

5.1. An experimentally determined value of the standard cell voltage $\mathcal{E}^0$ is 1.00 V at 25°C. Calculate the equilibrium cell voltage for the reaction quotients = 0.01, 0.1, 10, and 100. The number of electrons transferred (z) in the cell reaction is 2. At 25°C, RT/F = 25.7 mV.

5.2. Calculate the standard cell voltage $\mathcal{E}^0$ of the Daniel cell from the standard electrode potentials of Cu/Cu^{2+} and Zn/Zn^{2+} electrodes, which for Cu/Cu^2 is 0.34 V and for Zn/Zn^{2+} is -0.76 V. The Daniel cell reaction is

$$Zn + Cu^{2+} = Zn^{2+} + Cu$$

5.3. Calculate the equilibrium constant for the cell reaction (Daniel cell)

$$Zn + Cu^{2+} \rightleftharpoons Zn^{2+} + Cu$$

The standard potential $\mathcal{E}^0$ for the Daniel cell is 1.10 V at 25°C. Method: $\Delta G = -zF\mathcal{E}$, ΔG at equilibrium = 0.

5.4. Calculate the mean ionic activity coefficient $\gamma_{\pm}$ of a 0.1 M water solution of NaCl at 25°C in water, using the Debye–Hückel limiting law.

5.5. Calculate the mean ionic activity coefficient $\gamma_{\pm}$ of 0.1, 0.2, 0.3, 0.4, and 0.5 M solutions of KCl in water at 25°C using the Debye–Hückel limiting law. Compare the results with Figures 5.13 and 5.14.

6

Kinetics and Mechanism of Electrodeposition

6.1. INTRODUCTION

When an electrode is made a part of an electrochemical cell through which current is flowing, its potential will differ from the equilibrium potential. If the equilibrium potential of the electrode (potential in the absence of external current) is E and the potential of the same electrode as a result of external current flowing is $E(I)$, the different η between these two potentials,

$$\eta = E(I) - E \tag{6.1a}$$

is called *overpotential.* Or, in terms of $\Delta\phi$,

$$\eta = \Delta\phi(i) - \Delta\phi_{eq} \tag{6.1b}$$

The overpotential η is required to overcome hindrance of the overall electrode reaction, which is usually composed of a sequence of partial reactions. There are four possible partial reactions and thus four types of rate control: charge transfer, diffusion, chemical reaction, and crystallization. *Charge-transfer reaction* involves transfer of charge carriers, ions or electrons, across the double layer. This transfer occurs between the electrode and an ion or molecule. The charge-transfer reaction is the only partial reaction directly affected by the electrode potential. Thus, the rate of charge-transfer reaction is determined by the electrode potential. Pure charge-transfer overpotential η_{ct} exists only if the charge-transfer reaction is hindered and none of the other partial reactions is hindered. In this case the charge-transfer reaction is the rate-determining step.

Mass transport processes are involved in the overall reaction. In these processes the substances consumed or formed during the electrode reaction are transported from the bulk solution to the interphase (electrode surface) and from the interphase to the bulk

Fundamentals of Electrochemical Deposition, Second Edition.
By Milan Paunovic and Mordechay Schlesinger
Copyright © 2006 John Wiley & Sons, Inc.

solution. This mass transport takes place by *diffusion*. Pure diffusion overpotential η_d occurs if the mass transport is the slowest process among the partial processes involved in the overall electrode reaction. In this case diffusion is the rate-determining step.

Chemical reactions can be involved in the overall electrode process. They can be homogeneous reactions in the solution and heterogeneous reactions at the surface. The rate constant of chemical reactions is independent of potential. However, chemical reactions can be hindered, and thus the reaction overpotential η_r can hinder the current flow.

Processes at metal/metal-ion electrodes include *crystallization* partial reactions. These are processes by which atoms are either incorporated into or removed from the crystal lattice. Hindrance of these processes results in crystallization overpotential η_c. The slowest partial reaction is rate determining for the total overall reaction. However, several partial reactions can have low reaction rates and can be rate determining.

Thus, four different kinds of overpotential are distinguished and the total overpotential η can be considered to be composed of four components:

$$\eta = \eta_{ct} + \eta_d + \eta_r + \eta_c \tag{6.2}$$

where η_{ct}, η_d, η_r, and η_c are, as defined above, charge-transfer, diffusion, reaction, and crystallization overpotentials, respectively.

The term *overpotential* was introduced in 1899 by Caspari (1). The empirical relationship between current density i (in A/cm^2) and overpotential η was established in 1905 by Tafel (2):

$$\eta = a + b \log i \tag{6.3}$$

where a and b are constants. Erdey-Gruz and Volmer (3) derived the current–potential relationship using the Arrhenius equation (1889) for the rate constant:

$$k = A \exp\left(-\frac{\Delta G^{\ddagger}}{RT}\right) \tag{6.4}$$

where A is a constant and $\Delta G^{\ddagger}$ is the activation energy. They eliminated the need to know $\Delta G^{\ddagger}$ in order to use the Arrhenius equation by introducing the transfer coefficient α. The use of α is explained in the next section. However, work on the development of the modern theory of the activation overpotential started about 45 years later (from the Arrhenius equation, 1889) when Eyring (4) and Wynne-Jones and Eyring (5) formulated the absolute rate theory on the basis of statistical mechanics. This theory expressed the rate constant k of a chemical reaction in terms of the activation energy $\Delta G^{\ddagger}$, Boltzmann constant k_B, and Planck constant h:

$$k = \frac{k_B T}{h} \exp\left(-\frac{\Delta G^{\ddagger}}{RT}\right) \tag{6.5}$$

where R and T are the gas constant and absolute temperature, respectively.

In this chapter we derive the Butler–Volmer equation for the current–potential relationship, describe techniques for the study of electrode processes, discuss the influence of mass transport on electrode kinetics, and present atomistic aspects of electrodeposition of metals.

6.2. RELATIONSHIP BETWEEN CURRENT AND POTENTIAL: BUTLER–VOLMER EQUATION

Equations (6.4) and (6.5) can be used to derive the current–potential relationship for a general electrochemical equation:

$$Ox + ze \rightleftharpoons Red \tag{6.6}$$

where Ox is the oxidized form of a species and Red is its reduced form. For example, Ox is the metal ion $M(H_2O)_x^{z+}$ and Red is metallic M. Alternatively, Ox can be copper ion Cu^{2+} and Red, metallic copper, Cu. Also, Fe^{3+} and Fe^{2+} couple. Derivation of the current–potential relationship is done in two steps: (1) expressing the electrochemical rate v of reaction (6.6) in terms of current density; then (2) introducing the electrode potential E into the rate constant k given by Eq. (6.4) or (6.5). In this section we assume that the charge transfer is the slow process and that other processes (e.g., mass transport) are fast. Also, all steps that precede or follow reactions (6.6) are neglected.

Rate of Electrochemical Reaction in Terms of Current. In this part of the derivation we start with a definition of the rate of reaction and the definition of the electric current. The rate of the reduction reaction $\vec{v}$, reaction (6.6) from left to right, is defined as the number of moles m of Ox reacting per second and per unit area of the electrode surface:

$$\vec{v} = \frac{dm}{dt} \tag{6.7}$$

where dm/dt represents the change in number of moles m with time t. The rate $\vec{v}$ is given by

$$\vec{v} = \vec{k}[Ox] \tag{6.8}$$

where $\vec{k}$ is the rate constant of the reduction reaction and $[Ox]$ represents the activity of Ox. In this reaction one molecule of Ox is reduced by transfer of z electrons across the electrode–electrolyte interphase in the rate-determining step. The magnitude of the charge transferred per reaction event q is ze. The charge transferred per mole of events is

$$q_{m=1} = zeN \tag{6.9}$$

where N is Avogadro's number, the number of particles in 1 mol. Since eN is the Faraday constant (1 mol of electrons), the charge transferred per mole of events per unit area is

$$q_{m=1} = zF \qquad (6.10)$$

The charge transferred per m moles per unit area, q_m, is given as the product of $q_{m=1}$ and m:

$$q_m = mzF \qquad (6.11)$$

At this point we recall the definition of current, I. The *current* is the rate at which charge q passes a given area element in a conductor:

$$I = \frac{dq}{dt} \qquad (6.12)$$

Or, the average current, I_{av}, is equal to the net charge q that is passed in the time interval t:

$$I_{av} = \frac{q}{t} \qquad (6.13)$$

Combinating Eqs. (6.11) and (6.12) gives

$$I = \frac{d(mzF)}{dt} \qquad (6.14)$$

and since z and F are constants, we write

$$I = zF \frac{dm}{dt} \qquad (6.15)$$

Since dm/dt is the rate of reaction [Eq. (6.7)],

$$I = zFv \qquad (6.16)$$

For the reduction reaction of reaction (6.6), the reduction current density is

$$\vec{i} = zF\vec{v} \qquad (6.17)$$

and substituting the value of $\vec{v}$ given by Eq. (6.8) into Eq. (6.17) gives

$$\vec{i} = zF\vec{k}[\text{Ox}] \qquad (6.18)$$

The same reasoning applies to the reaction from right to left in the general equation (6.6). The rate of the oxidation reaction $\overleftarrow{v}$ is defined as the number of moles m of Red reacting per second and per unit area of the electrode surface. The rate, again, is

$$\overset{\leftarrow}{v} = \overset{\leftarrow}{k}[\text{Red}] \tag{6.19}$$

where $\overset{\leftarrow}{k}$ is the rate constant of the oxidation reaction, and the oxidation current density obtained from Eqs. (6.16) and (6.19) is

$$\overset{\leftarrow}{i} = zF\overset{\leftarrow}{k}\left[\text{Red}\right] \tag{6.20}$$

Equations (6.18) and (6.20) give the reaction rates of the general electrochemical reaction (6.6) in terms of current density.

The goal of the present section was to derive the relationship between current and potential for reaction (6.6). Thus, we have to introduce electrode potential, or overpotential, into Eqs. (6.18) and (6.20). This is achieved by expressing the rate constants $\vec{k}$ and $\overset{\leftarrow}{k}$ as a function of potential.

Rate Constant as a Function of Potential. Here we can start with either Eq. (6.4) or (6.5). For reasons of simplicity, we start with Eq. (6.4). Thus, the rate constant for the reduction reaction in the general reaction (6.6) can be written as

$$\vec{k} = \vec{B}\exp\left(-\frac{\Delta G^{\ddagger}}{RT}\right) \tag{6.21}$$

and for the oxidation reaction as

$$\overset{\leftarrow}{k} = \vec{B}\exp\left(-\frac{\Delta \overset{\leftarrow}{G}^{\ddagger}}{RT}\right) \tag{6.22}$$

For an electrochemical reaction the rate of reaction v and the rate constant k depend on potential E: specifically, the potential difference across electrode–solution interphase $\Delta\phi$ through the electrochemical activation energy $\Delta G_e^{\ddagger}$. Thus, the central problem here is to find the function

$$\Delta G_e^{\ddagger} = f(E) \tag{6.23}$$

It is not possible at present to evaluate this function theoretically. The system is too complicated. The complexity of theoretical calculations can best be explained by considering the definition of $\Delta G_e^{\ddagger}$ and processes involved in the activation process. $\Delta G_e^{\ddagger}$ is defined in Figure 6.1. Free energy of activation for the forward reaction is the free-energy difference between the free energy of the activated state, $G^{\ddagger}$, and the free energy of the initial state, G^{I}:

$$\Delta \vec{G}_e^{\ddagger} = G^{\ddagger} - G^{\text{I}} \tag{6.24}$$

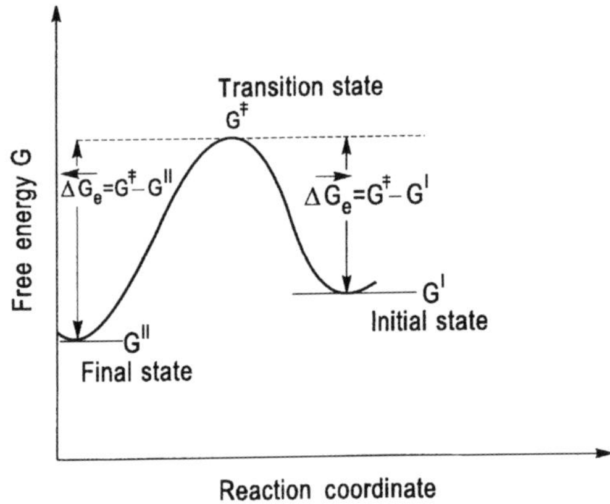

Figure 6.1. Free-energy change for the general electrochemical reaction, Eq. (6.6): initial state, Ox, in the bulk of the solution, outside the diffusion double layer; final state, Red, in the bulk of the solution outside the diffuse double layer.

Free energy of activation for the backward reaction is the free-energy difference between the free energy of the activated state, $G^{\ddagger}$, and the free energy of the final state, II:

$$\Delta \overleftarrow{G}_e^{\ddagger} = G^{\ddagger} - G^{II} \tag{6.25}$$

Thus, calculation of $\Delta G_e^{\ddagger}$ requires knowledge of the free energy of the transition state (activated complex) $G^{\ddagger}$ and of the activation process and a model of the transition state. The electrochemical activation process includes (1) stretching of bonds and changes in the configuration of the solution and ionic environment of the reacting species, and (2) changes in the electrical potential energy of the reacting species due to their transport through the electric field in the double layer. The first part of the activation process is very difficult to estimate theoretically, but the second part can be estimated by introducing simplifying assumptions that lead to an approximate solution. An approximate solution is indeed possible if we assume that $\Delta G_e^{\ddagger}$ can be separated into two parts. In the presence of $\Delta \phi$, the electrochemical activation energy $\Delta G_e^{\ddagger}$ may be written as

$$\Delta G_e^{\ddagger} = \Delta G_{\text{in}}^{\ddagger} + \Delta G_{\text{pd}}^{\ddagger} \tag{6.26}$$

where $\Delta G_{\text{in}}^{\ddagger}$ and $\Delta G_{\text{pd}}^{\ddagger}$ are the potential-independent and potential-dependent parts of the electrochemical activation energy, respectively. This separation of $\Delta G_e^{\ddagger}$ into potential-independent and potential-dependent parts is analogous to the separation of the electrochemical potential of species i ($\bar{\mu}_i$) into the chemical potential (μ) and the inner potential (ϕ):

$$\bar{\mu}_i = \mu_i + z_i F \phi \tag{6.27}$$

The potential-dependent part of the activation energy $\Delta G_{\text{pd}}^{\ddagger}$ can be estimated by introduction of the transfer coefficient α, which was introduced in 1930 by Erdey-Gruz and Volmer (3). The potential-dependent contribution $\Delta \overrightarrow{G}_{\text{pd}}^{\ddagger}$ to the free energy of activation

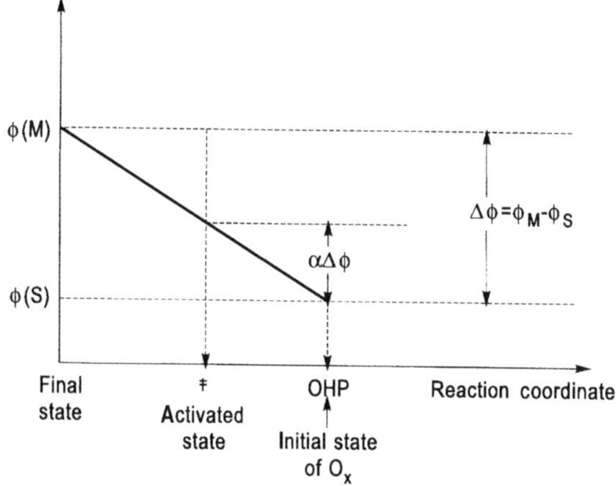

Figure 6.2. The electrical work of activation of Ox [general electrode reaction, Eq. (6.6)] in the forward direction is determined by the potential difference $\alpha\Delta\phi$.

for the reduction reaction can be evaluated in the following way. The electrical work W is given as a product of the charge Q and the potential difference ΔV this charge passed through:

$$W = Q\,\Delta V \tag{6.28}$$

In the case of the electrical work of activation, the charge Q is zF. The potential difference ΔV across which the ion has moved and that determines $\Delta G^{\ddagger}_{\mathrm{pd}}$ is part of the total $\Delta\phi(\Delta\phi = \phi_M - \phi_S)$. The part of $\Delta\phi$ that determines $\Delta G^{\ddagger}_{\mathrm{pd}}$ can be estimated considering Figure 6.2. The electrical work of activation of Ox [the general electrochemical reaction (6.6) in the forward direction] is determined by the potential difference $\alpha\Delta\phi$ across which the ion has moved to reach the top of the energy barrier (Fig. 6.2). The top of the barrier is usually halfway across the double layer. Thus, the electrical work of activation of Ox is $\alpha zF\,\Delta\phi$, where α is in the range 0 to 1 and is often 0.5 (symmetric energy barrier). Then the electrical work of activation of Red in the reverse reaction is $(1 - \alpha)zF\,\Delta\phi$. With these values for the electrical work of activation, $\Delta\vec{G}^{\ddagger}_{e}$ for the reduction process is

$$\Delta\vec{G}^{\ddagger}_{e} = \Delta\vec{G}^{\ddagger}_{\mathrm{in}} + \alpha zF\Delta\phi \tag{6.29}$$

and $\Delta\overleftarrow{G}^{\ddagger}_{e}$ for the oxidation process is

$$\Delta\overleftarrow{G}^{\ddagger}_{e} = \Delta\overleftarrow{G}^{\ddagger}_{\mathrm{in}} - (1 - \alpha)zF\Delta\phi \tag{6.30}$$

Introduction of Eq. (6.29) into (6.21) yields $\vec{k}$ as a function of $\Delta\phi$:

$$\vec{k} = \vec{B}\,\exp\left(-\frac{\Delta\vec{G}^{\ddagger}_{\mathrm{in}}}{RT}\right)\exp\left(-\frac{\alpha zF\Delta\phi}{RT}\right) \tag{6.31}$$

Introduction of Eq. 6.30 into 6.22 yields $\overleftarrow{k}$ as a function of $\Delta\phi$:

$$\overrightarrow{k} = \overrightarrow{B} \exp\left(-\frac{\Delta\overleftarrow{G}_{\text{in}}^{\ddagger}}{RT}\right) \exp\left(\frac{(1-\alpha)\,zF\Delta\phi}{RT}\right) \tag{6.32}$$

Thus, we have expressed the rate constant k as a function of potential difference $\Delta\phi$. This was the aim of the second part of the derivation of the Butler–Volmer equation.

Current–Potential Relationship for Partial Reactions. Partial $i = f(\Delta\phi)$ functions can be derived by joining equations expressing the rate of electrochemical reactions in terms of current [Eqs. (6.18) and (6.20)] and equations expressing the rate constant as a function of potential [Eqs. (6.31) and (6.32)]. Thus, the cathodic partial current density $\overrightarrow{i}$ is obtained from Eqs. (6.18) and (6.31) to yield

$$\overrightarrow{i} = zF[\text{Ox}]\overrightarrow{B} \exp\left(\frac{\Delta G_{\text{in}}^{\ddagger}}{RT}\right) \exp\left(-\frac{\alpha zF\Delta\phi}{RT}\right) \tag{6.33}$$

and the anodic partial current density $\overleftarrow{i}$ is obtained from Eqs. (6.20) and (6.32):

$$\overleftarrow{i} = zF[\text{Red}]\overleftarrow{B} \exp\left(-\frac{\Delta\overleftarrow{G}_{\text{in}}^{\ddagger}}{RT}\right) \exp\left(\frac{(1-\alpha)zF\,\Delta\phi}{RT}\right) \tag{6.34}$$

Exchange Current Density. When an electrode is at equilibrium, the equilibrium value of $\Delta\phi$ is $\Delta\phi_{\text{eq}}$ and the equilibrium partial current densities $\overrightarrow{i}$ and $\overleftarrow{i}$ are equal:

$$\overrightarrow{i} = \overleftarrow{i} \tag{6.35}$$

Equality of $\overrightarrow{i}$ and $\overleftarrow{i}$ on an atomic scale means that a constant exchange of charge carriers (electrons or ions) takes place process the metal–solution interphase. Figure 6.3

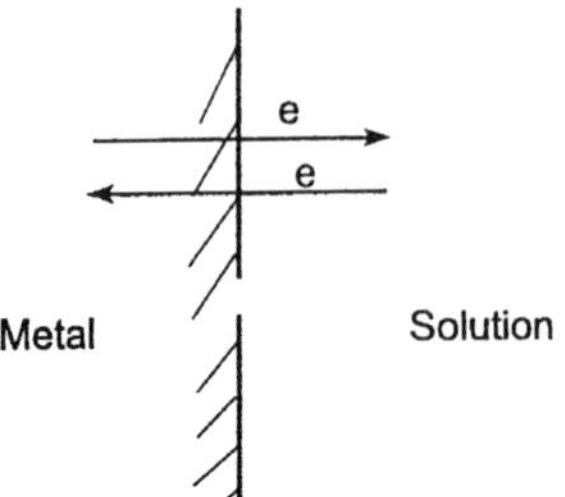

Figure 6.3. RedOx interphase at equilibrium: an equal number of electrons crossing in both directions across the metal–solution interphase.

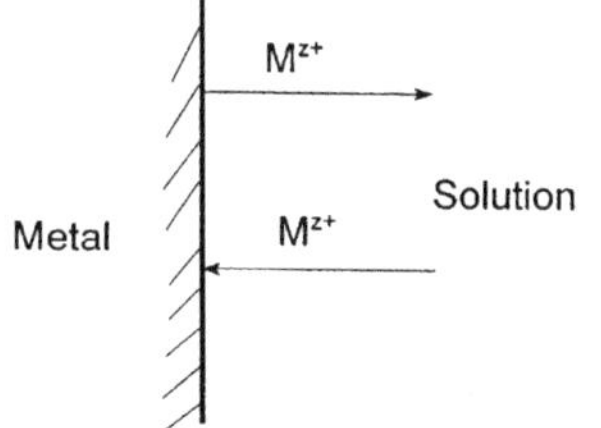

Figure 6.4. Metal/metal-ion interphase at equilibrium: an equal number of metal ions M^{z+} crossing in both directions across the metal–solution interphase.

illustrates a RedOx electrode at equilibrium. Figure 6.4 illustrates a metal/metal-ion electrode at equilibrium.

Since the two equilibrium current densities are equal, they will be designated by one symbol, i_0. Substituting for $\Delta\phi$ in Eqs. (6.33) and (6.34) by $\Delta\phi_{eq}$, one obtains

$$
i_0 = \begin{cases} zF[\text{Ox}]\vec{B}\exp\left(\dfrac{\Delta\vec{G}_{in}^{\ddagger}}{RT}\right)\exp\left(-\dfrac{\alpha zF\Delta\phi_{eq}}{RT}\right) & (6.36a) \\[3em] zF[\text{Red}]\overleftarrow{B}\exp\left(-\dfrac{\Delta\overleftarrow{G}_{in}^{\ddagger}}{RT}\right)\exp\left(\dfrac{(1-\alpha)zF\Delta\phi_{eq}}{RT}\right) & (6.36b) \end{cases}
$$

The exchange current density i_0 is one of the most important parameters of electrochemical kinetics. It defines the kinetic properties of the particular electrochemical reaction and the electrode material. For a homogeneous electrode material, a single crystal, one i_0 value is usually sufficient to characterize the electrochemical reaction and the electrode material. However, for a polycrystalline electrode material, each surface grain has its characteristic i_0. The exchange current densities of different crystallographic faces may be different. Moreover, different atomic sites (e.g., kink, surface) on the surface of a single crystal, or a grain of a polycrystalline material, will have different exchange current densities (e.g., $i_{0,\text{kink}}$, $i_{0,\text{surface}}$).

Thus, the overall exchange current density i_0 is composed of different partial exchange current densities:

$$ i_k,\ i_l,\ i_m \qquad (6.37) $$

and

$$ i = \sum_j i_j \qquad (6.38) $$

In Sections 6.5 and 6.8 we continue the discussion concerning exchange current density.

Potential Difference $\Delta\phi$ Departs from Equilibrium: Butler–Volmer Equation. When the interphase is not in equilibrium, a net current density i flows through the electrode (the double layer). It is given by the difference between the anodic partial current density $\overleftarrow{i}$ (a positive quantity) and the cathodic partial current density $\vec{i}$ (a negative quantity):

$$ i = \overleftarrow{i} - \vec{i} \qquad (6.39) $$

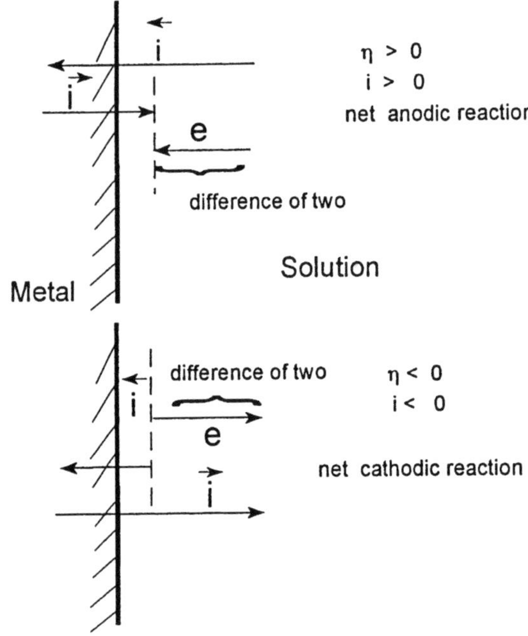

Figure 6.5. RedOx electrode deviates from the equilibrium potential; two currents flowing in opposite directions. The lengths of the arrows represent a measure of the magnitude of the process per unit area per unit time.

Figures 6.5 and 6.6 illustrate the RedOx and the metal/metal-ion electrodes in a nonequilibrium state, respectively. When current flows through an electrode, its potential $\Delta\phi(i)$ deviates from the equilibrium value $\Delta\phi_{eq}$ by the amount η, which was defined by Eq. (6.1a). Thus, the nonequilibrium potential difference $\Delta\phi$ in Eqs. (6.33) and (6.34) can be substituted by the nonequilibrium value $\Delta\phi(i)$:

$$\Delta\phi(i) = \Delta\phi_{eq} + \eta \tag{6.40}$$

This substitution results in the following equations:

$$\vec{i} = zF[\text{Ox}]\vec{B}\exp\left(\frac{\Delta\vec{G}_{in}^{\ddagger}}{RT}\right)\exp\left(-\frac{\alpha zF\Delta\phi_{eq}^{\ddagger}}{RT}\right)\exp\left(\frac{\alpha zF\eta}{RT}\right) \tag{6.41}$$

and

$$\overleftarrow{i} = zF[\text{Red}]\overleftarrow{B}\exp\left(-\frac{\Delta\overleftarrow{G}_{in}^{\ddagger}}{RT}\right)\exp\left(\frac{(1-\alpha)zF\Delta\phi_{eq}}{RT}\right)\exp\left(\frac{(1-\alpha)zF\eta}{RT}\right) \tag{6.42}$$

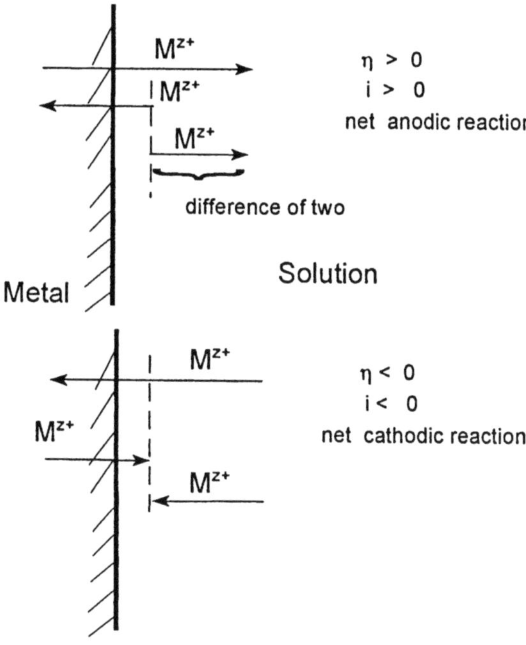

Figure 6.6. Metal/metal-ion electrode deviates from the equilibrium potential: two currents flowing in opposite directions.

Since the equilibrium current densities i_0 are equal [Eqs. (6.36a) and (6.36b)], from Eqs. (6.41) and (6.42) it follows that

$$\vec{i} = i_0 \exp\left(- \frac{\alpha z F \eta}{RT} \right) \tag{6.43}$$

$$\overleftarrow{i} = i_0 \exp\left(\frac{(1 - \alpha) z F \eta}{RT} \right) \tag{6.44}$$

Substituting these equations for $\vec{i}$ and $\overleftarrow{i}$ into Eq. (6.39), we get the Butler–Volmer equation:

$$i = i_0 \left[\exp\left(\frac{(1 - \alpha) z F \eta}{RT} \right) - \exp\left(- \frac{\alpha z F \eta}{RT} \right) \right] \tag{6.45}$$

This equation gives the relationship between the current density i and the charge-transfer overpotential η in terms of two parameters, the exchange current density i_0 and the transfer coefficient α. Figure 6.7 depicts the variation of the partial current densities and the net current density with overpotential. It can be seen that for large

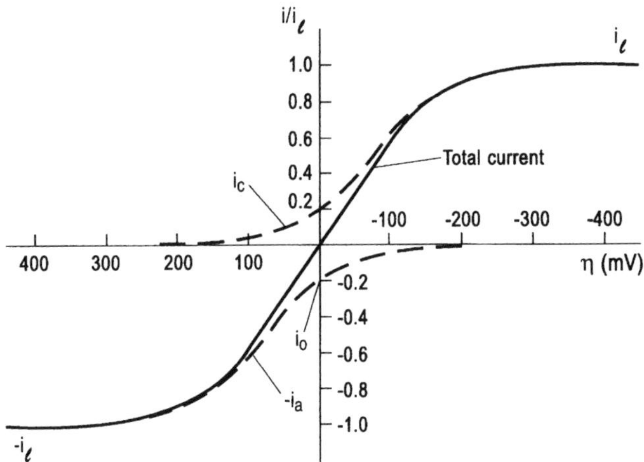

Figure 6.7. Variation of partial current densities (dashed line) and net current density (solid line) with overpotential η.

departures from the equilibrium potential, large $\pm\eta$ values, the partial currents approach the net current i. For large positive values of overpotential, $\overleftarrow{i} \approx i$; and for large negative values of overpotential, $\overrightarrow{i} \approx i$.

6.3. BUTLER–VOLMER EQUATION: HIGH-OVERPOTENTIAL APPROXIMATION

Large Cathodic Current. We have seen from Figure 6.7 that for the large negative values of overpotential η, the partial cathodic current density $\overrightarrow{i}$ approaches i, $i \approx \overrightarrow{i}$. For these conditions the Butler–Volmer equation (6.45) can be simplified. Analysis of Eq. (6.45) shows that when η becomes more negative, the first exponential term in the equation (corresponding to the anodic partial current) decreases, whereas the second exponential term (corresponding to the cathodic partial reaction) increases. Thus, under these conditions,

$$\exp\left(-\frac{\alpha zF\eta}{RT}\right) \gg \frac{(1-\alpha)zF\eta}{RT} \tag{6.46}$$

and the smaller term $\exp\{[(1 - \alpha)zF\eta]/RT\}$ in Eq. (6.45) can be neglected, yielding an approximate new form of the Bulter–Volmer equation:

$$i = -i_0 \exp\left(-\frac{\alpha zF\eta}{RT}\right) \tag{6.47}$$

This is essentially Eq. (6.43).

Large Anodic Current. When η has large positive values, the second exponential term in Eq. (6.45) (corresponding to the cathodic partial current density) decreases while the first term (corresponding to the anodic partial reaction) increases, resulting in $i \approx \overset{\rightharpoonup}{i}$. Thus, the smaller term,

$$\exp\left(-\frac{\alpha z F \eta}{RT}\right) \tag{6.48}$$

in Eq. (6.45) can be neglected, and the Butler–Volmer equation reduces to

$$i = i_0 \exp\left(\frac{(1-\alpha)zF\eta}{RT}\right) \tag{6.49}$$

This is essentially Eq. (6.44). Equations (6.47) and (6.49) show that there is an exponential relationship between i and high values of η, usually greater than 0.10 V.

Tafel Equation. So far we have expressed the current density i as a function of η, $i = f(\eta)$. Let us now consider the inverse:

$$\eta = f(i) \tag{6.50}$$

We will consider this relationship for the large cathodic and anodic values of η. For large cathodic current densities, we start with Eq. (6.47), omitting the minus sign and writing the absolute value of i:

$$|i| = i_0 \exp\left(-\frac{\alpha z F \eta}{RT}\right) \tag{6.51}$$

Taking the logarithm of both sides of this equation, we have

$$\ln|i| = \ln i_0 - \frac{\alpha z F \eta}{RT} \tag{6.52}$$

After separation of the i and η terms in this equation, we have

$$\frac{\alpha z F}{RT}\eta = \ln i_0 - \ln|i| \tag{6.53}$$

so we get $\eta = f(i)$:

$$\eta = \frac{RT}{\alpha z F}\ln i_0 - \frac{RT}{\alpha z F}\ln|i| \tag{6.54}$$

or with transformation of the natural logarithm (ln) into a decimal logarithm (log),

$$\eta = \frac{2.303RT}{\alpha zF} \log i_0 - \frac{2.303RT}{\alpha zF} \log |i| \tag{6.55}$$

This equation can be simplified if we note that the first term is a constant for a given electrode reaction (given i_0 at a given temperature),

$$\frac{2.303RT}{\alpha zF} \log i_0 = \vec{a} \tag{6.56}$$

and that $2.303RT/\alpha zF$ is a constant:

$$\frac{2.303RT}{\alpha zF} = \vec{b} \tag{6.57}$$

Introducing Eqs. (6.56) and (6.57) into Eq. (6.55), we get the Tafel line for the cathodic process:

$$\eta = \vec{a} - \vec{b} \log |i| \tag{6.58}$$

This equation shows that there is a linear relationship between η and $\log i$ when η has large cathodic values.

A similar linear relationship is obtained for the large anodic values of η. Starting from Eq. (6.49), taking the logarithm, and rearranging results in

$$\eta = -\frac{2.303RT}{(1-\alpha)zF} \log i_0 + \frac{2.303RT}{(1-\alpha)zF} \log i \tag{6.59}$$

and setting

$$-\frac{2.303RT}{(1-\alpha)zF} \log i_0 = \overleftarrow{a} \tag{6.60}$$

and

$$\frac{2.303RT}{(1-\alpha)zF} = \overleftarrow{b} \tag{6.61}$$

we get the Tafel line for the anodic process:

$$\eta = \overleftarrow{a} + \overleftarrow{b} \log i \tag{6.62a}$$

Or, emphasizing linearity for both cases, we can write

$$\eta = a \pm b \log |i| \tag{6.62b}$$

where the $\pm$ sign holds for anodic and cathodic process, respectively. It should be noted that the theoretical values of the constants a and b are different for the two processes. The linear relationship between η and $\log i$ was established experimentally by Tafel in 1905. The derivation given above shows that the Tafel equation is a special case of a general Butler–Volmer equation.

6.4. BUTLER–VOLMER EQUATION: LOW-OVERPOTENTIAL APPROXIMATION

When η is small we have

$$\frac{\alpha z F \eta}{RT} \ll 1 \tag{6.63a}$$

and

$$\frac{(1-\alpha)z F \eta}{RT} \ll 1 \tag{6.63b}$$

Under these conditions the expotentials in the Butler–Volmer equation (6.45) can be approximated using a power series:

$$e^x = 1 + x + \frac{x^2}{2!} + \frac{x^3}{3!} + \cdots \tag{6.64}$$

Applying this to the first exponential term in Eq. (6.45), $[(1-\alpha)z F \eta]/RT$, and taking only the first-order terms in η in Eq. (6.64), one gets the approximation

$$\exp\left(\frac{(1-\alpha)z F \eta}{RT}\right) \approx 1 + \frac{(1-\alpha)z F \eta}{RT} \tag{6.65}$$

For the second exponential term in Eq. (6.45) we get the approximation

$$\exp\left(-\frac{\alpha z F \eta}{RT}\right) \approx 1 - \frac{\alpha z F \eta}{RT} \tag{6.66}$$

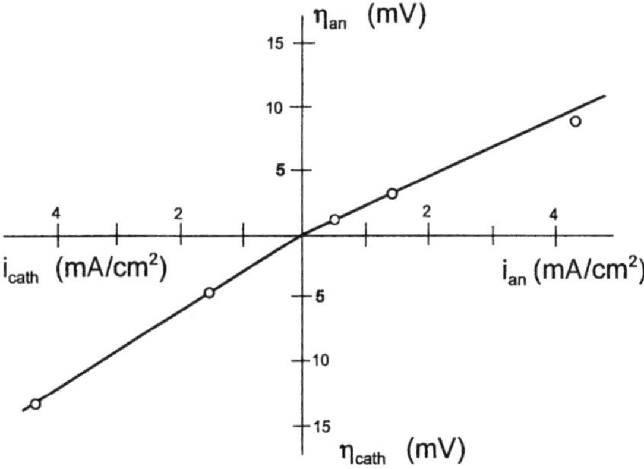

Figure 6.8. Linear relationship between i and η for electrodeposition of Cu from $CuSO_4-H_2SO_4$ solution. (From Ref. 10, with permission from the Royal Society of Chemistry.)

Substituting Eqs. (6.65) and (6.66) into the Butler–Volmer equation (6.45) results in

$$i = i_0 \left(\frac{zF\eta}{RT} - \frac{\alpha zF\eta}{RT} + \frac{\alpha zF\eta}{RT} \right) \tag{6.67}$$

$$i = i_0 \frac{zF}{RT} \eta \tag{6.68a}$$

or

$$i = i_0 f\eta \tag{6.68b}$$

where $f = F/RT$.

Thus, for small values of η, less than about 0.01 V, when the electrode potential is near the reversible potential, the current varies linearly with the overpotential. An example of the linear relationship is shown in Figure 6.8.

6.5. KINETIC DERIVATION OF NERNST EQUATION

The Nernst equation defines the equilibrium potential of an electrode. A simplified thermodynamic derivation of this equation is given in the Sections 5.3 to 5.5. Here we will give the kinetic derivation of this equation.

Equations (6.36a) and (6.36b) constitute the starting point of this derivation. Equations (6.36) define the equilibrium partial current densities $\vec{i}$ and $\overleftarrow{i}$, which are equal to the exchange current density i_0. Equations (6.36) can be simplified by introducing potential-independent constants k_{in}:

$$\vec{k}_{in} = \vec{B} \exp\left(\frac{\Delta \vec{G}_{in}^{\ddagger}}{RT}\right) \tag{6.69a}$$

$$\overleftarrow{k}_{in} = \overleftarrow{B} \exp\left(-\frac{\Delta \overleftarrow{G}_{in}^{\ddagger}}{RT}\right) \tag{6.69b}$$

Introducing these values into Eqs. (6.36), we obtain

$$zF[Ox]\vec{k}_{in} \exp\left(\frac{-\alpha zF\Delta\phi_{eq}}{RT}\right) = zF[Red]\overleftarrow{k}_{in} \exp\left(\frac{(1-\alpha)zF\Delta\phi_{eq}}{RT}\right) \tag{6.70}$$

The second exponential term in Eq. (6.70) can be resolved into two exponential terms after multiplication and remembering that $e^{x+y} = e^x e^y$:

$$\exp\left[\frac{(1-\alpha)zF\Delta\phi_{eq}}{RT}\right] = \exp\left(\frac{zF\Delta\phi_{eq}}{RT}\right)\exp\left(-\frac{\alpha zF\Delta\phi_{eq}}{RT}\right) \tag{6.71}$$

Substituting Eq. (6.71) into (6.70) and canceling equal terms on the left and right sides, one gets

$$\vec{k}_{in}[Ox] = \overleftarrow{k}_{in}[Red]\exp\left(\frac{zF\Delta\phi_{eq}}{RT}\right) \tag{6.72}$$

or

$$\frac{\vec{k}_{in}[Ox]}{\overleftarrow{k}_{in}[Red]} = \exp\left(\frac{zF\Delta\phi_{eq}}{RT}\right) \tag{6.73}$$

Taking the natural (ln) logarithm of both sides of the equation, we have

$$\ln\frac{\vec{k}_{in}}{\overleftarrow{k}_{in}} + \ln\frac{[Ox]}{[Red]} = \frac{zF\Delta\phi_{eq}}{RT} \tag{6.74}$$

Solving Eq. (6.74) for $\Delta\phi_{eq}$, we get the Nernst equation,

$$\Delta\phi_{eq} = \Delta\phi^0 + \frac{RT}{zF}\ln\frac{[Ox]}{[Red]} \tag{6.75}$$

where

$$\Delta\phi^0 = \frac{RT}{zF} \ln \frac{\vec{k}_{\text{in}}}{\overleftarrow{k}_{\text{in}}}$$

6.6. INFLUENCE OF MASS TRANSPORT ON ELECTRODE KINETICS

In Section 6.2 we derived the current–potential relationship for the case where the charge transfer is the slow process. In this case the rate of the electrode reaction is determined by the charge-transfer overpotential, η_{ct}. It is assumed here that other processes, such as mass transport, are fast. Figure 6.7 illustrates the $i-\eta_{\text{ct}}$ relationship, which has a linear range [low η; see Eq. (6.68)] followed by an exponential range (large η; see Eqs. (6.47) and (6.49)]. This relationship has a limit where the rate of reaction is limited by transport to the electrode. A general current-potential relationship is shown in Figure 6.9. In this section we analyze the effect of mass transport on the electrode kinetics.

Diffusion-Layer Model. Let us consider again the general electrochemical reaction (6.6). Initially, at time t_0, before electrolysis, the concentration of the solution is homogeneous at all distances x from the electrode, equal to the bulk concentration of reactant Ox. In a more rigorous consideration, one would say that the concentration of the solution is homogeneous up to the outer Helmholtz plane (OHP), that is, up to $x = x_{\text{OHP}}$. When a constant current is applied to the test electrodes and counterelectrodes such that the reaction

$$\text{Ox} + ze \rightarrow \text{Red} \tag{6.76}$$

occurs at the test electrode, the reactant Ox is consumed at the electrode, and its concentration at the interface decreases.

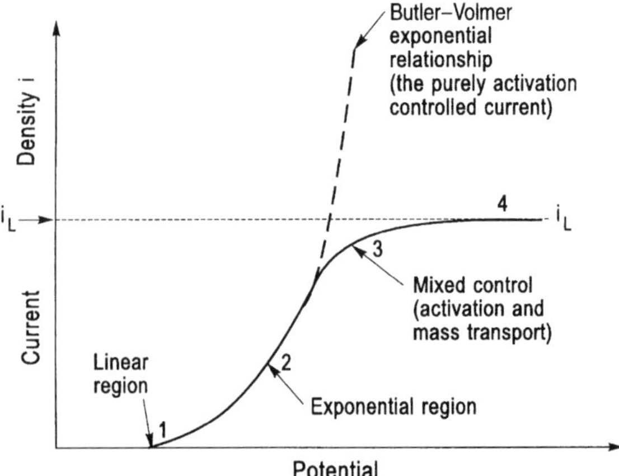

Figure 6.9. Four regions in the general current–overpotential relationship: 1, linear; 2, exponential; 3, mixed control; 4, limiting current density region.

As the electrolysis proceeds, there is a progressive depletion of the Ox species at the interface of the test electrode (cathode). The depletion extends farther and farther away into the solution as the electrolysis proceeds. Thus, during this non-steady-state electrolysis, the concentration of the reactant Ox is a function of the distance x from the electrode (cathode) and the time t, $[Ox] = f(x,t)$. Concurrently, concentration of the reaction product Red increases with time. For simplicity, the concentrations will be used instead of activities. Weber (19) and Sand (20) solved the differential equation expressing Fick's diffusion law (see Chapter 18) and obtained a function expressing the variation of the concentration of reactant Ox and product Red on switching on a constant current. Figure 6.10 shows this variation for the reactant.

In consideration of the M^{z+}/M electrode, Ox would be replaced by M^{z+}, and Figure 6.10 would represent variation of concentration of the M^{z+} ions with distance from the cathode in case of the non-steady-state deposition of metal M. The variation of the concentration of the reactant Ox at the electrode ($x = 0$), $c_{Ox}(x = 0,t)$, is given by the equation

$$C_{Ox}(0, t) = c^0 - \frac{2i\sqrt{t}}{nF\sqrt{\pi}\sqrt{D_{Ox}}}$$

(6.77)

where D_{Ox} is the diffusion coefficient of species Ox. According to Eq. (6.77), the concentration C_{Ox} at the electrode ($x = 0$, or better, $x = x_{OHP}$) is a function of the current density i and time t. It varies with the square root of t, or $i\sqrt{t}$. Thus, one result of the electrolysis is the development of a concentration gradient in the solution layer close to the electrode. This concentration gradient dc_{Ox}/dx changes with time and assumes a maximum value when the reactant is completely depleted at the electrode, $c_{Ox}(x = 0) = 0$. This maximum concentration gradient is $(dc/dx)_{x=0}$. Since the rate of an electrochemical reaction is as given by Eq. (6.14), and since dm/dt is the flux of the reactant at the electrode, which is given by Fick's law,

$$\frac{dm}{dt} = D_{Ox}\left(\frac{\partial C_{Ox}}{\partial x}\right)_{x=0}$$

(6.78)

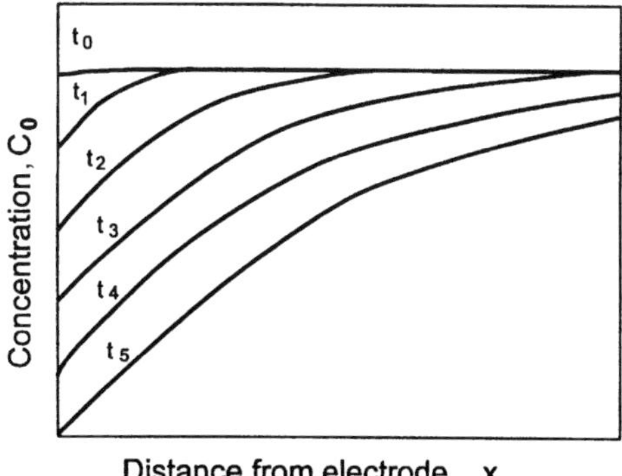

Figure 6.10. Variation of concentration of reactant during non-steady-state electrolysis. The number t_n on each curve is the time elapsed since the beginning of electrolysis, $t_5 > t_4 > \ldots t_1$. (From Ref. 23, with permission from Elsevier.)

the rate of the reaction at the surface is given by

$$i = zFD_{Ox}\left(\frac{\partial C_{Ox}}{\partial x}\right)_{x=0} \tag{6.79}$$

An exact treatment of the function $c_{Ox} = f(x,t)$ and evaluation of the gradient $(\partial c_{Ox}/\partial t)_{x=0}$ is complicated. Therefore, an approximate and a simplified method is proposed and presented below.

Nernst Diffusion-Layer Model. This model assumes that the concentration of Ox has a bulk concentration c_{Ox}^{b} up to a distance δ from the electrode surface and then falls off linearly to Ox ($x = 0$) at the electrode (neglecting the double-layer effect). The Nernst diffusion-layer model is illustrated in Figure 6.11.

In this model it is assumed that the liquid layer of thickness δ is practically stationary (quiescent). At distances greater than δ from the surface, the concentration of the reactant and other species is assumed to be equal to the bulk concentration. At these distances, $x > \delta$, stirring is efficient. The reacting species must diffuse through the diffusion layer to reach the electrode surface.

According to the Nernst model, the concentration gradient at the electrode is given by

$$\left(\frac{\partial c}{\partial x}\right)_{x=0} \approx \frac{c_b - c_{x=0}}{\delta} \tag{6.80}$$

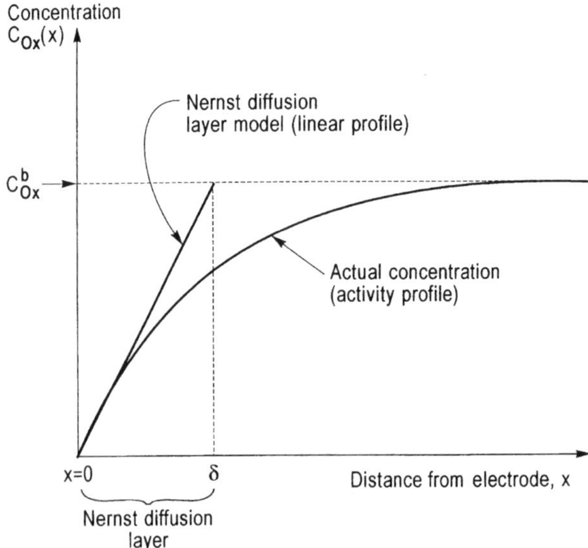

Figure 6.11. Variation of concentration of reactant during non-steady-state electrolysis; c_{Ox}^{b} is the concentration in the bulk; $c_{Ox}(x)$ is the concentration at the surface.

and the rate of reaction is given by

$$i = nFD_{Ox} \frac{c_b - c_{x=0}}{\delta}$$

(6.81)

The variation of the thickness of the diffusion layer δ with time can be evaluated from Eqs. (6.81) and (6.77). It follows from these two equations that

$$\delta = 2\sqrt{D_{Ox}} \sqrt{\frac{t}{\pi}}$$

(6.82)

for constant-current polarization. This equation shows that the diffusion-layer thickness increases with the square root of time.

Limiting Current Density.   According to Eq. (6.81), the current density i is a function of the concentration gradient $(c_b - c_{x=0})/\delta$. The maximum value of the concentration gradient is when $c_{x=0} = 0$, that is, when the concentration gradient value is c_b/δ. This is the steady-state value for the constant gradient and steady-state diffusion. Corresponding to this maximum concentration gradient is the maximum current density, called the *limiting diffusion current density* i_L, which is equal to [from (Eq. 6.81)]

$$i_L = \frac{nFD_{Ox}c_b}{\delta}$$

(6.83)

At the value of the limiting current density, the species Ox is reduced as soon as it reaches the electrode. At these conditions the concentration of the reactant Ox at the electrode is zero and the rate of reaction (6.76) is controlled by the rate of transport of the reactant Ox to the electrode.

   If an external current greater than the limiting current I_L is forced through the electrode, the double layer is charged further, and the potential of the electrode will change until some process other than reduction of Ox can occur. It will be shown later that the limiting current density is of great practical importance in metal deposition since the type and quality of metal deposits depend on the relative values of the deposition current and the limiting current. One extreme example is shown in Figure 6.12.

6.7. MULTISTEP REACTIONS

In previous sections we treated electrochemical processes that proceed according to a single electrochemical reaction [Eq. (6.6)]. However, many electrochemical processes proceed by a multitude of processes characterized by complex kinetic schemes. There are various types of complex mechanisms.

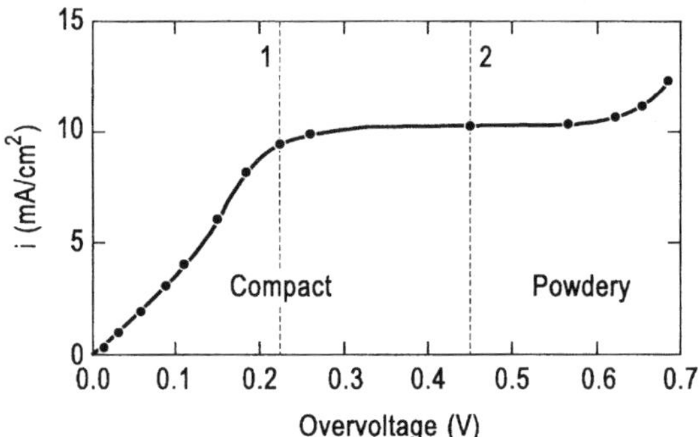

Figure 6.12. Overpotential characteristic of transition from compact to powdery deposit in electrodeposition of Cu from $CuSO_4$ (0.1 M) + H_2SO_4 (0.5 M) solutions. (From Ref. 22, with permission from Wiley.)

One important type of complex mechanism in electrode reactions is a series of consecutive reactions. One example of this type is electrochemical deposition from complexed ions. In this case the electrochemical reaction is preceded by a chemical reaction. Another example is that of inclusion of cathodic hydrogen evolution. We discuss these two cases next.

Electrochemical Reaction Preceded by a Chemical Reaction: Electrochemical Deposition from Complexed Ions. The kinetic scheme for such a complex reaction in an electrochemical deposition is

$$[ML_x]^{2+} \rightleftharpoons M^{2+} + xL \tag{6.84}$$

$$M^{2+} + 2e \rightarrow M \tag{6.85}$$

where L is a neutral ligand. We consider the case when the complexed ion $[ML_x]^{2+}$ is not reduced directly by the applied potential but is transformed into an electroactive form, M^{2+}, by a preceding chemical reaction, Eq. (6.84). Rate constants for the chemical reaction, Eq. (6.84), the rate constant for dissociation of the complex $[k_f$ (f, forward)] and the formation of the complex $[k_b$, (b, backward)], do not depend on the electrode potential. When the electroactive species is the complexed metal ion, the kinetic scheme involves a simple charge transfer.

In the kinetic scheme in Eqs. (6.84) and (6.85), the electrode reaction proceeds by three consecutive steps:

1. Transfer of $[ML_x]^{2+}$ and M^{2+} to and from the interphase
2. Chemical transformation according to Eq. (6.84)
3. Transfer of electrons to the electrochemically active species M^{2+}

The concentration of reactant M^{2+} at the interphase is controlled by diffusion and the chemical reaction (6.84). Either of these three steps can be slow and thus can be the rate-determining step (RDS). Dissociation of a complexed metal ion prior to charge transfer has been studied electrochemically by the constant-current (chronopotentiometry) and potential sweep methods (these methods are described in Section 6.9). If a constant current is used, proper selection of an intermediate current density is important for the following reasons. At low current densities there is a slow consumption of the reactant M^{2+} at the electrode. This allows sufficient production of M^{2+} by complex dissociation, and the system behaves as if there were no preceding chemical reaction. At very high current densities there is not enough time for chemical reaction (6.84) to occur. The system again behaves as if there were no preceding chemical reaction. Thus, the presence of a prior chemical reaction in the kinetic scheme (6.84)–(6.85) can be detected and studied quantitatively by using intermediate current densities for a given set of dissociation (k_d) and formation (k_f) rate constants.

Cathodic Evolution of Hydrogen. The cathodic evolution of hydrogen is of great scientific and technological importance. Technological importance stems from the fact that electrodeposition of some metals, such as Ni and Cr, is accompanied by simultaneous hydrogen evolution.

The overall reaction of hydrogen evolution in an acid solution is

$$2H^+ + 2e \rightarrow H_2 \tag{6.86}$$

In a water solution, a proton (hydrogen ion) is hydrated, forming the hydroxonium ion H_3O^+. For the sake of simplicity we write H^+ instead of H_3O^+. For further simplicity, we assume that the diffusion of protons from the bulk of the solution to the electrode [outer Helmholtz plane (OHP)],

$$H^+(\text{solution}) \rightarrow H^+(\text{OHP}) \tag{6.87}$$

and the hydrogen molecule evolution,

$$H_2(\text{OHP}) \rightarrow H_2(\text{atmosphere}) \tag{6.88}$$

are fast processes. In this case the mechanism of the cathodic evolution of hydrogen consists of two steps:

Step 1: Charge transfer:

$$H^+(\text{OHP}) + e(M) \rightarrow MH \tag{6.89}$$

Step 2: Hydrogen combination:

$$2MH \rightarrow 2M + H_2(\text{electrode}) \tag{6.90}$$

where MH denotes hydrogen adsorbed or (H_{ads}) at the electrode M. This mechanism is further complicated by the fact that the combination step (step 2) can proceed in either of two ways.

Step 2a: Chemical desorption step (CD step) or atom–atom combination step:

$$H_{ads} + H_{ads} \rightarrow H_2(\text{electrode}) \tag{6.91}$$

In this reaction path two adsorbed hydrogen atoms diffuse on the electrode surface, collide, and react to form an H_2 molecule (Tafel recombination).

Step 2b: Electrochemical desorption step (ED step) or ion–atom combination step:

$$H_{ads} + H^+ + e \rightarrow H_2 \tag{6.92}$$

Thus, there are two kinetic paths for the hydrogen evolution. The first path consists of charge transfer (CT) followed by chemical desorption (CD): path CT–CD. The second path consists of charge transfer (CT) followed by electrochemical desorption (ED): path CT–ED. Within each path, either of the consecutive steps can be slow and thus can be the rate-determining step (RDS). Each of these paths has two possible mechanisms.

Thus, there are four possible mechanisms for the hydrogen evolution in an acid solution: (1) CT is RDS, CD fast; (2) CD is RDS, CT is fast; (3) CT is RDS, ED is fast; and (4) ED is RDS, CT is fast. Different paths and different mechanisms have different Tafel slopes. Readers are referred to Refs. 11, 15, 21–23, and 26 for determination of the reaction mechanisms.

The simplified analysis of cathodic evolution of hydrogen presented above shows how an apparently simple reaction may have a rather complicated mechanism.

6.8. ATOMISTIC ASPECTS OF ELECTRODEPOSITION OF METALS

In the electrodeposition of metals, a metal ion M^{n+} is transferred from solution into the ionic metal lattice. A simplified atomistic representation of this process is

$$M^{n+}(\text{solution}) \rightarrow M^{n+}(\text{lattice}) \tag{6.93}$$

This reaction is accompanied by the transfer of n electrons from the external electron source (power supply) to the electron gas of the metal M.

Atomic processes that constitute the electrodeposition process, Eq. (6.93), can be seen by presenting the structure of the initial, M^{n+} (solution), and the final state, M^{n+} (lattice). Since metal ions in the aqueous solution are hydrated, the structure of the initial state in Eq. (6.93) is represented by $[M(H_2O)_x]^{n+}$. The structure of the final state is the M adion (adatom) at the kink site (Fig. 6.13), since it is generally assumed that atoms (ions) are attached to the crystal via a kink site (3). Thus, the final step of the overall reaction, Eq. (6.93), is the incorporation of the M^{n+} adion into the kink site.

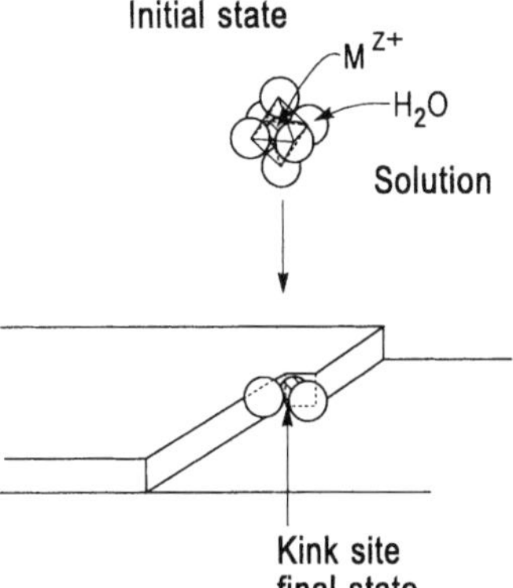

Figure 6.13. Initial and final states in metal deposition.

Because of surface inhomogeneity, the transition from the initial state $[M(H_2O)_x]^{n+}$ (solution) to the final state M^{n+} (kink),

$$[M(H_2O)_x]^{n+}(\text{solution}) \rightarrow M^{n+}(\text{kink}) \tag{6.94}$$

can proceed via either of two mechanisms: (1) step-edge site ion-transfer mechanism or (2) terrace site ion-transfer mechanism.

Step-Edge Ion-Transfer Mechanism. The step-edge site ion-transfer, or direct transfer mechanism, is illustrated in Figure 6.14. It shows that in this mechanism, ion transfer from the solution (OHP) takes place on a kink site of a step edge or on any other site on the step edge. In both cases the result of the ion transfer is an M adion in the metal crystal lattice. In the first case, a direct transfer to the kink site, the M adion is in the half-crystal position, where it is bonded to the crystal lattice with one half of the bonding energy of the bulk ion. Thus, the M adion belongs to the bulk crystal. However,

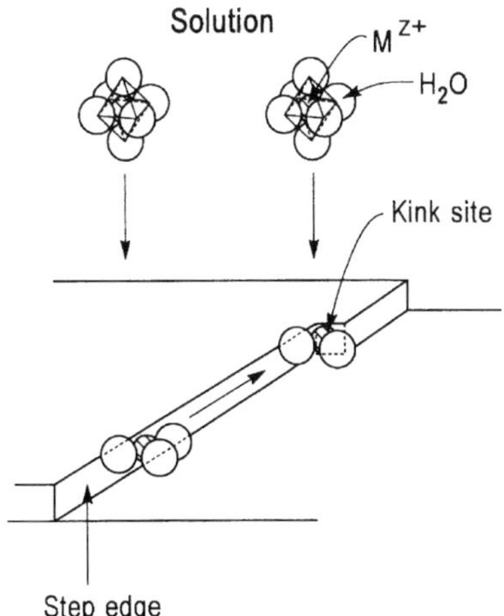

Figure 6.14. Step-edge ion-transfer mechanism.

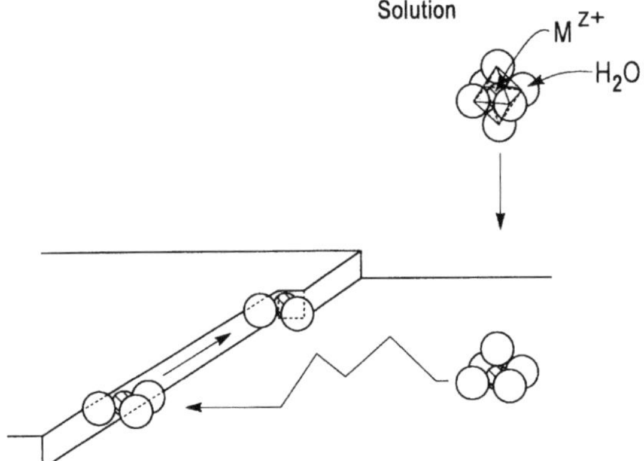

Figure 6.15. Ion transfer to the terrace site, surface diffusion, and incorporation at kink site.

it still has some water of hydration (Fig. 6.14). In the second case, direct transfer to the step-edge site other than kink, the metal ion transferred diffuses along the step edge until it finds a kink site (Fig. 6.14). Thus, in a step-edge siteion-transfer mechanism there are two possible paths: direct transfer to a kink site and the step-edge diffusion path.

Terrace Ion-Transfer Mechanism. In the terrace siteion-transfer mechanism a metal ion is transferred from the solution (OHP) to the flat face of the terrace region (Fig. 6.15). At this position the metal ion is in the adion (adsorbed-like) state, having most of its water of hydration. It is weakly bound to the crystal lattice. From this position it diffuses on the surface, seeking a position of lower energy. The final position is a kink site.

In view of these two mechanisms, step edge and terrace ion transfer, the overall current density i is considered to be composed of two components,

$$i = i_{se} + i_{te} \tag{6.95}$$

where i_{se} and i_{te} are the step-edge and terrace site current density components, respectively.

The initial theoretical treatment of these mechanisms of deposition was given by Lorenz (31–34). The initial experimental studies on surface diffusion were published by Mehl and Bockris (35, 38). Conway and Bockris (36, 40) calculated activation energies for the ion-transfer process at various surface sites. The simulation of crystal growth with surface diffusion was discussed by Gilmer and Bennema (43).

6.9. TECHNIQUES FOR STUDY OF ELECTRODE PROCESSES

A general Red/Ox electrochemical reaction, Eq. (6.6), Ox $+ ze \rightleftharpoons$ Red, proceeds in at least five steps:

1. Transfer of Ox from the bulk solution to the interphase
2. Adsorption of Ox onto the electrode surface

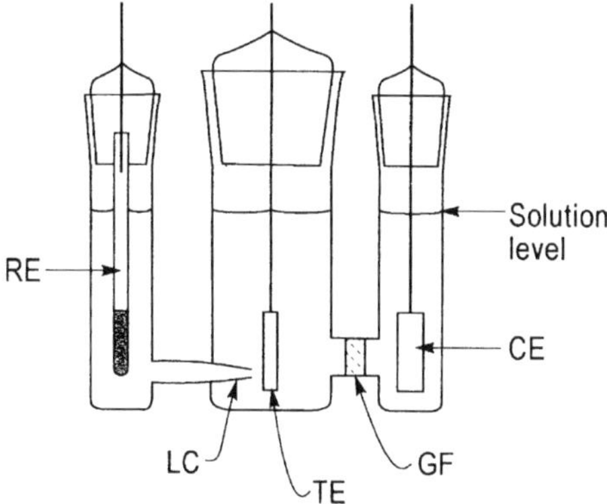

Figure 6.16. Three-compartment three-electrode electrochemical cell; RE, reference electrode; LC, Lugin capillary; TE, test electrode; GF, glass frit; CE, counter-electrode.

3. Charge transfer at the electrode to form Red
4. Desorption of Red from the surface
5. Transport of Red from the interphase into the bulk of the solution

For the M/M^{z+} electrode, steps 4 and 5 are absent since the product of the charge transfer M remains incorporated in the metal electrode. Steps 1 and 5 are mass-transport processes.

Study of the charge-transfer processes (step 3 above), free of the effects of mass transport, is possible by the use of transient techniques. In the transient techniques the interface at equilibrium is changed from an equilibrium state to a steady state characterized by a new potential difference $\Delta\phi$. Analysis of the time dependence of this transition is the basis of transient electrochemical techniques. We will discuss galvanostatic and potentiostatic transient techniques; for other techniques [e.g., alternating current (ac)], the reader is referred to Refs. 50 to 55.

Study of transient processes utilizes the three-electrode cell design. One type of three-compartment electrochemical cell is shown in Figure 6.16. Other designs can be found in the literature (e.g., Ref. 54).

Galvanostatic Transient Technique. In the galvanostatic technique the current between the test electrode and the auxiliary (counter-) electrode is held constant with a current source (galvanostat), and the potential between the test electrode and the reference electrode is determined as a function of time. The potential is the dependent variable, which is recorded with suitable recording systems, such as recorders or oscilloscopes (Fig. 6.17).

The input signal and the response to the signal are compared in Figure 6.18. The response function, $E_i = f(t)$, shows that a certain time (t_1) is necessary to reach a potential E_i when the electrode reaction begins at the measurable rate. The duration of this time can be estimated by considering a simplified equivalent circuit to the single-electrode reaction (Fig. 6.19). When a constant current is applied to the system shown in Figure 6.17, the current (electron flow) is used for (1) charging the double-layer capacitance C_{dl} up to the potential at which the electrode reaction can

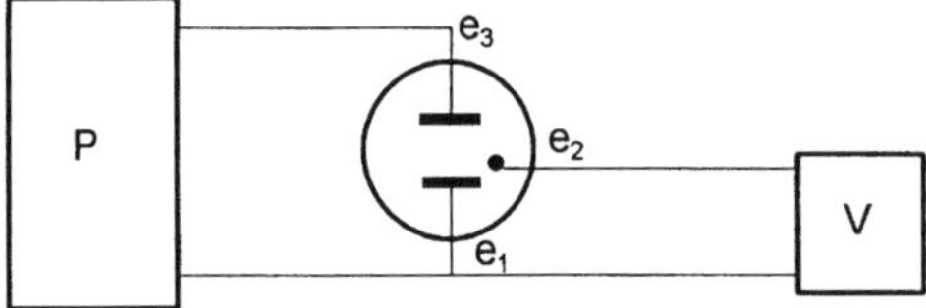

Figure 6.17. Schematic diagram of apparatus for galvanostatic measurements; P, constant current power supply; e_t, test electrode; e_2, reference electrode; e_3, counter (auxiliary)-electrode; V, potential–time recording instrument.

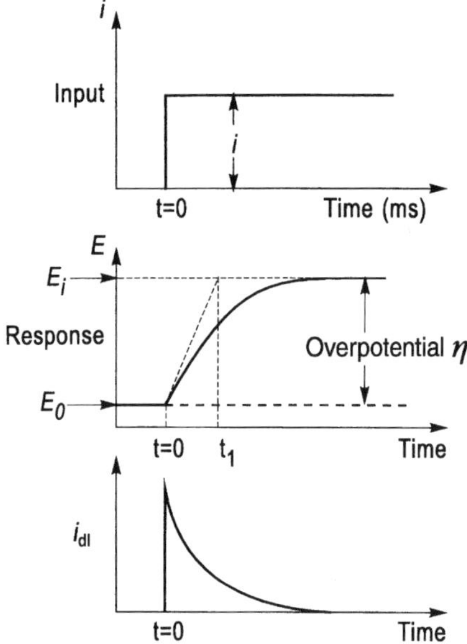

Figure 6.18. Variation of potential of the test electrode, E, with time during galvanostatic electrolysis (milisecond range); E_0, equilibrium potential; E_i, potential of the test electrode at beginning of electrolysis at constant current density i.

proceed with a measurable velocity and (2) electrode reaction (charge transfer). Thus, the total galvanostatic current density i_g is given by

$$i_g = i_{dl} + i_{ct} \tag{6.96}$$

where i_{dl} is the capacitive and i_{ct} is the faradaic current (charge transfer). The first process, after applying a current to the system (Fig. 6.19), involves charging the double-layer capacitance, C_{dl}, from the reversible potential E up to the potential E_i when the electrode reaction begins at a measurable rate. The time necessary to charge capacitor C in an RC (resistance–capacitance) circuit to 99.0% of the imposed voltage is

$$t_{V=0.99\,V}(C) = 4.6RC \tag{6.97}$$

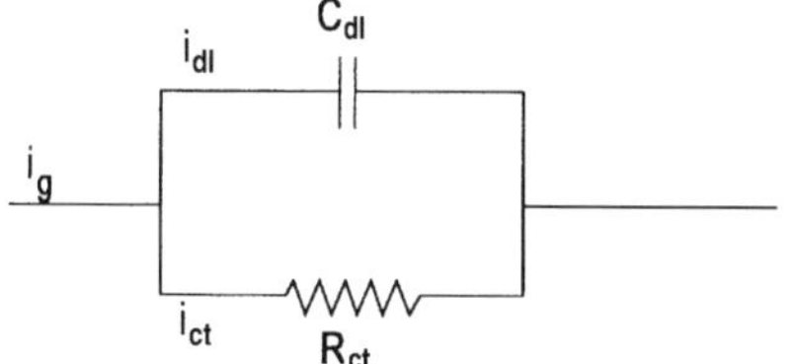

Figure 6.19. Simplified equivalent circuit for single-electrode reaction [e.g., Eq. (6.6)]; C_{dl}, double-layer capacitance of test electrode; R_{ct}, charge-transfer resistance of electrode reaction.

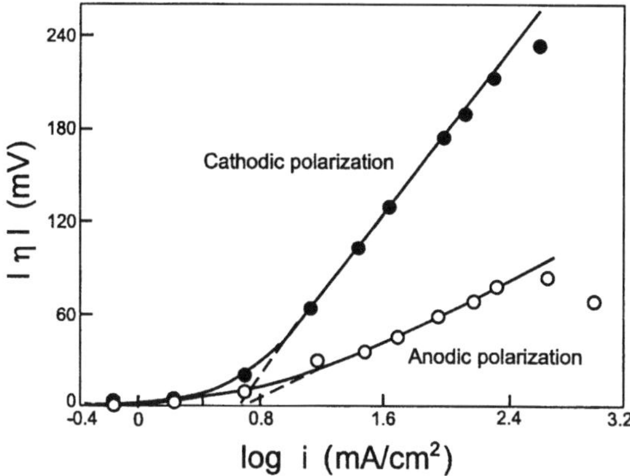

Figure 6.20. Current–potential relationship for electrodeposition of copper from acid $CuSO_4$ solution. (From Ref. 53, with permission from the Electrochemical Society.)

For example, if $C_{dl} = 50\,\mu F/cm^2$ and $R = 2\,\Omega$, $t = 4.6 \times 10^{-4}$ s (0.46 ms). Thus, in the galvanostatic transient technique, the duration of the input current density pulse is on the order of milliseconds. From a series of measurements of E_i for a set of i values, one can construct the current–potential relationship for an electrochemical process. For example, Figure 6.20 shows the current–potential relationship for the electrodeposition of copper from acid $CuSO_4$ solution.

Potentiostatic Transient Technique. In the potentiostatic technique the potential of the test electrode is controlled, while the current, the dependent variable, is measured as a function of time. The potential difference between the test electrode and the reference electrode is controlled by a potentiostat (Fig. 6.21). The input function, a constant potential, and the response function, $i = f(t)$, are shown in Figure 6.22.

Potential Sweep Method. In the transient techniques described above, a set of measurements of the potential for a given current or the current for a given potential is measured in order to construct the current–potential function, $i = f(E)$. For example, the Tafel lines shown in Figure 6.20 were constructed from a set of galvanostatic transients of the type shown in Figure 6.18. In the potential sweep technique, $i = f(E)$, curves are recorded directly in a single experiment. This is achieved by sweeping the potential with time. In linear sweep voltammetry, the potential of the test electrode is varied linearly with time (Fig. 6.23a). If the sweep rate is

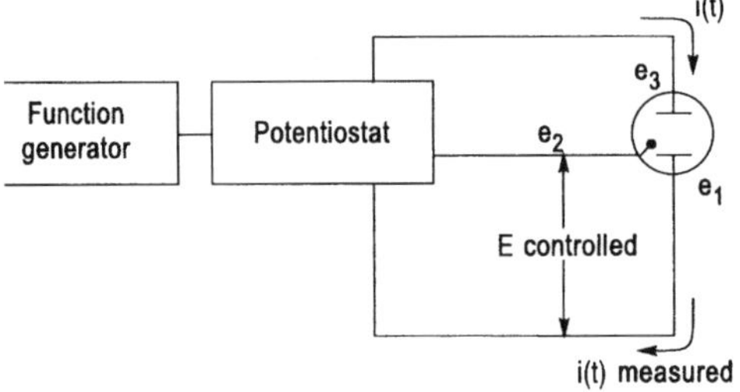

Figure 6.21. Schematic diagram of apparatus for potentiostatic measurements; E, controlled potential; e_1, test electrode; e_2, reference electrode; e_3, counter (auxiliary)-electrode.

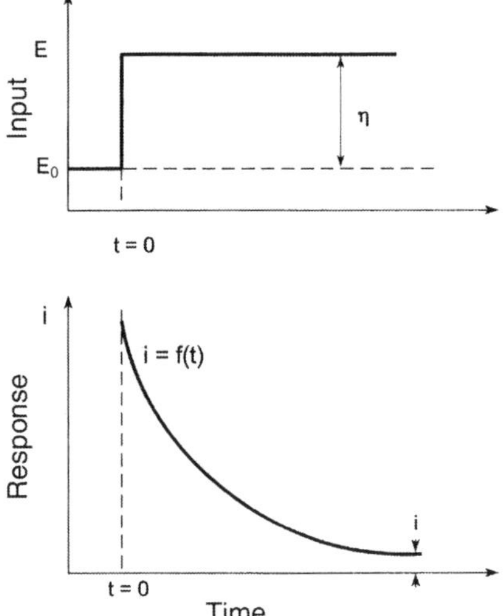

Figure 6.22. Variation of current with time during potentiostatic electrolysis.

v (mV/s), the potential E_t of the test electrode at time t in the cathodic polarization is given by

$$E_t = E_{t=0} - vt \tag{6.98}$$

where $E_{t=0}$ is the initial potential. The sweep rate is usually in the range 1 to about 1000 mV/s (depending on application). A typical current–potential response curve is shown in Figure 6.23*b*.

Rotating-Disk Electrode. Study of the kinetics and mechanism of electrode processes under well-defined mass transport conditions is possible through use of methods of the rotating-disk electrode (RDE). The RDE consists of a disk of metal embedded in a cylindrical insulator (e.g., Teflon) holder (see Fig. 6.24). It is rotated about its center. Only the bottom end of the metal disk is exposed to the solution.

For an RDE the diffusion layer thickness depends on the angular speed of rotation ω according to

$$\delta = 1.61 D^{1/3} \, \nu^{1/6} \, \omega^{-1/2} \tag{6.99}$$

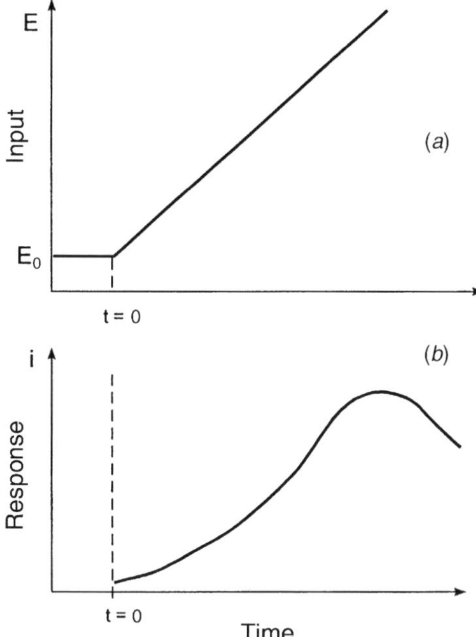

Figure 6.23. Linear potential sweep voltammetry: (a) input function; (b) response function.

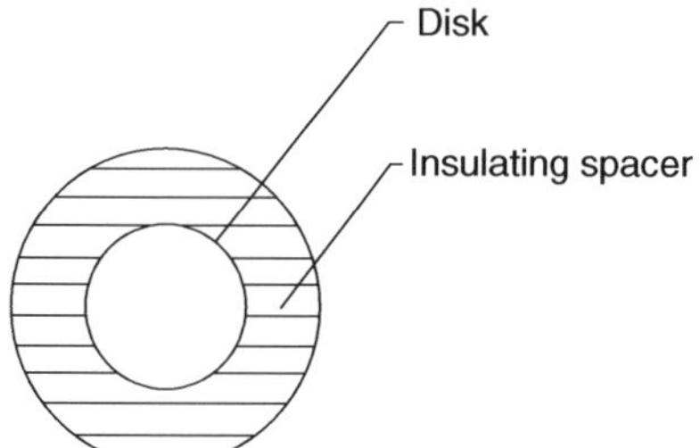

Figure 6.24. Schematic representation of the bottom of a rotating-disk electrode.

and the limiting current density depends on ω according to

$$i_L = 0.62\, nFAD^{2/3}\, c\nu^{-1/6}\, \omega^{-1/2} \tag{6.100}$$

where D is the diffusion coefficient, ν is the kinematic viscosity (the coefficient of viscosity/density of solution), ω is the rotation angular speed (in s^{-1}), c is the concentration of the solution, A is the disk surface area, and n is the number of electrons involved in the reaction (33, 34). Equations (6.99) and (6.100) may be presented in a more practical form, replacing the rotation speed ω in rad/s by $\omega = 2\pi N$, where N is the number of rotations per second (rps). For example, for Eq. (6.100) we obtain

$$i_L = 0.20\, nFAD^{2/3}\, cv^{-1/6}(\text{rpm})^{1/2} \qquad (6.101)$$

where rpm is the number of rotations per minute. We use Eq. (6.101) in the following example.

Example 6.1. Consider an RDE immersed in a solution of a metal ion M^{2+} where the concentration of M^{2+} is $10^{-5}\,\text{mol/cm}^3$ ($10^{-2}\,\text{mol/L}$). The diffusion coefficient D of M^{2+} is $2 \times 10^{-5}\,\text{cm}^2/\text{s}$, $v = 10^{-2}\,\text{cm}^2/\text{s}$. Calculate the limiting current density i_L for the deposition of a metal ion M^{2+} at the RDE with a surface area $A = 1\,\text{cm}^2$ and a rotation speed of 300 rpm. Substitution of the values given above for A, D, v, and c into Eq. (6.101) yields

$$i_L = 1.91 \times 10^2\, c\omega^{1/2} = 1.07 \times 10^{-2}\,\text{A/cm}^2 = 10.7\,\text{mA/cm}^2$$

This value should be compared with the i_L value for a quiescent (unstirred) solution in Example 6.2. $\square$

Example 6.2. Calculate the diffusion limiting current density i_L for the deposition of a metal ion M^{2+} at a cathode in a quiescent (unstirred) solution assuming a diffusion layer thickness δ of 0.05 cm. The concentration of M^{2+} ions in the bulk (c_b) is $10^{-2}\,\text{mol/L}$ ($10^{-5}\,\text{mol/cm}^3$), the same as in Example 6.1. The diffusion coefficient D of M^{2+} in the unstirred solution is $2 \times 10^{-5}\,\text{cm}^2/\text{s}$. Using Eq. (6.83), we calculate that the limiting diffusion current density for this case is

$$i_L = 0.72\,\text{mA/cm}^2$$

A comparison of this value of i_L with the value of i_L in Example 6.1 shows that there is a more than tenfold increase in i_L for the RDE at 300 rpm. $\square$

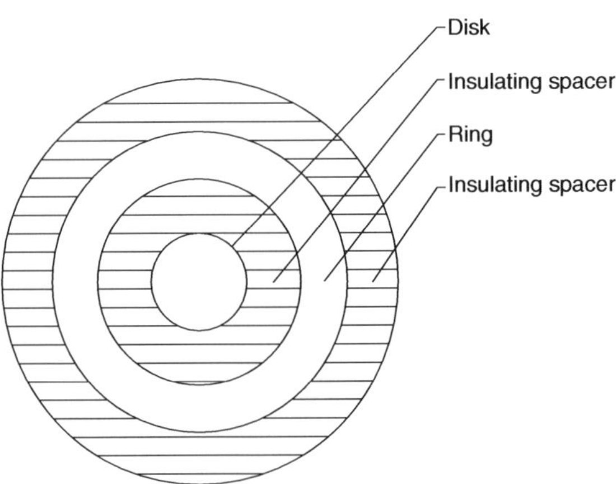

Figure 6.25. Schematic representation of the bottom of a rotating ring-disk electrode.

Rotating Ring-Disk Electrode. The rotating ring-disk electrode (RRDE) consists of two electrodes: a central disk electrode and a ring electrode (Fig. 6.25). The two electrodes are separated by a thin insulating spacer (e.g., Teflon). The electrodes and the spacer are both in the same plane. The entire system is rotated about its center. Each electrode is an independent entity. The potential (or current) of each electrode is controlled and monitored independently. The importance of the RRDE technique is that it allows identifying intermediate products (radicals) in a multistep electrode reaction occurring on the central disk electrode. The ring electrode is maintained at the potential characteristic of the intermediates produced at the disk electrode reaction.

6.10. DETERMINATION OF KINETIC PARAMETERS α AND i_0

Data presented in Figure 6.20 can be used to evaluate the Tafel slope b [Eq. (6.62a)] for the deposition and dissolution of copper. From the cathodic polarization curve we obtain the Tafel slope $\vec{b}$ as

$$\vec{b} = \frac{\partial \eta}{\partial \log |i|} = \frac{\Delta \eta}{\Delta \log |i|} = 125 \tag{6.102}$$

and the anodic polarization curve as

$$\overleftarrow{b} = \frac{\partial \eta}{\partial \log |i|} = \frac{\Delta \eta}{\Delta \log |i|} = 50 \tag{6.103}$$

The transfer coefficient α for the cathodic process may be determined with the help of Eqs. (6.102) and (6.57) and that for the anodic process using Eqs. (6.103) and (6.61). From these equations we obtain $\alpha_{\text{cathodic}} = 0.25$ and $\alpha_{\text{anodic}} = 0.59$ for the case presented in Fig. 6.20. The exchange current density i_0 may be obtained from the intercepts for $\eta = 0$ (i.e., intersection of the Tafel lines for high cathodic and anodic overpotentials). Setting $\eta = 0$ in Eq. (6.55) or Eq. (6.59), we obtain

$$\log |i| = \log i_0 \qquad \text{for } \eta = 0$$

and

$$i_{\eta=0} = i_0 = 4.79 \, \text{mA/cm}^2$$

REFERENCES AND FURTHER READING

1. W. A. Caspari, *Z. Phys. Chem.* **30**, 89 (1899).
2. J. Tafel, *Z. Phys. Chem.* **50**, 641 (1905).
3. T. Erdey-Gruz and M. Volmer, *Z. Phys. Chem.* **A150**, 203 (1930).
4. H. Eyring, *J. Chem. Phys.* **3**, 107 (1935).

5. W. F. K. Wynne-Jones and H. Eyring, *J. Chem. Phys.* **3**, 492 (1935).

6. J. Horiuti and M. Polanyi, *Acta Physicochim. URSS* **2**, 505 (1935).

7. H. Eyring, S. Glasstone, and K. L. Laidler, *J. Chem. Phys.* **7**, 1053 (1939).

8. G. E. Kimball, *J. Chem. Phys.* **8**, 199 (1940).

9. S. Glasstone, K. J. Laidler, and H. Eyring, *The Theory of Rate Processes,* McGraw-Hill, New York, 1941.

10. E. Mattson and J. O'M. Bockris, *Trans. Faraday Soc.* **55**, 1586 (1959).

11. B. E. Conway, *Theory and Principles of Electrode Processes*, Ronald Press, New York, 1965.

12. P. Delahay, *Double Layer and Electrode Kinetics*, Interscience, New York, 1965.

13. M. H. Back and K. J. Laidler, eds., *Selected Readings in Chemical Kinetics*, Pergamon Press, Oxford, 1967.

14. K. J. Vetter, *Electrochemical Kinetics: Theoretical and Experimental Aspects*, Academic Press, New York, 1967.

15. T. N. Andersen and H. Eyring, in *Physical Chemistry: An Advanced Treatise*, Vol. 9A, *Electrochemistry*, H. Eyring, ed., Academic Press, New York, 1970.

16. J. Goodisman, *Electrochemistry: Theoretical Foundations*, Wiley, New York, 1987.

17. J. O'M. Bockris and S. U. M. Khan, *Surface Electrochemistry*, Plenum Press, New York, 1993.

18. E. Gileadi, *Electrode Kinetics*, VCH, New York, 1993.

19. H. F. Weber, *Wied, Ann.* **7**, 536 (1879).

20. H. J. Sand, *Philos. Mag.* **1**, 45 (1901).

21. P. Delahay, *New Instrumental Methods in Electrochemistry*, Wiley, New York, 1954.

22. N. Ibl, in *Advances in Electrochemistry and Electrochemical Engineering*, Vol. 2, C. W. Tobias, ed., Wiley, New York, 1962.

23. M. Paunovic, *J. Electroanal. Chem.* **14**, 447 (1967).

24. D. D. Macdonald, *Transient Techniques in Electrochemistry*, Plenum Press, New York, 1977.

25. A. J. Bard and L. R. Faulkner, *Electrochemical Methods*, Wiley, New York, 1980.

26. M. Paunovic, *J. Electrochem. Soc.* **124**, 349 (1977).

27. W. Kossel, *Nachr. Ges. Wiss. Goettingen* **1927**, 135.

28. I. Stranski, *Z. Phys. Chem.* **136**, 259 (1928).

29. W. K. Burton, N. Cabrera, and F. C. Frank, *Philos. Trans. R. Soc. London* **A243**, 299 (1951).

30. W. J. Lorenz, *Z. Naturforsch.* **7a**, 750 (1952).

31. W. Lorenz, *Z. Phys. Chem.* **B202**, 275 (1953).

32. W. Lorenz, *Z. Elektrochem.* **57**, 382 (1953).

33. W. Lorenz, *Naturwissenschaften* **40**, 576 (1953).

34. W. J. Lorenz, *Z. Naturforsch.* **9a**, 716 (1954).

35. W. Mehl and J. O'M. Bockris, *J. Chem. Phys.* **27**, 817 (1957).

36. B. E. Conway and J. O'M. Bockris, *Proc. R. Soc. London* **A248**, 394 (1958).

37. E. Bauer, *Z. Kristallogr.* **110**, 372 (1958).

38. W. Mehl and J. O'M. Bockris, *Can. J. Chem.* **37**, 190 (1959).

39. H. Gerischer, *Electrochim. Acta* **2**, 1 (1960).

40. B. E. Conway and J. O'M. Bockris, *Electrochim. Acta* **3**, 340 (1961).

41. N. F. Mott and R. J. Watts-Tobin, *Electrochim. Acta* **4**, 79 (1961).

42. J. O'M. Bockris and G. A. Razumney, *Fundamental Aspects of Electrocrystallization*, Plenum Press, New York, 1967.

43. G. H. Gilmer and P. Bennema, *J. Appl. Phys.* **43**, 1347 (1972).

44. T. Vitanov, A. Popov, and E. Budevski, *J. Electrochem. Soc.* **121**, 207 (1974).

45. M. J. Jaycock and G. P. Parfitt, *Chemistry of Interfaces*, Wiley, New York, 1981.

46. M. Boudart and G. Djega-Mariadasson, *Kinetics of Heterogeneous Catalytic Reactions*, Princeton University Press, Princeton, NJ, 1984.

47. J. Lapujoulade, in *Interaction of Atoms and Molecules with Solid Surfaces*, V. Bortolani, N. H. March, and M. P. Tosi, eds., Plenum Press, New York, 1990.

48. G. Ehrlich, *Surf. Sci.* **331–333**, 865 (1995).

49. E. Budevski, G. Staikov, and W. J. Lorenz, *Electrochemical Phase Formation and Growth*, VCH, New York, 1996.

50. H. Gerischer, *Z. Elektrochem.* **62**, 256 (1958).

51. P. Delahay, in *Advances in Electrochemistry and Electrochemical Engineering*, Vol. 1, P. Delahay and C. W. Tobias, eds., Interscience, New York, 1961.

52. W. Lorenz and G. Salie, *J. Electroanal. Chem.* **80**, 1 (1977).

53. J. O'M. Bockris, in *Transactions of the Symposium on Electrode Processes*, E. Yeager, ed., Wiley, New York, 1961.

54. E. Yeager, J. O'M. Bockris, B. E. Conway, and S. Sarangapani, eds., *Comprehensive Treatise of Electrochemistry*, Vol. 9, Plenum Press, New York, 1984.

55. J. O'M. Bockris, A. K. N. Reddy, and M. E. Gamboa-Adelco, *Modern Electrochemistry*, Vol. 2, Plenum Press, New York, 2001.

PROBLEMS

6.1. Calculate the cathodic current density i for the deposition of a metal ion as a function of the overpotential η for $\eta = -10, -50, -100$, and $200\,\text{mV}$ using (a) the Butler–Volmer equation and, (b) the Butler–Volmer equation for high overpotential approximation. The exchange current density i_0 is $2 \times 10^{-3}\,\text{A/cm}^2$, the transfer coefficient α is 0.5, and the temperature is 25°C. The electrode process is a single-electron process ($z = 1$). Compare the results for parts (a) and (b) and calculate the relative error (%) in current density resulting from the use of the high-overpotential approximation. State, in addition, for what values of overpotential the high overpotential is an acceptable approximation. Use $F = 96{,}487\,\text{C/mol}$, $R = 8.3144\,\text{J/mol} \cdot \text{K}$ (1 joule = 1 V.C), and $F/RT = f = 38.9\,\text{V}$.

6.2. Calculate the cathodic current density i for the deposition of a metal ion as a function of overpotential η for $\eta = -2, -5, -10, -20, -50$, and $-100\,\text{mV}$ using (a) the Butler–Volmer equation, and (b) the Butler–Volmer equation for low-overpotential approximation. The exchange current density i_0 is $2 \times 10^{-3}\,\text{A/cm}^2$, the transfer coefficient α is 0.5, and the temperature is 25°C. The electrode process is a single-electron transfer process ($z = 1$). Compare results for parts (a) and (b) and calculate the relative error (%) in current density resulting from the use of low-overpotential approximation. State for what values of overpotential the low-overpotential approximation is acceptable. $F/RT = f = 38.9$ V.

6.3. Calculate the diffusion-limiting current density i_L for the deposition of a metal ion M^{2+} at a cathode in a quiescent (unstirred) solution assuming a diffusion layer thickness δ of 0.05 cm. The concentration of M^{2+} in the bulk (c_b) is 2×10^{-5} mol/cm^3 (2×10^{-2} mol/L) and the diffusion coefficient D of M^{2+} in the unstirred solution is 2×10^{-5} cm^2/s.

6.4. Determine the diffusion layer thickness δ for the deposition of a metal ion M^{2+} at a cathode in a quiescent (unstirred) solution using the experimentally determined diffusion limiting current density $i_L = 5 \times 10^{-3}$ A/cm^2. The concentration of M^{2+} ions in the bulk of the solution (c_b) is 1×10^{-5} mol/cm^3 and the diffusion coefficient D of the M^{2+} ions in the unstirred solution is 2×10^{-5} cm^2/s.

6.5. Determine the diffusion layer thickness δ as a function of time t for constant-current deposition of a metal ion at a cathode in a quiescent (unstirred) solution at 25°C. Use time t values of 20, 40, 80, 100, 200, and 300 s. Plot the function $\delta = f(t)$ using the δ values calculated.

7

Nucleation and Growth Models

7.1. INTRODUCTION

The first theoretical interpretation of the electrochemical crystal growth in terms of atomic models considered the substrate as a perfect crystal surface. That perfect surfaces do not have sites for growth, and nucleation has to be the first step in the deposition process. Erdey-Gruz and Volmer (1,2) formulated such a nucleation model of electrochemical crystal growth in 1930, and we describe this model. Subsequently, Frank (4) and Burton et al. (74) realized that real crystal surfaces (substrates) have imperfections and a variety of growth sites. This consideration introduced a major change in the theoretical interpretation of the deposition process and resulted in a series of new models. Further major contributions to an understanding of the electrochemical crystal growth stem from experimental applications of in situ surface analytical methods, including scanning tunneling microscopy (STM) and atomic force microscopy (AFM). In this chapter we discuss basic models and show how significant contributions were made in understanding the initial stages in electrochemical deposition, which moved deposition from "art" to "science."

7.2. FORMATION AND GROWTH OF ADION CLUSTERS

In the formation and growth of adion clusters, two processes are of fundamental importance: (1) the arrival and adsorption of ions (atoms) at the surface, and (2) the motion of these adsorbed ions (adions, adatoms) on the surface. An adion deposited on the surface of a perfect crystal stays on the surface as an adion only temporarily since its binding energy to the crystal is small. It is not a stable entity on the surface, but it can increase its stability by the formation of clusters. The free energy of formation of a cluster of N ions, $\Delta G(N)$, has two components (terms):

$$\Delta G(N) = -Nze|\eta| + \phi(N) \tag{7.1}$$

Fundamentals of Electrochemical Deposition, Second Edition.
By Milan Paunovic and Mordechay Schlesinger
Copyright © 2006 John Wiley & Sons, Inc.

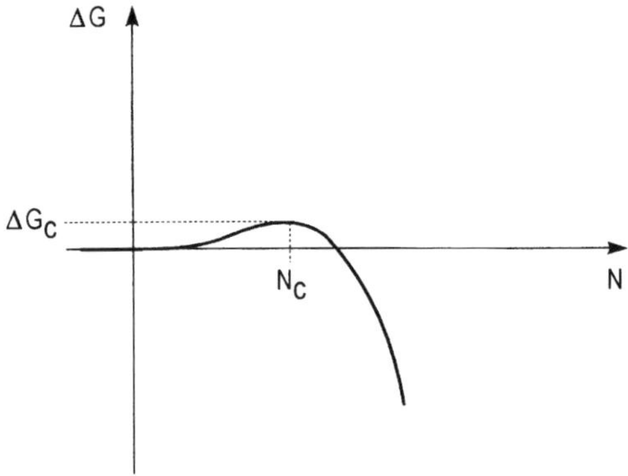

Figure 7.1. Free energy of formation of a cluster as a function of size N (a cluster of N atoms); N_c, the size of the critical cluster (nucleus).

where the first term is related to the transfer of N ions from solution to the crystal phase and the second term is related to the increase of the surface energy due to creation of the surface of a cluster. This increase in surface energy, or this excess energy, is equal to the difference of the binding energies of N bulk ions and N ions as arranged on the surface of the crystal.

Both terms in Eq. (7.1) are functions of the size of the cluster N. The first term increases linearly with N, and the second increases as $N^{2/3}$. Dependence of the energy of formation of a cluster $\Delta G(N)$ on the number of adions N in a two-dimensional (2D) cluster is shown in Figure 7.1. It is seen from the figure that ΔG initially increases, reaches a maximum, and then decreases with increasing N. At the maximum, the cluster size is N_c. The size of the critical nucleus N_c (the number of atoms in the cluster) in 2D nucleation is given by

$$N_c = \frac{bs\varepsilon^2}{(ze\eta)^2} \tag{7.2a}$$

where b is the factor relating the surface area S of the nucleus to its perimeter P ($b = P^2/4S$; $b = \pi$ for a circular nucleus), s is the area occupied by one atom on the surface of the nucleus, and ε is the edge energy. For Ag nuclei, using Eq. (7.2a), one calculates that N_c at $\eta = 5$, 10, and 25 mV is 128, 32, and 5, respectively ($s = 6.55 \times 10^{-16}\,\text{cm}^2$, $\varepsilon = 2 \times 10^{-13}\,\text{J/cm}$, $z = 1$, and $e = 1.602 \times 10^{-19}\,\text{C}$). Thus, N_c depends strongly on the overpotential; it is inversely proportional to η^2. The critical radius of the surface nucleus r_c is a function of the overpotential:

$$r_c = \frac{s\varepsilon}{ze\eta} \tag{7.2b}$$

For Ag ($r = 1.444\,\text{Å}$) the critical radius r_c at $\eta = 5$, 10, and 25 mV is 16.35, 8.18, and 3.27 Å, respectively.

The spontaneous growth of clusters is possible after the maximum in ΔG is reached. This critical cluster is the nucleus of the new phase and is characterized by equal probability for growth and dissolution. The growth of clusters before a maximum is reached, and when average ΔG is increasing, is due to statistical energy fluctuations that allow local higher-value ΔG, beyond the maximum (8).

In some theoretical treatments of the growth–decay process of clusters, growth is considered to proceed by the gain or loss of single adions, A_1. Thus, allowed reactions for the model system are

$$A_1 + A_1 \rightleftharpoons A_2$$
$$A_2 + A_1 \rightleftharpoons A_3$$
$$A_{m-1} + A_1 \rightleftharpoons A_m$$

where m is a number slightly larger than the size of the cluster whose Gibbs free energy of formation is maximum.

7.3. NUCLEATION OF SURFACE NUCLEI

The nucleation law for a uniform probability with time t of conversion of a site on the metal electrode into nuclei is given by

$$N = N_0(1 - \exp(-At)) \tag{7.3a}$$

where N_0 is the total number of sites (the maximum possible number of nuclei on the unit surface) and A is the nucleation rate constant. This equation represents the first-order kinetic model of nucleation. The rate of 2D nucleation J is given by

$$J = k_1 \exp\left(-\frac{bs\varepsilon^2}{zekT\eta}\right) \tag{7.3b}$$

were k_1 is the rate constant, b is a geometric factor depending on the shape of the 2D cluster [Eq. (7.2a)], s is the area occupied by one atom on the surface of the cluster, ε is the specific edge energy, and k is the Boltzmann constant; the other symbols have their usual meaning.

There are two limiting cases for Eq. (7.3) for the initial stages of nucleation (low t value). First, for large nucleation constant A, Eq. (7.3) reduces to

$$N \simeq N_0 \tag{7.4}$$

indicating that all electrode sites are converted to nuclei instantaneously. Thus, this is referred to as *instantaneous nucleation*. Second, for small A and small t, Eq. (7.3) reduces to

$$N \simeq AN_0t \tag{7.5}$$

since the exponential term in Eq. (7.3) may be represented as a linear approximation $(-e^{-At} \simeq -1 - At)$. In this case the number of nuclei N is a function of time t and the nucleation is termed *progressive*. It is possible to distinguish between these two modes of nucleation experimentally, such as by the use of potentiostatic current–time transients (discussed in Section 7.7).

7.4. GROWTH OF SURFACE NUCLEI

When the charge-transfer step in an electrodeposition reaction is fast, the rate of growth of nuclei (crystallites) is determined by either of two steps: (1) the lattice incorporation step or (2) the diffusion of electrodepositing ions into the nucleus (diffusion in the solution). We start with the first case. Four simple models of nuclei are usually considered: (*a*) a two-dimensional (2D) cylinder, (*b*) a three-dimensional (3D) hemisphere, (*c*) a right-circular cone, and (*d*) a truncated four-sided pyramid (Fig. 7.2).

Growth of Independent Nuclei. In the initial stages of growth of the nuclei it can be assumed that nuclei grow independent of each other. In this case the rate of growth of a single (free) 2D cylindrical nucleus is given by

$$i = \frac{nFk^2 \cdot 2\pi hM}{\rho} t \tag{7.6}$$

where k is the rate constant of 2D nucleus growth (in mol/cm$^2 \cdot$ s), h is the height of a cylindrical nucleus (a height of a monolayer), M is the molecular weight, and ρ is the density of deposit (mass/volume). A 2D nucleus grows only laterally.

The rate of growth of a single 3D hemispherical nucleus is given by

$$i = \frac{nFk^3 \cdot 2\pi M^2}{\rho^2} t^2 \tag{7.7}$$

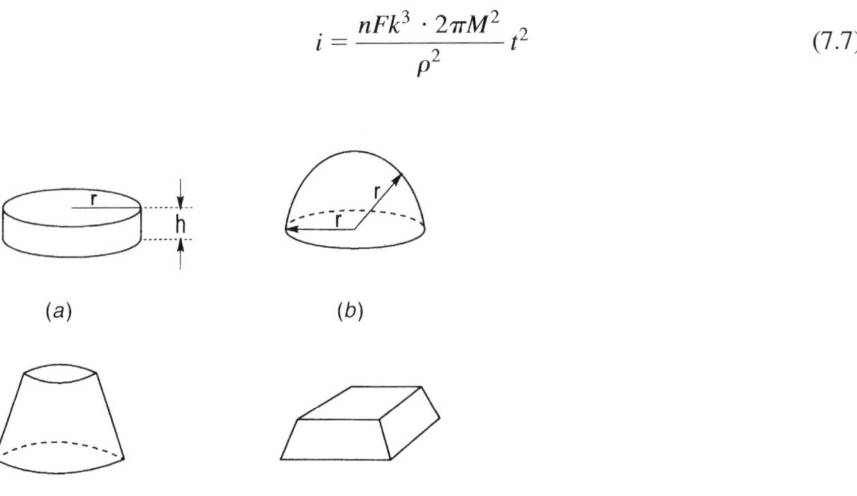

Figure 7.2. Models of surface nuclei.

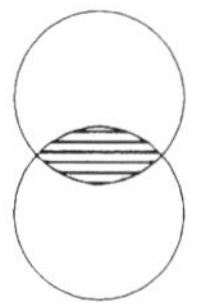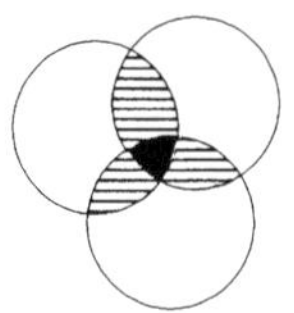

Figure 7.3. Overlap of diffusion zones of cylindrical nuclei growing on a surface. Shaded regions indicate two zones overlapping; black region, three zones overlapping.

Growth of Interacting Nuclei. The main assumption made in deriving Eqs. (7.6) and (7.7) is that nuclei are independent of each other. This is valid only for the initial stages of growth. However, in succeeding stages it is necessary to consider the effect of the overlap between diffusion fields around growing nuclei (Fig. 7.3). The result of this overlap is the development of local concentration and overpotential distribution in the neighborhood of the growing nuclei (clusters). Overlap areas are zones of reduced concentration and reduced nucleation rate. Growing nuclei cannot grow freely in all directions since they will impinge on each other. The growth will stop at the point of contact, resulting in a limitation in the size of the growth center. For further discussion, see Section 7.10.

7.5. SIMULTANEOUS NUCLEATION AND GROWTH OF NUCLEI

The overall current–time relationships for the simultaneous nucleation and growth of nuclei are of the form $i \propto t^{\beta}$, where β is a variable depending primarily on the model of nuclei (2D, 3D) and the type of nucleation (instantaneous, progressive).

Without Overlap. The equations of growth for a single nucleus, such as Eqs. (7.6) and (7.7), can be combined with the equations of nucleation, such as Eqs. (7.3) to (7.5), to give the overall current–time relationship. Thus, for simultaneous two-dimensional (2D) cylindrical growth and instantaneous nucleation,

$$i = \frac{2nF\pi MhN_0 K^2}{\rho} t \tag{7.8}$$

and 2D growth and progressive nucleation,

$$i = \frac{nF\pi MhAN_0 K^2}{\rho} t^2 \tag{7.9}$$

For simultaneous three-dimensional (3D) hemispherical growth and instantaneous nucleation,

$$i = \frac{2nF\pi M^2 N_0 K^3}{\rho^2} t^2 \tag{7.10}$$

and 3D growth and progressive nucleation,

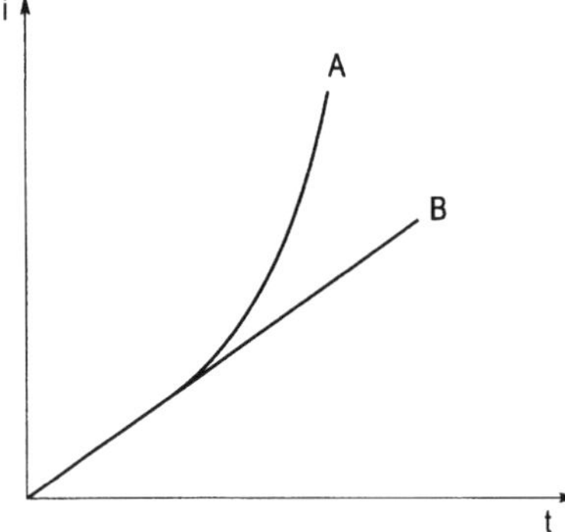

Figure 7.4. Overall current–time relationship for the simultaneous nucleation and growth of nuclei: (A) without overlap; (B) with overlap.

$$i = \frac{2nF\pi M^2 A N_0 K^3}{3\rho^2} t^3 \tag{7.11}$$

It can be seen that Eqs. (7.9) and (7.10) represent the same type of current–time transient, $i \propto t^2$. Thus, to distinguish between 2D growth (progressive nucleation) and 3D growth (instantaneous nucleation), it is necessary to perform additional optical microscopic or electron microscopic experiments. These experiments can provide information enabling one to distinguish between progressive nucleation [Eq. (7.9)] and instantaneous nucleation [Eq. (7.10)].

Equations (7.8) to (7.11) show that the exponent β in the $i \propto t^\beta$ function may vary from 1 to 3. Therefore, all models presented here predict that the current density i increases for all time (Fig. 7.4). This is not what is observed experimentally and thus is unacceptable. A new model that takes into account the effect of overlap between diffusion fields around nuclei (Fig. 7.3) is required. This is the subject of the next section.

With Overlap. For two-dimensional cylindrical growth and instantaneous nucleation,

$$i = \frac{2nF\pi h M N_0 k^2}{\rho} t \exp\left(-\frac{\pi M^2 N_0 k^2 t^2}{\rho^2} \right) \tag{7.12}$$

and for progressive nucleation,

$$i = \frac{nF\pi h M A k^2}{\rho} t^2 \exp\left(-\frac{\pi M^2 A k^2 t^3}{3\rho^2} \right) \tag{7.13}$$

A comparison between Eqs. (7.8)–(7.9) and (7.12)–(7.13) shows that the exponential terms in Eqs. (7.12) and (7.13) represent the overlap effect (overlap correction). This new model of simultaneous nucleation and growth of nuclei, Eqs. (7.12) and (7.13),

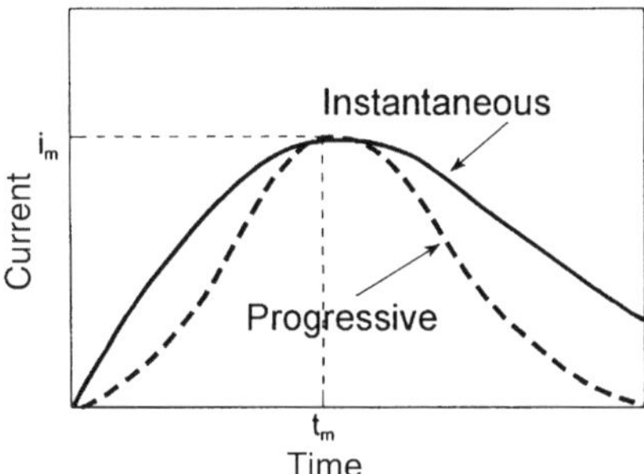

Figure 7.5. Theoretical current–time transients for instantaneous and progressive nucleation.

predicts an initial increase in current, reaching a maximum current, and a decrease in current. In the case of progressive nucleation, the current decreases asymptotically to zero (Fig. 7.5). This model of nucleation and growth is in agreement with experimental determination of $i = f(t)$, as shown in the next section.

7.6. FORMATION OF A SINGLE MONOLAYER

There are two mechanisms for formation of a monolayer: (1) the instantaneous nucleation mechanism according to Eq. (7.12)—in this case the monolayer is spreading out on the substrate from nuclei formed at time $t = 0$; and (2) the progressive nucleation mechanism, in which, according to Eq. (7.13), nuclei appear randomly in space and time. The current–time relationships for these two mechanisms are shown in Figure 7.5. In both cases the current passes through a maximum.

7.7. FORMATION OF MULTILAYERS

Two different mechanisms of multilayer growth on a perfect or quasiperfect surface can be distinguished: (1) mononuclear layer-by-layer growth and (2) multinuclear mutilayer growth. The first mechanism proceeds at low overpotentials, that is, overpotentials slightly above the critical overpotential. In this case the nucleation rate is lower than the rate of nucleus growth, and each nucleus spreads over the entire surface before the next nucleus is formed. Thus, each layer is formed by one nucleus only. Figure 7.6 shows that in this case the current fluctuates. A time integral under the i–t pulse (fluctuation) is equal to the deposition of one monoatomic layer.

At higher overpotentials the nucleation rate increases faster than the step (Chapter 3) propagation rate, and the deposition of each layer proceeds with the formation of a large number of nuclei. This is the multinuclear multilayer growth. Armstrong and Harrison (13) have shown that initially, the theoretical current–time transient for the two-dimensional nucleation (Fig. 7.7) has a rising section, then passes through several damped oscillations, and finally, levels out to a steady state.

The initial rise section in Figure 7.7, which follows the quadratic law (i–t^2) given by the preexponential term in Eq. (7.13), corresponds to formation of the first monolayer.

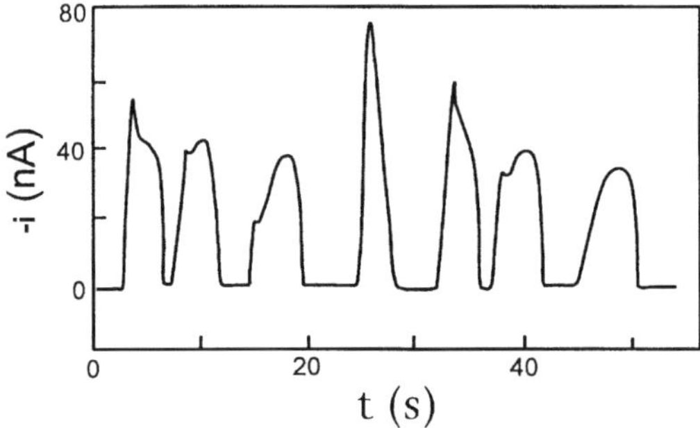

Figure 7.6. Part of a current–time record during a mononuclear layer-by-layer growth of a quasiperfect Ag(100) crystal face with a circular form in the standard system Ag(100)/6 M AgNO$_3$ at $\eta = -6\,$mV and $T = 318\,$K. Surface area A $= 3 \times 10^{-4}\,$cm^{-2}. Current density $i = 1\,$mA/cm^2. The current spikes indicate the formation, growth, and decay of new layers. (From Ref. 34, with permission from the Electrochemical Society.)

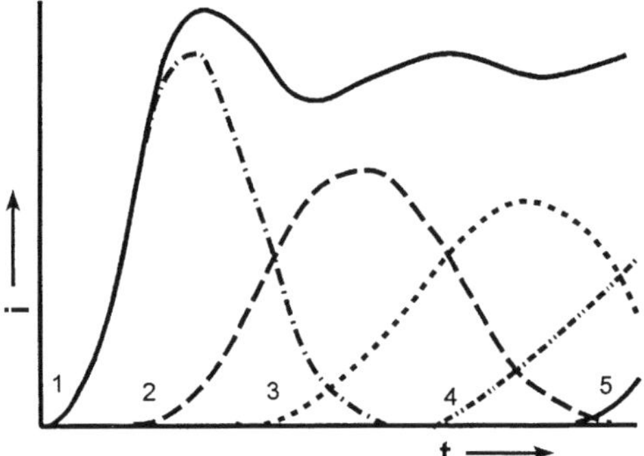

Figure 7.7. Potentiostatic current–time transient for the metal deposition together with theoretical currents for individual layers (1–5). Two-dimensional progressive nucleation taking overlap into account. (From Ref. 13, with permission from the Electrochemical Society.)

Figure 7.7 also shows the theoretical i–t transients for the formation of successive layers under conditions of progressive nucleation. The theoretical current–time transient for three-dimensional nucleation is shown in Figure 7.8. The difference between 2D and 3D nucleation (Fig. 7.7 and 7.8) is in the absence of damped oscillations in the latter case. A comparison between the theoretical and experimental transients for the 2D polynuclear multilayer growth is shown in Figure 7.9.

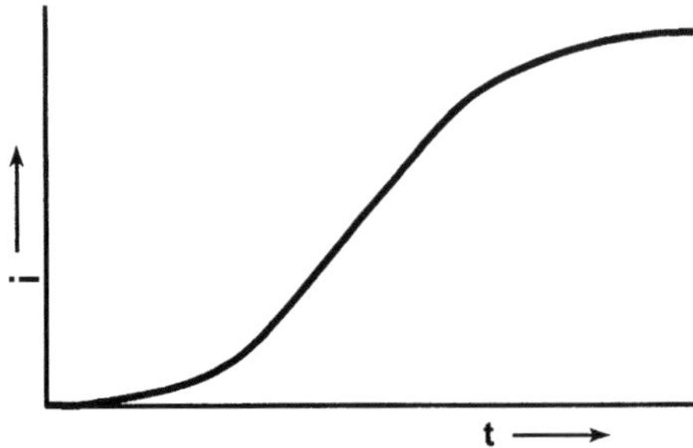

Figure 7.8. Current–time transient for three-dimensional nucleation taking overlap into account. (From Ref. 13, with permission from the Electrochemical Society.)

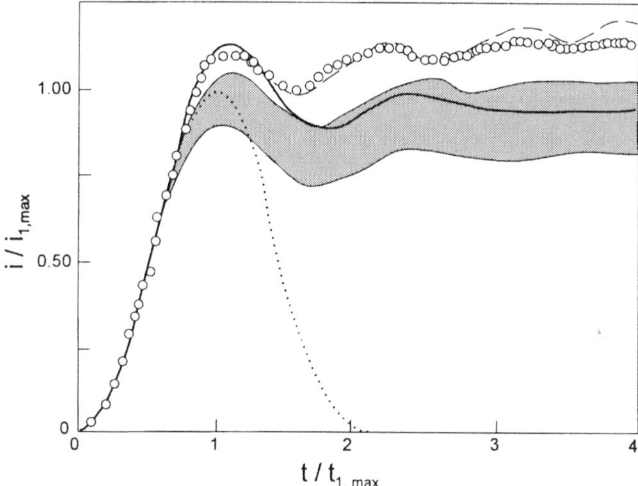

Figure 7.9. Experimental and theoretical (open circles, Monte Carlo simulation) current transients for polynuclear multilayer growth. (From Ref. 22, with permission from Annual Reviews, Inc.)

Potentiostatic Current–Time Transients. The theoretical potentiostatic current–time transient, including the effect of overlap, is shown in Figure 7.10. The potentiostatic transient can be divided into three time intervals. At the beginning, the first time interval, the current decays during the process of nucleation and growth. This is the double-layer charging current, i_{dl}. In the second time interval the current increases. This increase can be due to the growth of either independent nuclei alone or independent nuclei and simultaneous increase in number of nuclei. This is the deposition current without the overlapping effect, i_{free}. If the nucleation is instantaneous, the current i_{free} increases linearly with time. If nucleation is progressive, the current i_{free} increases as t^2. In the third time interval there are two opposing effects: growth of independent nuclei and

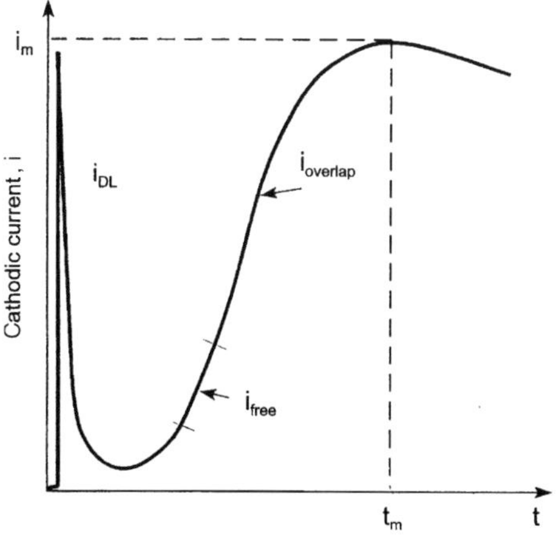

Figure 7.10. Theoretical potentiostatic current–time transient, including the effect of overlap. (From Ref. 27, with permission from Elsevier.)

overlap. The result of these opposing effects is an initial increase in the current, a maximum, and a decrease in the current in the third time interval, $i_{overlap}$. The current increases until the growth centers begin to overlap. A decrease in the current starts when the diffusion zones around the nuclei overlap and the growth centers impinge on each other. Growth stops at the point of contact. At that time the current decreases as a result of a decrease in the effective electrode surface area and a change from the hemispherical to the linear mass transfer and to an effectively planar surface.

Diagnostic Relationships Between Current and Time. Theoretical diagnostic relationships between current and time for 2D nucleation are given in the following equations. For the instantaneous nucleation case,

$$\ln \frac{i}{t} = a - bt^2 \tag{7.14}$$

and for the progressive nucleation,

$$\ln \frac{i}{t^2} = c - dt^3 \tag{7.15}$$

where a, b, c, and d are constants (which can be derived theoretically).

Diagnostic Relationships Between Current, Maximum Current, and Time. Scharifker and Hills (26) developed a theory that deals with the potentiostatic current transients for 3D nucleation with diffusion-controlled growth. According to this theory, the theoretical diagnostic relationship in a nondimensional form is given by

$$\left(\frac{i}{i_m} \right)^2 = \frac{3.8181}{t/t_m} \left\{ 1 - \exp\left[-1.2564 \left(\frac{t}{t_m} \right) \right] \right\}^2 \tag{7.16}$$

for instantaneous nucleation, and

$$\left(\frac{i}{i_m} \right)^2 = \frac{1.2254}{t/t_m} \left\{ 1 - \exp\left[-2.3367 \left(\frac{t}{t_m} \right)^2 \right] \right\}^2 \tag{7.17}$$

for progressive nucleation. Theoretical current transients (i/i_m) according to Eqs. (7.16) and (7.17) have the same shape as those shown in Figure 7.5.

Equations (7.16) and (7.17) are used in an analysis of experimental data. For example, Rynders and Alkire (32) used these equations to analyze copper electrodeposition on platinum. They concluded that at the intermediate overpotentials (120 and 170 mV), the dimensionless current transients are consistent with the theoretical predictions for progressive nucleation, Eq. (7.17). At overpotentials higher than 220 mV, nucleation shifted to the instantaneous nucleation theory.

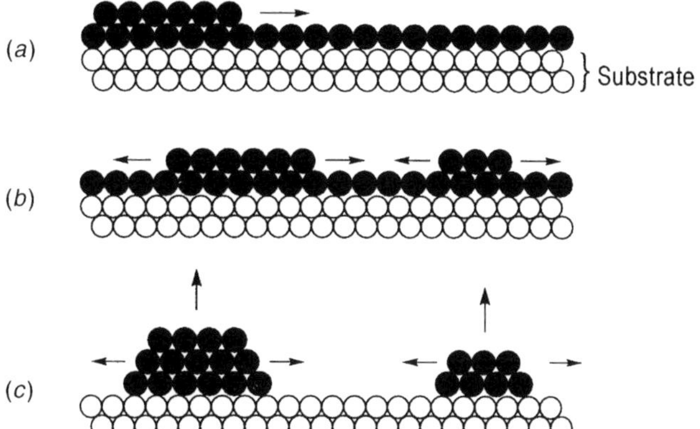

Figure 7.11. Schematic representation of layer growth (*a,b*) and the nucleation–coalescence mechanism (*c*).

7.8. FORMATION OF COHERENT DEPOSIT

Up to this point we have considered the mechanism of deposition of a single monoatomic layer (Section 7.6) and multilayers composed of a few monoatomic layers (Section 7.7). In Sections 7.8 to 7.12 we discuss how coherent electrodeposits develop.

There are two basic mechanisms for formation of a coherent deposit: layer growth and 3D crystallite growth (or nucleation–coalescence growth). A schematic presentation of these two mechanisms is shown in Figure 7.11. The layer growth mechanism is discussed in Section 7.9. In this mechanism a crystal grows by the lateral spreading of discrete layers (steps), one after another across the surface. In this case a growth layer, a step, is a structure component of a coherent deposit.

The 3D crystallites growth mechanism is discussed in Section 7.10. In this case the structure components are 3D crystallites, and a coherent deposit is built by coalescence (joining) of these crystallites.

7.9. LAYER GROWTH MECHANISM

Steps, or growth layers, are structure components for construction of a variety of growth forms in the electrodeposition of metals (e.g., columnar crystals, whiskers, fiber textures). We can distinguish between monoatomic steps, polyatomic microsteps, and polyatomic macrosteps. Only the propagation of polyatomic steps can be observed directly, in situ.

Microsteps. There is in general a tendency for a large number of thin steps to bunch into a system of a few thick steps. Many monoatomic steps can unite (bunch, coalesce) to form a polyatomic step. Frank (4) proposed a bunching mechanism to explain this process. Bunching of steps is illustrated schematically in Figure 7.12.

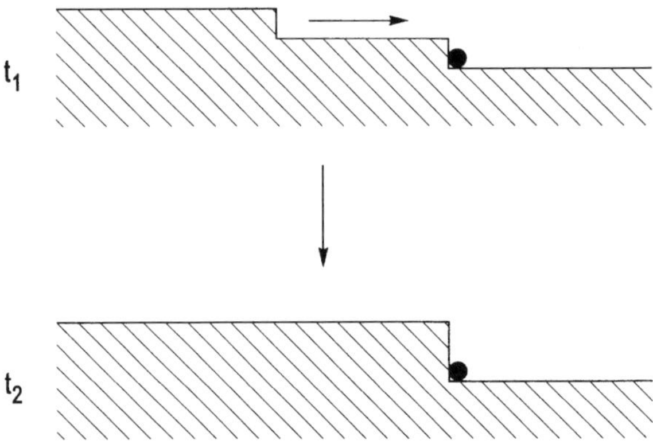

Figure 7.12. Bunching of steps.

Propagation of copper microsteps on real surfaces of Pt was observed by Rynders and Alkire (32) using in situ atomic force microscopy (AFM). They used a Pt single crystal as the substrate, cut not exactly parallel to the (100) plane but with misorientation of about 2°. This misorientation in cutting resulted in microsteps 25 to 50 Å in height and terraces approximately 1 μm in length (Fig. 7.13a). Copper was electrodeposited from a $CuSO_4/H_2SO_4$ electrolyte. A sequence of potential (or current) pulses (10 ms to 10 s) was used to obtain a sequence of data that represent the initial growth processes. Figure 7.13 shows AFM images which illustrate that the growth of copper on Pt (100) at 1 mA/cm^2 was initiated along the largest of the steps and that the deposition rate was uniform along the step. At an intermediate current density, 10 mA/cm^2, depositions grew on all steps. At current densities above 15 mA/cm^2, deposition occurred on both step and terrace sites.

Propagation of microsteps with a height of 30 to 100 Å (15 to 50 monoatomic layers) on a quasi-ideal surface of Ag was observed directly by Bostanov et al. (60) using the Nomarski differential contrast technique.

Macrosteps. Polyatomic macrosteps originate either from screw dislocations or from 3D nucleation. In the former case, steps are self-perpetuating, and in the latter case, they are nucleation dependent. Macrosteps can be formed by bunching of 1000 or more microsteps.

Using an optical microscope, Wranglen (43) observed that many layers originate, one after the other, at growth centers. At low current densities, a new layer does not start until the former has reached the edge of the crystal. At higher current densities, new layers are developed before the forerunners have reached their final size at the edge of the crystal. In this case many layers propagate simultaneously. Damjanovic et al. (47) observed in situ lateral and vertical growth of macrosteps using optical microscopy. They found that the average velocity of the step motion is about 2×10^{-6} cm/s in the

Figure 7.13. Propagation of copper microsteps on real surfaces of Pt. AFM images ($z = 100$ nm/div) of copper electrodeposition at $i = 1$ mA/cm^2 for a total charge of (a) 0.0, (b) 0.8, and (c) 1.6 mC. (From Ref. 32, with permission from the Electrochemical Society.)

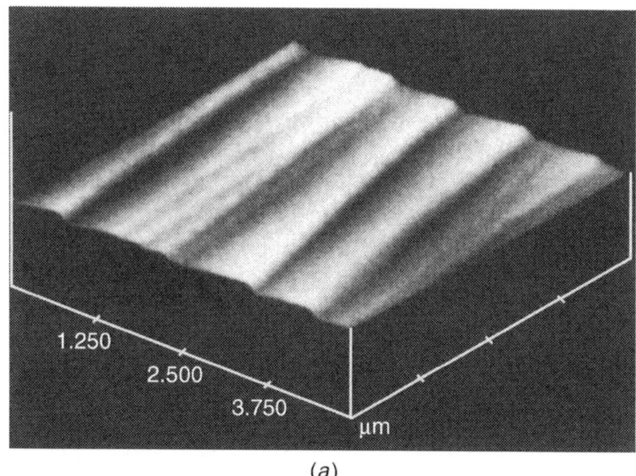

(a)

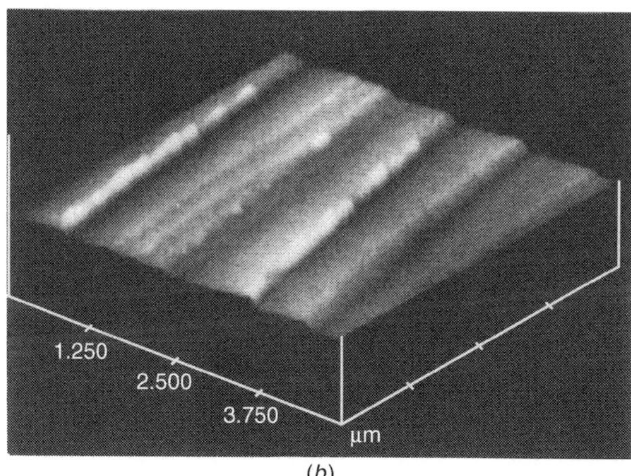

(b)

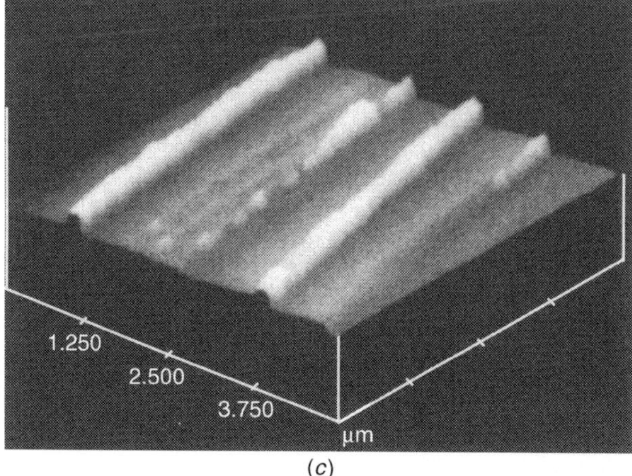

(c)

case of current density of $5\,mA/cm^2$ for electrodeposition of Cu on a Cu(100) single-crystal face from a $CuSO_4/H_2SO_4$ solution. The average height of steps varies linearly with the thickness of the deposit (2 to 12×10^{-6} cm at $5\,mA/cm^2$ for the time of deposition of 9 to 35 m).

Lateral Merging of Steps.   Bertochi and Bertochi (53) found that the principal mechanism for the formation of the defect clusters in the electrodeposition of Cu on single crystals is the interaction between steps laterally merging and annihilating. Defect concentration tends to increase when the misorientation is larger than a few degrees.

7.10. NUCLEATION–COALESCENCE GROWTH MECHANISM

As an example of the nucleation–coalescence growth mechanism, we describe an example of the nucleation and growth of electrodeposited gold on (111) surfaces of silver by means of electron microscopy (46). They found that growth from $HAuCl_4-KCl$ solutions occurs by a nucleation mechanism and that in the first stage of deposition, gold is in the form of very thin platelike isolated islands [or 3D crystallites (TDCs)]. At this stage the isolated nuclei or TDCs had a population density of $\simeq 4 \times 10^{10}\,cm^{-2}$, and the total surface coverage was about 22%. After this stage of isolated nuclei, further deposition leads to the coalescence of some of the TDCs into the form of elongated crystallites, with a reduction in population to $\simeq 1 \times 10^{10}\,cm^{-2}$. In the next stage a linked network structure is formed corresponding to a surface coverage of about 58%. The next structure can be described as a continuous deposit containing holes and channels. Further deposition leads to the filling of these holes and channels, and a completely hole-free film is formed at a thickness of 80 to 100 Å.

Thus, the sequence of growth of electrodeposited gold by the nucleation–coalescence mechanism has four stages: (1) formation of isolated nuclei and their growth to TDCs, (2) coalescence of TDCs, (3) formation of a linked network, and (4) formation of a continuous deposit. This sequence of growth stages is shown schematically in Figure 7.14.

It is interesting to note that it was concluded in this work that the first few angstroms of deposit form an alloy layer. Another interesting and important conclusion in this work is that the sequence of growth of electrodeposits (at 20°C) is very similar to that of the evaporated deposits (at 250 to 300°C).

Tanabe and Kamasaki (52) observed the nucleation growth mechanism in the deposition of Au on Fe(001) and Fe(110) single crystals. The population of nuclei (TDCs) of Au electrodeposited in the initial stages of deposition was $3 \times 10^{11}\,cm^{-2}$. In further deposition, micro-TDCs were connected one to another forming a network structure. Stable coherent deposits of Au were formed when the surface coverage was about 80%.

Three-dimensional epitaxial crystallites (TECs) were observed in the first stages of electrodeposition of copper (51) and nickel (58) on copper substrates. TECs of nickel formed on copper–film substrate from nickel sulfate solutions in low concentration are shaped rectangularly with edges averaging 1300 Å in length. The coherent

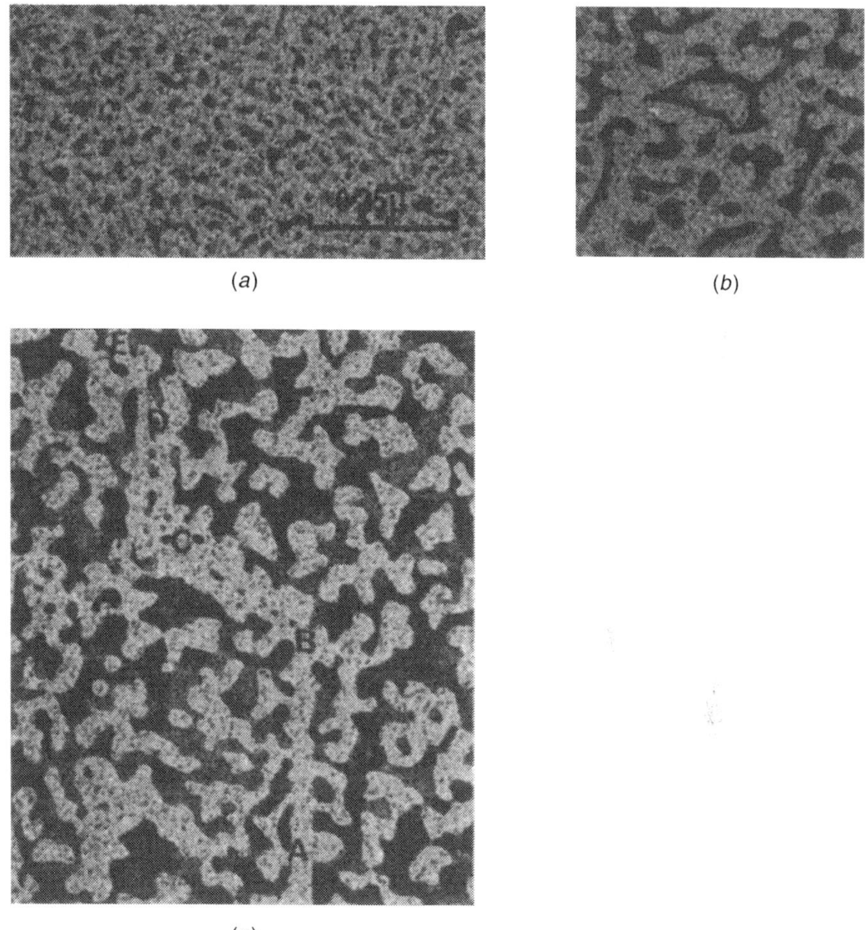

Figure 7.14. Sequence of growth of electrodeposited gold: (*a*) 20 Å; (*b*) 25 Å; (*c*) 40 Å; $I = 200\,\mu A$, $A = 20\,mm^2$. (From Ref. 46, with permission from Taylor & Francis.)

deposit was formed by growth and coalescence of TECs. Continuous deposition was observed when the average thickness was 20 to 50 Å.

Coalescence-Induced Defects. The two most significant coalescence-induced defects are voids and dislocations. Nakahara (61) has shown by transmission electron microscopy that microscopic voids (<50 Å) are generated in electrodeposited Cu, Au, and Ni–P films at the boundaries between 3D crystallites during their coalescence. When the coalescing crystallites are misoriented with respect to each other, the coalescing boundaries are grain boundaries. Weil and Wu (57) have shown that misorientation (misalignment) between neighboring Ni crystallites generates dislocations.

7.11. DEVELOPMENT OF TEXTURE

Here we examine models that try to explain how textures develop during deposition on oriented (single-crystal), textured, polycrystalline, and amorphous substrates. We select electrodeposition of nickel as a model system.

Single-Crystal Substrate. Single crystals are selected as substrates for two main reasons: (1) to eliminate the effects of grain boundaries, and (2) to eliminate heterogeneity of the substrate due to the presence of grains of various orientations and thus various exchange current densities. Single-crystal substrates exert a strong epitaxial influence on the growth process. Amblart et al. (64) have shown that Ni epitaxial growth from Watts solution starts with the formation of epitaxial crystallites (TECs). Three-dimensional epitaxial crystallites then coalesce to give a continuous epitaxial layer. Further growth is complex and involves two different nucleation processes: 3D epitaxial nucleation and independent nucleation. The first process includes epitaxial growth, and the second includes nonepitaxial growth. The balance between these competitive processes depends on the substrate orientation and the deposition conditions. For example, on a (100) Cu face, epitaxial growth continues beyond Ni deposit thickness of 10 μm. In this case, further growth via the independent nucleation is predominant and a transition from epitaxial growth to polycrystalline growth occurs.

Polycrystalline Randomly Oriented Substrate. Electrodeposition on a randomly oriented polycrystalline substrate can result in the development of preferred orientation, or texture, in thicker deposits. In a polycrystalline material, crystallographic axes of the individual grains (individual crystallites) that constitute the material are randomly oriented with respect to the axes of a fixed reference system (Fig. 7.15a). If one or more crystallographic axes of grains constituting the polycrystal are fixed (have the same orientation) with respect to the axes of the reference system, the polycrystalline material exhibits preferred orientation or texture (Fig. 7.15b). The development of texture can occur during deposition or during postdeposition processing.

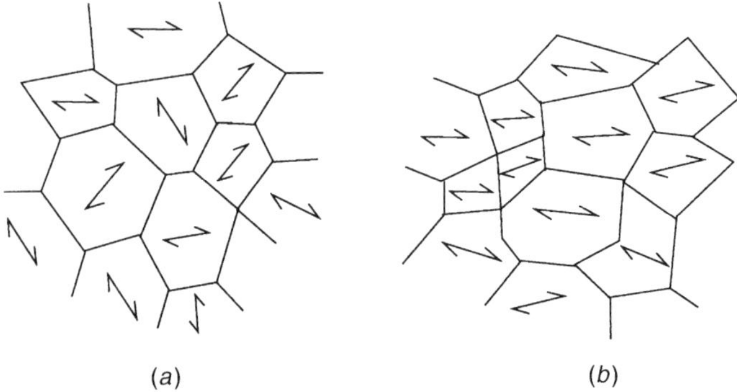

(a)   (b)

Figure 7.15. Schematic representation of polycrystalline randomly oriented substrate (a) and substrate with preferred orientation (texture) (b).

The competitive growth model of development of texture during deposition is based on the idea that different crystal faces have different rates of growth. Thus, there is growth rate competition between crystallites of various orientations. Crystallites of various orientations could be generated either during a preferential nucleation process or during the competitive growth mechanism subsequent to the stage of coalescence. The type of texture depends on the composition of electrolyte and substrate, the overpotential, and other parameters.

Amorphous Substrate.   An amorphous substrate (e.g., vitreous carbon) is without epitaxial influence; it is inert with respect to the growth process of the deposit. Amblart et al. (64) have shown that in the initial stages of Ni growth from Watts solution on vitreous carbon, substrate orientation of individual 3D nuclei is random. They have also shown that a newly coalesced compact deposit has perfectly random orientation. From these observations they concluded that the texture of thicker Ni deposits in this case is the result of a competitive growth mechanism occurring in a stage of growth subsequent to the coalescence stage.

7.12. DEVELOPMENT OF COLUMNAR MICROSTRUCTURE

Columnar microstructure, perpendicular to the substrate surface, is shown schematically in Figure 7.16. This microstructure is composed of relatively fine grains near the substrate, which then changes to the columnar microstructure with much coarser grains at greater distances from the substrate. The development of columnar microstructure can be interpreted on the basis of growth competition between adjacent grains similar to the method used for the development of texture (Section 7.11). The low-surface-energy grains grow faster than do the high-energy grains. This rapid growth of low-surface-energy grains at the expense of high-energy grains results in an increase in mean grain size with increased thickness of deposit and transition from a fine grain size near the substrate to a coarse columnar grain size. Columnar structures develop in deposits prepared by electrodeposition and evaporation. Srolovitz et al. (66) developed a theoretical model for the growth of columnar microstructure in vapor-deposited films. We also find this work interesting for electrodeposition. From this work we present Figure 7.17, which shows that the microstructure of the evaporated film, parallel to the substrate surface, has a relatively uniform microstructure composed of fine grains near the substrate surface and a bimodal grain size distribution at intermediate depths.

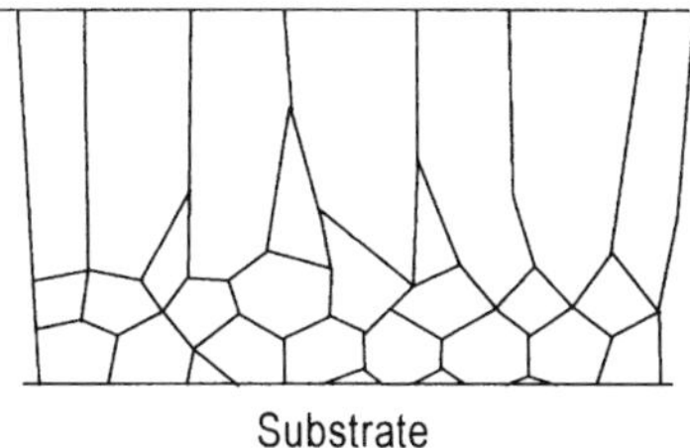

Figure 7.16. Schematic cross section (perpendicular to the substrate) of the columnar deposit.

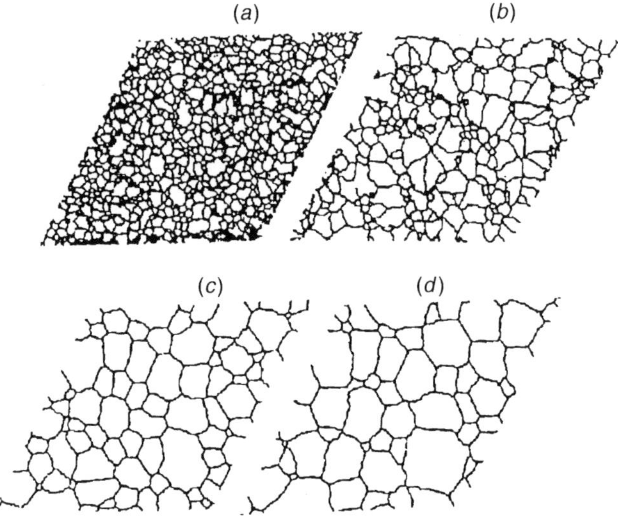

Figure 7.17. Theoretical model of the microstructure of columnar film, parallel to the substrate, at different depths, $t_d > t_c > t_b > t_a$ (t, time of deposition). (From Ref. 66, with permission from the American Institute of Physics.)

7.13. OVERPOTENTIAL DEPENDENCE OF GROWTH FORMS

The dependence of growth forms on overpotentials originates from the potential dependence of nucleation and growth processes. Competition between nucleation

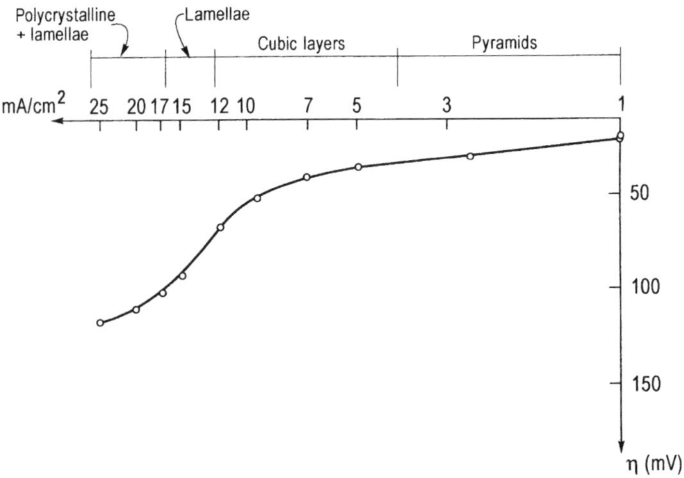

Figure 7.18. Current–potential curve showing the correlation between overpotential η and growth forms of electrodeposited copper from N CuSO$_4$ and N H$_2$SO$_4$ at 25°C. (From Ref. 40, with permission from Elsevier.)

and growth processes is strongly influenced by the potential of the cathode. Thus, major factors determining growth forms are as follows:

1. Structure of the double layer, including concentration of different species present in the solution. The potential dependence of additive adsorption and their effects on growth forms are discussed in Chapter 10.
2. Concentration on adions at the surface.
3. The radius and size of the critical nucleus [Eqs. (7.2a) and (7.2b)].
4. Rate of nucleation [Eq. (7.3a)].
5. Lateral and vertical growth rate of the crystal grains.

Seiter et al. (40) found a correlation between overpotential η and growth forms of electrodeposited copper on copper sheet substrate with (100) texture, shown in Figure 7.18.

Barnes et al. (44) observed similar results on copper single-crystal surfaces near the (100) face: below 10-mV ridges, 40- to 70-mV platelets, 70- to 100-mV blocks, and fine platelets; and above 100 mV, polycrystalline deposit. The four basic structural forms are shown in Figure 7.19. Less frequently observed growth forms are pyramids, spirals, whiskers, and dendrites. The structure of deposits is discussed further in Chapter 16.

7.14. DEPOSITION OF A METAL ON A FOREIGN METALLIC SUBSTRATE

In industrial applications of metal deposition a metal M is deposited either on the native metal substrate M or on a foreign metal substrate S. As an example of the former, Cu is electrodeposited on a Cu substrate formed by electroless Cu deposition on an activated nonconductor in the fabrication of printed circuit boards. As an example of the latter, Ni is electrodeposited on Cu in the fabrication of contact pads in the electronics industry.

The mechanism of growth of metal M on a foreign metallic substrate S is determined by the two most important parameters: (1) the binding energy between M and M and M and S, and (2) the crystallographic misfit between M and S. Three mechanisms are possible in this case, and they are related to these two parameters. The first mechanism is the TDC nucleation and growth, or the Volmer–Weber mechanism. This mechanism is operative when the binding energy between adions of metal M (M_{adi}) and the substrate M, $\psi(M_{adi} - M)$, is larger than that between M_{adi} and metal S, $\psi(M_{adi} - S)$:

$$\psi(M_{adi} - M) \gg \psi(M_{adi} - S)$$

In this case the crystallographic misfit does not have any effect. The crystallographic misfit (mf) is defined as

$$mf = \frac{a_S - a_M}{a_M} \tag{7.18}$$

where a_S and a_M are the lattice spacings for the substrate S and deposit M, respectively.

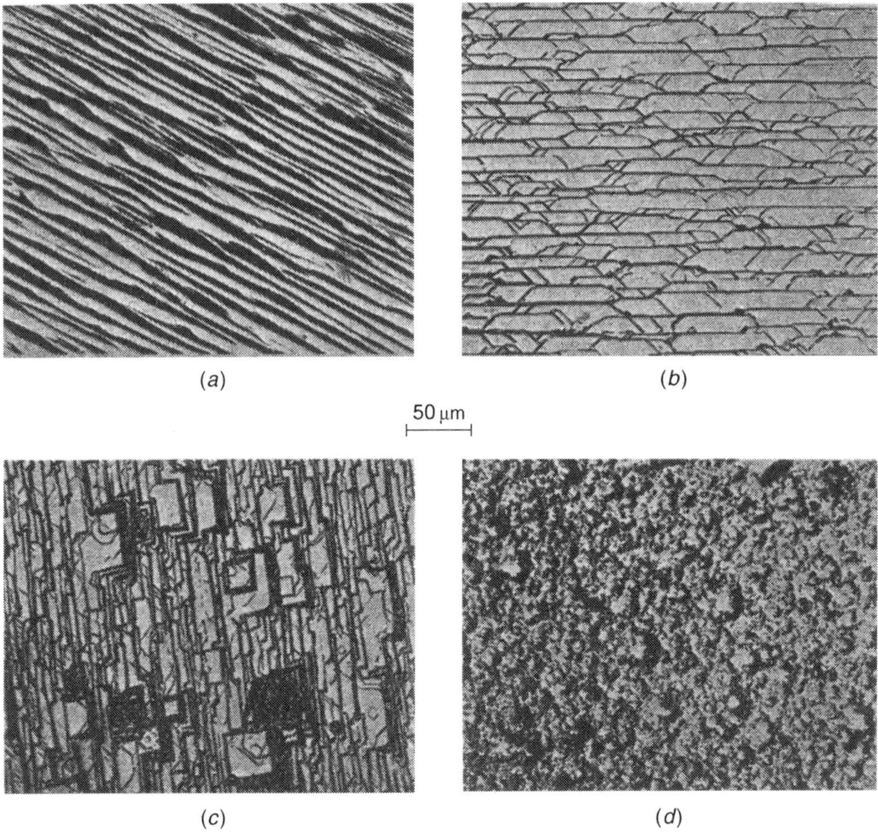

Figure 7.19. The four basic deposit structural forms of Cu deposited on Cu(100) face from acid copper solutions: (*a*) ridge; (*b*) platelet; (*c*) block; (*d*) polycrystalline. (From Ref. 44, with permission from Elsevier.)

One common characteristic of mechanisms 2 and 3 is the relationship between binding energies, and the difference is in the misfit. Thus, if

$$\psi(M_{adi} - S) \gg \psi(M_{adi} - M)$$

we distinguish two types of mechanisms:

1. If $a_S \approx a_M$, the misfit is zero and the deposition mechanism is layer by layer or a Frank–Van der Merve growth mechanism.
2. If $a_S \neq a_M$, misfit is present, positive or negative, and growth proceeds by the Stranski–Krastanov mechanism, which is composed of two steps. In the first step a 2D overlayer of M_{adi} on S is formed, and in the second step 3D crystallites grow on top of this predeposited overlayer (Fig. 7.20).

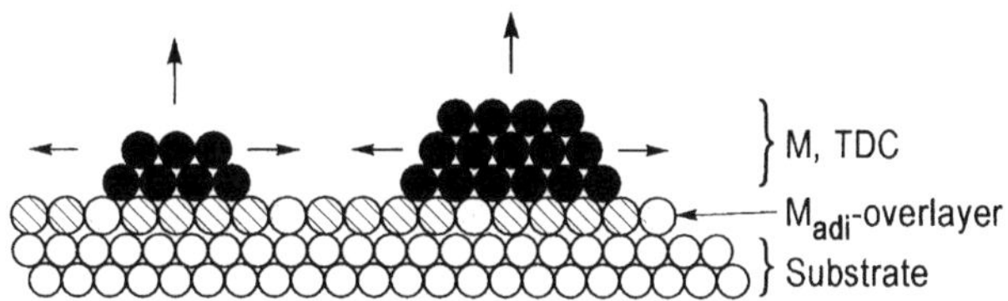

Figure 7.20. Stranski–Krastanov mechanism.

7.15. UNDERPOTENTIAL DEPOSITION

In Chapter 6 we have seen that metal M will be deposited on the cathode from the solution of M^{n+} ions if the electrode potential E is more negative than the Nernst potential of the electrode M/M^{n+}. However, it is known that in many cases metal M can be deposited on a foreign substrate S from a solution of M^{n+} ions at potentials more positive than the Nernst potential of M/M^{n+}. This electrodeposition of metals is termed *underpotential deposition*. Thus, in terms of the actual electrode potential E during deposition and the Nernst equilibrium potential $E(M/M^{n+})$ and their difference, $\Delta E = E - E(M/M^{n+})$, we distinguish two types of electrodeposition:

1. Overpotential deposition (OPD):

$$E < E(M/M^{n+}), \quad \Delta E < 0$$

2. Underpotential deposition (UPD):

$$E > E(M/M^{n+}), \quad \Delta E > 0$$

One interesting example of UPD deposition is deposition of Cu on Au(111) substrate. Figure 7.21b shows a cyclic voltammogram of Cu–UPD on Au(111) substrate, and Figure 7.21a shows a cyclic voltammogram of Au(111) electrode in the absence of Cu^{2+} ions. A comparison of Figure 7.21a with b shows that UPD deposition occurs in the potential range between 300 and 650 mV versus NHE, where two Cu adsorption peaks (A_1, A_2) and two desorption peaks (D_1, D_2) appear. In this potential region the gold electrode behaves as a quasi-ideal polarizable electrode. Peaks P_o and P_r represent gold oxide formation and reduction, respectively. Peak P_e corresponds to oxygen evolution.

The Cu adsorbate structure was studied using STM and EXAFS (extended x-ray absorption fine structure) techniques, but it is not yet well understood. UPD–OPD transition is in the range -82 to -71 mV. Bulk fcc Cu spacing is reached after deposition of about 10 Cu monolayers. Holzle et al. (72) have shown that UPD Cu deposition on Au(111) is a combined adsorption–nucleation and growth process.

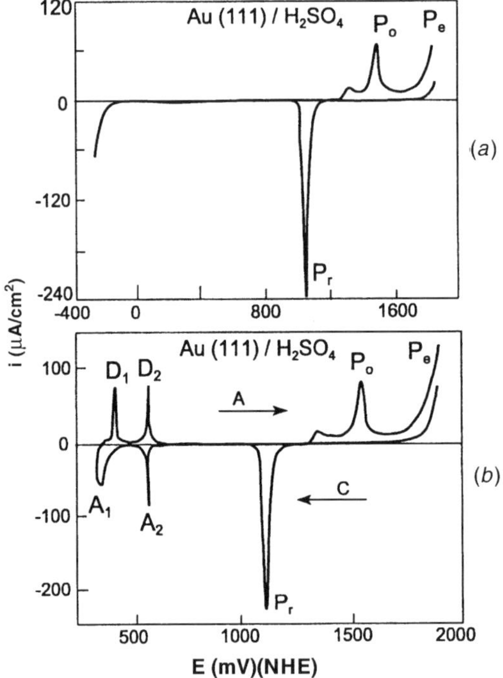

Figure 7.21. (Top) Cyclic voltammograms of Au(111) electrode in 9×10^{-2}M H$_2$SO$_4$; (bottom) absence of Cu^{2+} ions and presence of Cu^{2+} ions, 10^{-2}M CuSO$_4$. (From Ref. 71, with permission from the Electrochemical Society.)

7.16. FORMATION OF NANOCLUSTERS USING SCANNING TUNNELING MICROSCOPY

Scanning tunneling microscopy (STM) as a technique for surface analysis is described in Section 13.3. STM can also be used to produce metal nanoclusters (75–81). The process of the formation of nanoclusters via STM techniques involves two basic steps:

1. Electrochemical deposition of a metal from the solution onto the uncovered part of the STM tip.
2. Transfer of the electrochemically deposited metal from the STM tip to the substrate. This transfer (known as *jump to contact*) consists of two steps:
   a. Approach of the tip to the surface, resulting in formation of a connective neck between the tip and the substrate.
   b. Withdrawal of the tip, resulting in breaking of the connective neck and formation of a nanocluster.

The procedure is depicted schematically in Figure 7.22. For example, uniform-size Cu clusters were produced and distributed evenly over an atomically flat Au(111) electrode surface (76). This example is of the type of deposition of a metal on a foreign metallic substrate and involves UPD (defined in Section 7.15) (80). The procedure was made fully automated through the use of microprocessors in controlling the x-, y-, and z-movements of the tip. Arrays of up to 400 individual clusters were made at a rate of 50 clusters per second (76). The average height of the Cu clusters was 0.8 nm (75,80).

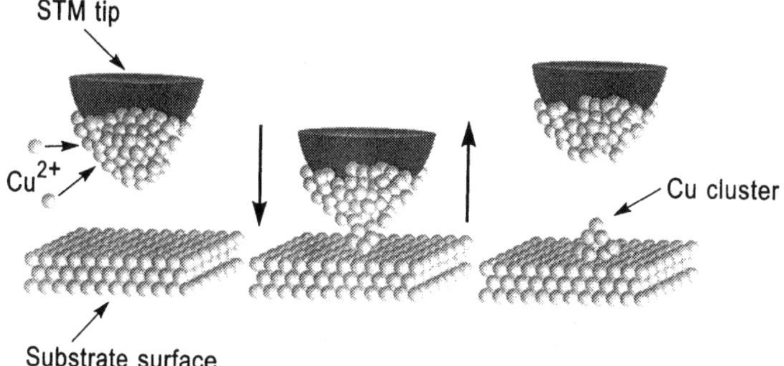

Figure 7.22. Schematic diagram for the transfer of electrodeposited Cu from a Cu-covered STM tip to the Au substrate during a tip approach. (From Ref. 78, with permission from Elsevier Science.)

7.17. FLUCTUATION PHENOMENA IN AN EQUILIBRIUM MICROSTRUCTURE ON METAL SURFACES

Analysis of the equilibrium step fluctuations on a Cu(100) electrode in an aqueous electrolyte was also done using a scanning tunneling microscope (81). Specifically, STM analysis of the fluctuating steps on metal surfaces in thermal equilibrium can be used to determine diffusion coefficients and energy barriers for atomic hopping. The motion of atoms involves (1) rapid exchange of atoms between steps, (2) atomic hopping along the step edges, and (3) exchange of atoms with adjacent terraces. The equilibrium step fluctuations are analyzed theoretically by means of a time correlation function as described in Ref. 81.

REFERENCES AND FURTHER READING

1. T. Erdey-Gruz and M. Volmer, *Z. Phys. Chem.* **A150**, 201 (1930).
2. T. Erdey-Gruz and M. Volmer, *Z. Phys. Chem.* **A157**, 165 (1931).
3. B. Becker and W. Doering, *Ann. Phys. Leipzig* **24**, 719 (1935).
4. F. C. Frank, *Discuss. Faraday Soc.* **5**, 48 (1949).
5. D. A. Vermilyea, *J. Chem. Phys.* **25**, 1254 (1954).
6. M. Fleischmann and H. R. Thirsk, *Electrochim. Acta* **1**, 146 (1959).
7. A. Bewick, M. Fleischmann, and H. R. Thirsk, *Trans. Faraday Soc.* **58**, 2200 (1962).
8. T. L. Hill, *Statistical Thermodynamics*, Addison-Wesley, Reading, MA, 1962.
9. M. Fleischmann and H. R. Thirsk, in *Advances in Electrochemistry and Electrochemical Engineering*, Vol. 3, P. Delahay, ed., Interscience, New York, 1963.
10. M. Fleischmann and H. R. Thirsk, *Electrochim. Acta,* **9**, 757 (1964).
11. M. Fleischmann, J. Pattison, and H. R. Thirsk, *Trans. Faraday Soc.* **61**, 1256 (1965).
12. R. D. Armstrong, M. Fleischmann, and H. R. Thirsk, *J. Electroanal. Chem.* **11**, 208 (1966).
13. R. D. Armstrong and J. A. Harrison, *J. Electrochem. Soc.* **116**, 328 (1969).

14. J. A. Harrison and H. R. Thirsk, in *Electroanalytical Chemistry*, Vol. 5, A. Bard, ed., Marcel Dekker, New York, 1971.

15. U. Bertocci, *J. Electrochem. Soc.* **119**, 822 (1972).

16. G. J. Hills, D. J. Schiffrin, and J. Thompson, *Electrochim. Acta* **19**, 657 (1974).

17. R. D. Armstrong and A. A. Metcalk, *J. Electroanal. Chem.* **63**, 19 (1975).

18. D. D. Macdonald, *Transient Techniques in Electrochemistry*, Plenum Press, New York, 1977, Chap. 8.

19. B. Lewis and J. C. Anderson, *Nucleation and Growth of Thin Films*, Academic Press, New York, 1978.

20. A. Bewick and M. Fleischmann, in *Topics in Surface Chemistry*, E. Kay and P. S. Bagus, eds., Plenum Press, New York, 1978.

21. G. H. Gilmer, *J. Cryst. Growth* **49**, 465 (1980).

22. E. Budevski, V. Bostanov, and G. Staikov, *Annu. Rev. Mater. Sci.* **10**, 85 (1980).

23. V. Bostanov, W. Obretenov, G. Staikov, D. Roe, and E. Budevski, *J. Cryst. Growth* **52**, 761 (1981).

24. M. Y. Abyaneh and M. Fleischmann, *J. Electroanal. Chem.* **119**, 187 (1981).

25. W. Obretenov, V. Bostanov, and V. Popov, *J. Electroanal. Chem.* **132**, 273 (1982).

26. B. Scharifker and G. Hills, *Electrochim. Acta* **28**, 879 (1983).

27. H. Bort, K. Juttner, W. J. Lorenz, G. Staikov, and E. Budevski, *Electrochim. Acta* **28**, 985 (1983).

28. E. B. Budevski, in *Comprehensive Treatise of Electrochemistry*, Vol. 7, B. E. Conway, J. O'M. Bockris, E. Yeager, S. U. M. Khan, and R. E. White, eds., Plenum Press, New York, 1983.

29. J. O'M. Bockris and S. U. M. Khan, *Surface Electrochemistry*, Plenum Press, New York, 1993.

30. A. Milchev, W. S. Kruijt, M. Sluyters-Rehbach, and J. H. Sluyters, *J. Electroanal. Chem.* **362**, 21 (1993).

31. W. S. Kruijt, M. Sluyters-Rehbach, J. H. Sluyters, and A. Milchev, *J. Electroanal. Chem.* **371**, 13 (1994).

32. R. M. Rynders and R. C. Alkire, *J. Electrochem. Soc.* **141**, 1166 (1994).

33. E. Budevski, G. Staikov, and W. J. Lorenz, *Electrochemical Phase Formation and Growth*, VCH, New York, 1996.

34. G. Staikov and W. J. Lorenz, in *Electrochemically Deposited Thin Films III*, M. Paunovic and D. A. Scherson, eds., *Proceedings*, Vol. 96-19, Electrochemical Society, Pennington, NJ, 1996.

35. W. K. Burton and N. Cabrera, *Discuss. Faraday Soc.* **5**, 33 (1949).

36. C. W. Bunn and H. Emmett, *Discuss. Faraday Soc.* **5**, 119 (1949).

37. H. Fischer, *Elektrolytische Abscheidung und Elektrokristallisation von Metallen*, Springer-Verlag, Berlin, 1954.

38. H. Fischer, *Z. Elektrochem.* **59**, 612 (1955).

39. F. C. Frank, in *Growth and Perfection of Crystals*, R. H. Doremus, B. W. Roberts, and D. Turnbull, eds., Wiley, New York, 1958.

40. H. Seiter, H. Fischer, and L. Albert, *Electrochim. Acta* **2**, 97 (1960).

41. H. J. Pick, G. G. Storey, and T. B. Vaughan, *Electrochim. Acta* **2**, 165 (1960).

42. T. B. Vaughan and H. J. Pick, *Electrochim. Acta* **2**, 179 (1960).

43. G. Wranglen, *Electrochim. Acta* **2**, 130 (1960).

44. S. C. Barnes, G. G. Storey, and H. J. Pick, *Electrochim. Acta* **2**, 195 (1960).

45. D. Shanefield and P. E. Lighty, *J. Electrochem. Soc.* **110**, 973 (1963).

46. E. W. Dickson, M. H. Jacobs, and D. W. Pashley, *Philos. Mag.* **11**, 575 (1965).

47. A. Damjanovic, M. Paunovic, and J. O'M. Bockris, *J. Electroanal. Chem.* **9**, 93 (1965).

48. N. Ibl, in *Advances of Electrochemistry and Electrochemical Engineering*, Vol. 2, C. W. Tobias, ed., Interscience, New York, 1966.

49. H. Fischer, *Plating* **56**, 1229 (1969).

50. E. R. Thompson and K. R. Lawless, *Electrochim. Acta* **14**, 269 (1969).

51. R. Sard and R. Weil, *Electrochim. Acta* **15**, 1977 (1970).

52. Y. Tanabe and S. Kamasaki, *J. Met. Finish. Soc. Jpn.* **22**, 54 (1971).

53. U. Bertocci and C. Bertocci, *J. Electrochem. Soc.* **118**, 1287 (1971).

54. S. K. Verma and H. Wilman, *J. Phys. D Appl. Phys.* **4**, 152 (1971).

55. E. Schmidt, P. Beutler, and W. J. Lorenz, *Ber. Bunsenges. Phys. Chem.* **75**, 71 (1971).

56. H. R. Thirsk and J. A. Harrison, *Electrode Kinetics*, Academic Press, New York, 1972.

57. R. Weil and J. B. C. Wu, *Plating* **60**, 622 (1973).

58. S. Nakahara and R. Weil, *J. Electrochem. Soc.* **120**, 1462 (1973).

59. W. J. Lorenz, H. D. Herman, N. Wuthrich, and F. Hilbert, *J. Electrochem. Soc.* **121**, 1167 (1974).

60. V. Bostanov, G. Staikov, and D. K. Roe, *J. Electrochem. Soc.* **122**, 1301 (1975).

61. S. Nakahara, *Thin Solid Films* **45**, 421 (1977).

62. D. M. Kolb, in *Advances in Electrochemistry and Electrochemical Engineering*, Vol. 11, H. Gerischer and C. W. Tobias, eds., Wiley, New York, 1978.

63. K. Juttner and W. J. Lorenz, *Z. Phys. Chem. N.F.* **122**, 163 (1980).

64. J. Amblart, M. Froment, G. Maurin, N. Spyrellis, and E. T. Trevisan-Souteyrand, *Electrochim. Acta* **28**, 909 (1983).

65. B. Yang, B. L. Walden, R. Messier, and W. B. White, *Proc. SPIE* **821**, 68 (1987).

66. D. J. Srolovitz, A. Mazor, and G. G. Bukiet, *J. Vac. Sci. Technol.* **A6**, 2371 (1988).

67. J. A. Giordmaine and J. B. Wachtman, Annual Reviews, Palo Alto, CA, 1990.

68. T. Hachinga, H. Houbo, and K. Itaya, *J. Electroanal. Chem.* **315**, 257 (1991).

69. H. J. Frost, in *Mater. Res. Soc. Symp. Proc.*, Vol. 202, C. V. Thompson, J. Y. Tsao, and D. J. Srolovitz, eds., Materials Research Society, Pittsburgh, PA, 1991.

70. K.-N. Tu, J. W. Mayer, and L. C. Feldman, *Electronic Thin Film Science*, Macmillan, New York, 1992.

71. I. H. Omar, H. J. Pauling, and K. Juttner, *J. Electrochem. Soc.* **140**, 2187 (1993).

72. M. H. Holzle, U. Retter, and D. M. Kolb, *J. Electroanal. Chem.* **371**, 101 (1994).

73. M. H. Holzle, V. Zwing, and D. M. Kolb, *Electrochim. Acta* **40**, 1237 (1995).

74. W. K. Burton, N. Cabrera, and F. C. Frank, *Philos. Trans. R. Soc. London* **A243**, 299 (1951).

75. R. Ullmann, T. Will, and D. M. Kolb, *Chem. Phys. Lett.* **209**, 238 (1993).

76. D. M. Kolb, R. Ullmann, and T. Will, *Science* **275**, 1097 (1997).

77. G. E. Engelmann, J. C. Ziegler, and D. M. Kolb, *Surf. Sci.* **401**, L420 (1998).

78. D. M. Kolb, R. Ullmann, and J. C. Ziegler, *Electrochim. Acta* **43**, 2751 (1998).

79. M. A. Schneeweiss and D. M. Kolb, *Phys. Status Solidi* **A173**, 51 (1999).

80. D. M. Kolb and M. A. Schneeweiss, *Interface*, Spring (1999).

81. M. Giesen, R. Randler, S. Baier, H. Ibach, and D. M. Kolb, *Electrochim. Acta* **45**, 527 (1999).

PROBLEMS

7.1. The size of the critical nucleus N_c in 2D nucleation as a function of the overpotential η is given by Eq. (7.2a).

    (a) Calculate the factor b in Eq. (7.2a) for (1) a circular nucleus and (2) a square nucleus.

    (b) Calculate the size of the Ag critical nuclei as a function of the overpotential η for the values $\eta = 2, 10, 20, 30,$ and $40 \times 10^{-3}\,\text{V}$.

    (c) Calculate $N_c = f(\eta)$ for (1) circular and (2) square nuclei. The edge energy $\varepsilon = 2 \times 10^{-13}\,\text{J/cm}$. Express $N_c = f(\eta)$ in the form $N_c = K_{2D}/\eta^2$ for (1) circular and (2) square nuclei; K_{2D} includes all the constants in this case. Present your results in tabular form.

7.2. The size of the spherical critical nucleus N_c in 3D nucleation as a function of the overpotential η is given as

$$N_c = \frac{32\pi\sigma^3 v^2}{3(z\eta e)^3}$$

where σ, v, z, and e are the surface energy, volume of an atom, number of electrons involved in the electrodeposition equation, and the electron charge, respectively.

Calculate $N_c = f(\eta)$ for the electrodeposition of Ag. The overpotential η values are 10 and $50 \times 10^{-3}\,\text{V}$. Express $N_c = f(\eta)$ in the form $N_c = K_{3D}/\eta^3$, where K_{3D} includes all the constants in this case. For Ag(100)/AgNO$_3$, H$_2$O, $\sigma = 1 \times 10^{-5}\,\text{J/cm}^2$. The radius of the Ag atom is $1.444\,\text{Å}$, $z = 1$, and $e = 1.602 \times 10^{-19}\,\text{C}$.

7.3. The free energy of formation of a cluster as a function of the size of a cluster of N atoms is shown in Figure 7.1. The figure shows that $\Delta G(N)$ increases initially, reaches a maximum ΔG_{max}, and then decreases with increasing N [Eq. (7.1)]. $\Delta G = f(N,\eta)$ for a three-dimensional spherical nucleation of Ag is given by

$$\Delta G = \Delta G_1 + \Delta G_2$$

where $\Delta G_1 = -Nze\eta$ and $\Delta G_2 = \Phi(N) = \sigma H N^{2/3}$; H is the shape factor (9) with $H = (4\pi)^{1/3}(3v)^{2/3}$; and v is the volume of the Ag atom. The radius of the Ag atom is $1.444 \times 10^{-8}\,\text{cm}$.

    Plot ΔG, ΔG_1, and ΔG_2 as a function of N for $\eta = 50 \times 10^{-3}\,\text{V}$. Determine the size of the critical nucleus N_c from ΔG_{max} in the graph. The surface energy for Ag(100)/Ag(NO$_3$) and H$_2$O is $\sigma = 1 \times 10^{-5}\,\text{J/cm}^2$.

8

Electroless Deposition

8.1. INTRODUCTION

The basic components of an electrolytic cell for electrodeposition of metals from an aqueous solution are, as shown in Figure 2.1, power supply, two metal electrodes (M_1 and M_2), water containing the dissolved ions, and two metal–solution interfaces; M_1–solution and M_2–solution. An electrolytic cell for electroless deposition is shown in Figure 8.1. A comparison of Figures 2.1 and 8.1 shows that in electroless deposition there is no power supply and the system has only one electrode. However, the solution is more complex. It contains water, a metal salt MA (M^{z+}; A^{z-}), and a reducing agent Red as basic components.

The overall reactions of electrodeposition and electroless deposition may be used to compare these two processes. The process of electrodeposition of metal M is represented by

$$M_{solution}^{z+} + ze \xrightarrow{\text{electrode}} M_{lattice} \tag{8.1}$$

In this process z electrons are supplied by an external power supply (Fig. 2.1). The overall reaction of electroless metal deposition is

$$M_{solution}^{z+} + Red_{solution} \xrightarrow{\text{catalytic surface}} M_{lattice} + Ox_{solution} \tag{8.2}$$

where Ox is the oxidation product of the reducing agent Red. The catalytic surface may be the substrate S itself or catalytic nuclei of metal M' dispersed on a noncatalytic substrate surface. In the electroless deposition process, a reducing agent Red in the solution is the electron source; the electron-donating species Red gives electrons to the catalytic surface and metal ions M^{z+} at the surface. The reaction represented by Eq. (8.2) must be conducted in such a way that a homogeneous reaction between M^{z+} and Red, in the bulk of the solution, is suppressed.

Fundamentals of Electrochemical Deposition, Second Edition.
By Milan Paunovic and Mordechay Schlesinger
Copyright © 2006 John Wiley & Sons, Inc.

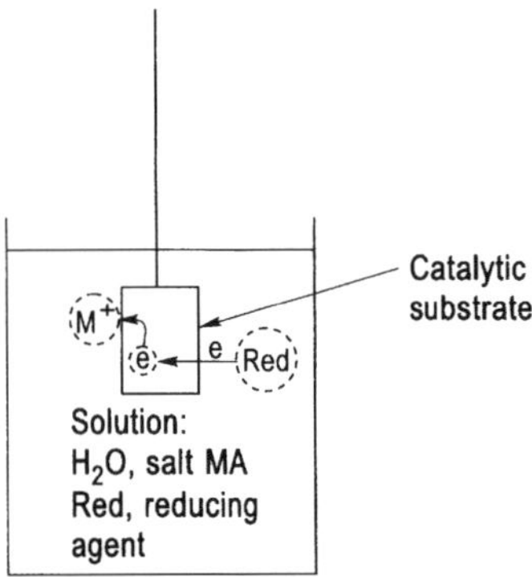

Figure 8.1. Electrolytic cell for electroless deposition of metal M from an aqueous solution of metal salt MA and a reducing agent Red.

Another comparison between Figures 2.1 and 8.1 points to a more basic difference between electrodeposition and electroless deposition. There are two electrodes in Figure 2.1: a cathode and an anode. Here two separate electron-transfer reactions occur at two spatially separated electrode–electrolyte interfaces. At the cathode a reduction reaction occurs [Eq. (8.1)], and at the anode an oxidation reaction occurs: for example,

$$M_{lattice} \xrightarrow{\text{anode}} M^{z+}_{solution} + ze \tag{8.3}$$

In electroless deposition the two electrochemical reactions, reduction of $M^{z+}_{solution}$ and oxidation of $Red_{solution}$, occur at the same electrode, at the same electrode–electrolyte interface [Eq. (8.2) and Fig. 8.1]. Thus, in electroless deposition there is a statistical division of the catalytic sites on the substrate into anodic and cathodic sites. Since these catalytic sites are part of the same piece of metal (substrate), there is a flow of electrons between these sites.

In this chapter we discuss the electrochemical model of electroless deposition (Sections 8.2 and 8.3), kinetics and mechanism of partial reactions (Sections 8.4 and 8.5), activation of noncatalytic surfaces (Section 8.6), kinetics of electroless deposition (Section 8.7), the mechanism of electroless crystallization (Section 8.8), and unique properties of some deposits (Section 8.9).

8.2. ELECTROCHEMICAL MODEL: MIXED-POTENTIAL THEORY

An electrochemical model for the process of electroless metal deposition was suggested by Paunovic (10) and Saito (8) on the basis of the Wagner–Traud (1) mixed-potential theory of corrosion processes. According to the mixed-potential theory of electroless deposition, the overall reaction given by Eq. (8.2) can be decomposed into one reduction reaction, the cathodic partial reaction,

$$M^{z+}_{solution} + ze \xrightarrow{\text{catalytic surface}} M_{lattice} \tag{8.4}$$

and one oxidation reaction, the anodic partial reaction,

$$\text{Red}_{\text{solution}} \xrightarrow{\text{catalytic surface}} \text{Ox}_{\text{solution}} + me \qquad (8.5)$$

Thus, the overall reaction [Eq. (8.2)] is the outcome of the combination of two different partial reactions, Eqs. (8.4) and (8.5). As mentioned above, these two partial reactions, however, occur at one electrode, the same metal–solution interphase. The equilibrium (rest) potential of the reducing agent, $E_{\text{eq,Red}}$ [Eq. (8.5)] must be more negative than that of the metal electrode, $E_{\text{eq,M}}$ [Eq. (8.4)], so that the reducing agent Red can function as an electron donor and M^{z+} as an electron acceptor. This is in accord with the discussion in Section 5.7 on standard electrode potentials.

Wagner–Traud Diagram. According to the mixed-potential theory, the overall reaction of the electroless deposition, [Eq. (8.2)] can be described electrochemically in terms of three current–potential (i–E) curves, as shown schematically in Figure 8.2. First, there are two current–potential curves for the partial reactions (solid curves): (1) $i_c = f(E)$, the current–potential curve for the reduction of M^{z+} ions, recorded from the rest potential $E_{\text{eq,M}}$, in the absence of the reducing agent Red (when the activity of M^{z+} is equal to 1, $E_{\text{eq,M}} = E_M^0$); and (2) $i_a = f(E)$, the current–potential curve for oxidation of the reducing agent Red, recorded from the rest potential $E_{\text{eq,Red}}$, in the absence of M^{z+} ions (when the activity of Red is equal to 1, $E_{\text{eq,Red}} = E_{\text{Red}}^0$). Then the third curve, $i_{\text{total}} = f(E)$, the dashed curve in Figure 8.2, is the current–potential curve for the overall reaction.

The two major characteristics of this system of curves are:

1. The $i_{\text{total}} = f(E)$ curve intersects the potential axis. At this intersection the current is zero and

$$i_c = i_a \qquad (8.6)$$

for $i_{\text{total}} = 0$. The potential where Eq. (8.6) holds is the mixed potential, E_{mp}.

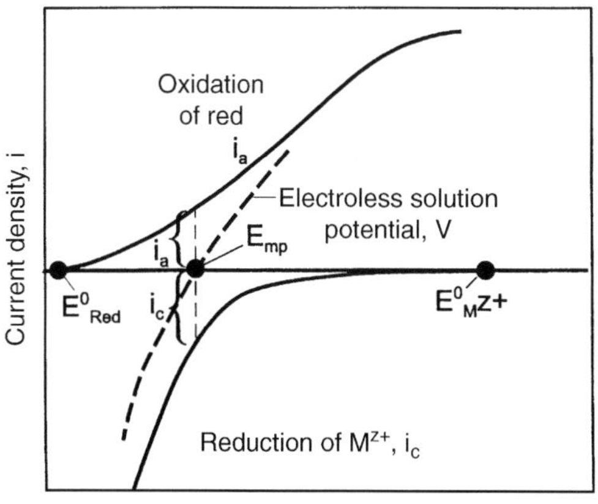

Figure 8.2. Wagner–Traud diagram for the total (i_{total}) and component current potential curves (i_a, i_c) for the overall reaction of electroless deposition.

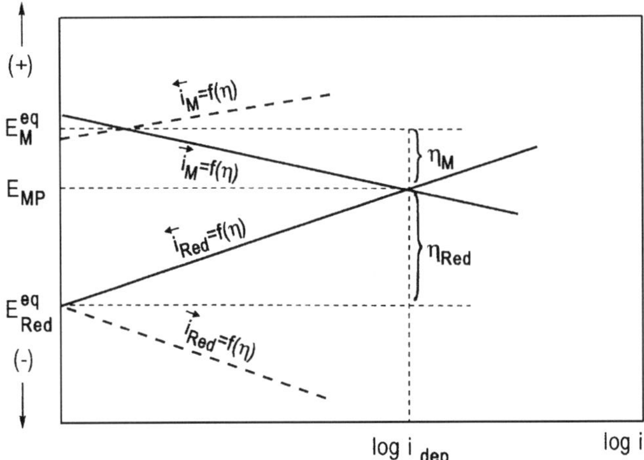

Figure 8.3. Evans diagram of current–potential curves for a system with two different simultaneous electrochemical reactions. Kinetic scheme: Eqs. (8.4) and (8.5).

2. At any point of the $i_{total} = f(E)$ curve,

$$i_{total} = i_a + i_c \tag{8.7}$$

Thus, the total current density, i_{total}, is composed of two components. It is a result of the addition of current densities of the two partial processes.

Evans Diagram. An alternative method of presenting the current–potential curves for electroless metal deposition is the Evans diagram. In this method the sign of the current density is suppressed. Figure 8.3 shows a general Evans diagram with current–potential functions $i = f(E)$ for the individual electrode processes, Eqs. (8.4) and (8.5). According to this presentation of the mixed-potential theory, the current–potential curves for individual processes, $i_c = i_M = f(E)$ and $i_a = i_{Red} = f(E)$, intersect. The coordinates of this intersection have the following meaning: (1) the abscissa, the current density of the intersection, is the deposition current density i_{dep} (i.e., log i_{dep}), that is, the rate of electroless deposition in terms of mA/cm^2; and (2) the ordinate, the potential of the intersection, is the mixed potential, E_{mp}.

Mixed Potential, E_{mp}. When a catalytic surface S is introduced into an aqueous solution containing M^{z+} ions and a reducing agent, the partial reaction of reduction [Eq. (8.4)] and the partial reaction of oxidation [Eq. (8.5)] occur simultaneously. Each of these partial reactions strives to establish its own equilibrium, E_{eq}. The result of these processes is the creation of a steady state with a compromised potential called the *steady-state mixed potential, E_{mp}*. The result of this mixed potential is that the potential of the redox couple Red/Ox [Eq. (8.5)] is raised anodically from the reversible value $E_{eq,Red}$ (Fig. 8.3), and the potential of the metal electrode M/M^{z+} [Eq. (8.4)] is depressed cathodically from its reversible value, $E_{eq,M}$, down to the mixed potential, E_{mp} (Fig. 8.3).

Thus, the basic four characteristics of the steady-state mixed potential are:

1. Both redox systems are shifted from their own characteristic equilibrium potentials by the amout η (overpotential):

$$\eta_M = E_{mp} - E_{eq,M} \tag{8.8}$$

$$\eta_{Red} = E_{mp} - E_{eq,Red} \tag{8.9}$$

2. A net electrochemical reaction occurs in each redox system since both reactions, Eqs. (8.4) and (8.5), are shifted from their equilibria by introduction of the mixed potential.

3. The criterion for a steady state is that the rate of reduction of M^{z+}, the cathodic current density i_M, is equal to the rate of oxidation of the reducing agent Red, the anodic current density i_{Red}:

$$i_{M,deposition} = (i_M)_{E_{mp}} = (i_{Red})_{E_{mp}} \tag{8.10}$$

since a net current cannot flow in an isolated system.

4. A system at the steady-state mixed potential is not in equilibrium since a net overall reaction does occur, and therefore the free energy change is not zero, which is the requirement for thermodynamic equilibrium.

8.3. TEST OF MIXED-POTENTIAL THEORY

Electroless Deposition of Copper. The basic ideas of the mixed-potential theory were tested by Paunovic (10) for the case of electroless copper deposition from a cupric sulfate solution containing ethylenediaminetetraacetic acid (EDTA) as a complexing agent and formaldehyde (HCHO) as the reducing agent (Red). The test involved a comparison between direct experimental values for E_{mp} and the rate of deposition with those derived theoretically from the current–potential curves for partial reactions on the basis of the mixed-potential theory.

The average rate of electroless deposition of copper determined gravimetrically (by weighing before and after deposition) at 24°C (±0.5) from a solution that contains 0.1 M $CuSO_4 \cdot 5H_2O$, 0.175 M EDTA, 0.05 M CH_2O, NaOH to pH 12.5, is 1.8 mg/h·cm^2. This value is obtained if the time of deposition is counted from the instant of immersion of the copper plate (substrate) into the solution. If the time of deposition is counted from the instant the mixed potential is reached (about 4 min after immersion of the Cu substrate), the deposition rate is 1.9 mg/h·cm^2. The average value of the mixed potential during copper deposition is -0.65 V versus SCE.

The rate of deposition and the mixed potential are determined on the basis of the mixed-potential theory using the Evans diagram. First, the current–potential curve for the reduction of cupric ions in a solution containing H_2O, 0.1 M $CuSO_4$, 0.175 M EDTA, and NaOH to pH 12.50 (no CH_2O present) at 24°C (±0.5) is determined using the galvanostatic technique. At this electrode only one reaction occurs, reduction of Cu^{2+}. An electrode with only one electrode process is called a *single electrode* here. The result is shown as $i_M = f(E)$ in Figure 8.4.

Second, the current–potential curve for the oxidation of formaldehyde at the single electrode was determined using the galvanostatic technique. The solution in this case contained H_2O, 0.05 M CH_2O, 0.075 M EDTA (excess of EDTA used in the solution for the single cathodic reaction), and NaON to pH 12.50 (no $CuSO_4$ was present in this solution). The temperature was 24°C (±0.5). The result is shown as $i_{Red} = f(E)$ in Figure 8.4, where it is seen that these two polarization curves, $i_{Red} = f(E)$ and $i_M = f(E)$, intersect. The coordinates of intersection are (1) abscissa, $i = 1.9 \times 10^{-3}\,A/cm^2$, and (2) ordinate, $E = -0.65\,V$ versus SCE. The current density $i = 1.9 \times 10^{-3}\,A/cm^2$ is the rate of electroless deposition of copper expressed in terms of A/cm^2. The potential $E = -0.65\,V$ versus SCE is the mixed potential of the electroless copper system studied. The rate of deposition expressed in mg/h·cm² is calculated on the basis of Faraday's law using the equation

$$w = i \times 1.18 \text{ mg/h} \cdot cm^2 \tag{8.11}$$

where i is given in mA/cm². For $i = 1.9 \times 10^{-3}\,A/cm^2$, it is 2.2 mg/h·cm².

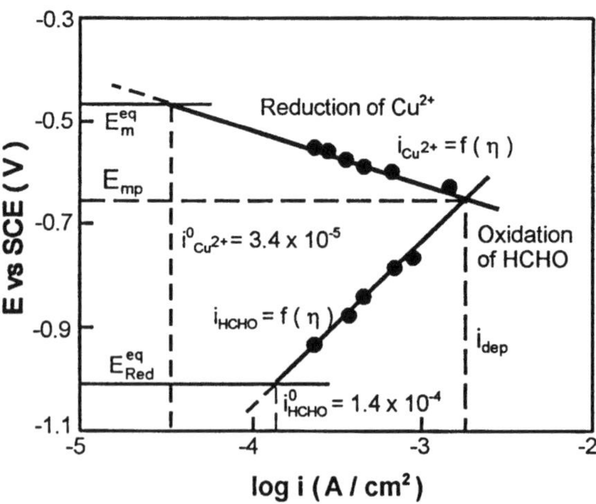

Figure 8.4. Current–potential curves for the reduction of Cu^{2+} ions and the oxidation of reducing agent Red, formaldehyde, combined into one graph (an Evans diagram). Solution for the Tafel line for the reduction of Cu^{2+} ions: 0.1 M $CuSO_4$, 0.175 M EDTA, pH 12.50, E_{eq} (Cu/Cu^{2+}) = $-0.47\,V$ versus SCE; for the oxidation of formaldehyde: 0.05 M HCHO and 0.075 M EDTA, pH 12.50, E_{eq} (HCHO) = $-1.0\,V$ versus SCE; temperature 25 $\pm$ 0.5°C. (From Ref. 10, with permission from the American Electroplaters and Surface Finishers Society.)

Now we compare the direct experimental values for the rate and E_{mp} with the theoretical values obtained on the basis of mixed-potential theory using an Evans diagram. First is the rate of deposition: (1) the experimental (gravimetric) value is 1.8 mg/h·cm^2 when the time of deposition is counted from the instant of immersion of the copper plate substrate into the solution, and 1.9 mg/h·cm^2 when the time is counted from the instant the mixed potential is reached; (2) the theoretical value, from the Evans diagram (Fig. 8.4), is 2.2 mg/h·cm^2. A comparison of these two values, 2.2 (theoretical) and 1.9 (experimental, gravimetric), shows that there is relatively good agreement between the theoretical and the experimental (gravimetric) data. Second is the mixed potential E_{mp}: (1) the experimental value (measured during deposition in the gravimetric measurements) is -0.65 V versus SCE; (2) the theoretical value, from the Evans diagram (Fig. 8.4), is -0.65 V versus SCE. Comparison of these two values shows that there is excellent agreement between the theoretical and the directly measured data.

Thus, one concludes that the mixed-potential theory is essentially verified for the case of electroless copper deposition. These conclusions were later confirmed by Donahue (15), Molenaar et al. (25), and El-Raghy and Abo-Salama (33). The mixed-potential theory has been verified for electroless copper deposition as well using hypophosphite as the reducing agent (72).

Electroless Deposition of Nickel. The mixed-potential theory was also tested and verified for the case of electroless nickel deposition using the potentiodynamic method and Wagner–Traud and Evans diagrams (43). The electroless Ni solution used for these studies contained $NiSO_4$ 39.4 g/L, sodium citrate 20 g/L, 85% lactic acid 10 g/L, and dimethylamine borane (DMAB) 2 g/L, as the reducing agent. The temperature was held at $40 \pm 0.5°C$ and pH 7.00 (measured at 25°C), adjusted by NH_4OH. The rate of deposition determined gravimetrically is 2.4 mg/h·cm^2, and the mixed potential is -0.85 V versus SCE. Current–potential curves for the partial cathodic reaction (absence of DMAB) and the partial anodic reaction (absence of $NiSO_4$) are shown in Figure 8.5 in the form of a Wagner–Traud diagram.

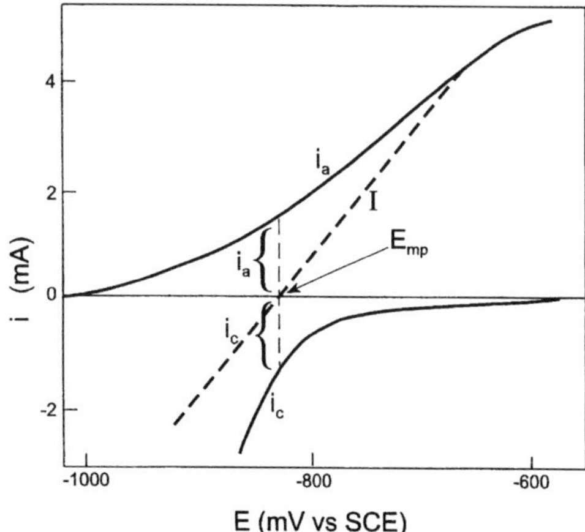

Figure 8.5. Wagner–Traud diagram for electroless Ni(B) deposition: $E_{mp} = -840$ mV versus SCE. Electrode area 0.68 cm^2. (From Ref. 43, with permission from the American Electroplaters and Surface Finishers Society.)

The current–potential curve for the overall reaction, $I = f(E)$ in Figure 8.5, is constructed from the partial currents I_c and I_a using the Wagner–Traud principle of additive combination of partial currents [Eq. (8.7)], where the partial cathodic current I_c has a negative sign and the partial anodic current I_a has a positive sign. The potential where $I = 0$ is the mixed potential E_{mp}, and it has a value of $-840\,$mV versus SCE. At the mixed potential E_{mp}, the values of the partial currents I_c and I_a are equal to the deposition rate itself expressed in terms of current. According to Figure 8.5, the deposition rate is $1.38\,$mA or $2.03\,$mA/cm^2. On the basis of Faraday's law, this deposition rate r may be expressed in mg/h·cm^2 using the equation

$$r = i \times 1.09 \ \text{mg/h} \cdot \text{cm}^2 \tag{8.12}$$

where i is given in mA/cm^2. From Eq. (8.12) the deposition current density of $2.03\,$mA/cm^2 corresponds to the rate of deposition $r = 2.2\,$mg/h·cm^2. The same result was obtained using the Evans diagram method. The coordinates of the intersection are: (1) abscissa, $i = 2.03\,$mA/cm^2, and (2) ordinate, $E_{mp} = -840\,$mV versus SCE. The gravimetrically determined rate of deposition is $2.4\,$mg/h·cm^2 and E_{mp} measured directly during deposition is $-850\,$mV versus SCE.

Comparison between the values of the mixed potential and the rate of deposition via direct determination with those derived from the mixed-potential theory is very good. Thus, the mixed-potential theory was verified for this case of electroless Ni deposition.

Electroless Deposition of Gold. Okinaka (21) verified the mixed-potential theory for the case of electroless gold deposition. Figure 8.6 shows that the partial cathodic

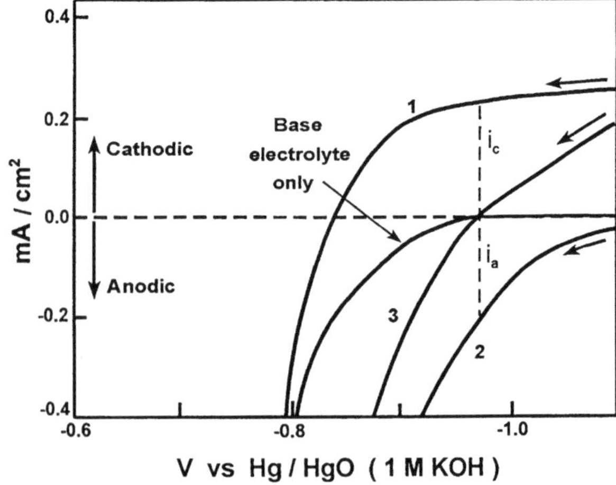

Figure 8.6. Current–potential curves for a gold electrode at 75°C. Base electrolytes, KOH and KCN. Curve 1, $2 \times 10^{-4}\,$M KAu(CN)$_2$ without KBH$_4$; curve 2, $0.1\,$M KBH$_4$ without KAu(CN)$_2$; curve 3, $2 \times 10^{-4}\,$M KAu(CN)$_2$ and $0.1\,$M KBH$_4$. Potential scanned at $5.56\,$mV/s. (From Ref. 21, with permission from the Electrochemical Society.)

current density of $[Au(CN)_2]^-$ reduction measured at E_{mp} in the absence of the reducing agent BH_3OH^- is equal to the partial anodic current density of oxidation of BH_3OH^- measured at the mixed potential in the absence of $[Au(CN)_2]^-$. Thus, the validity of the mixed-potential theory is shown in this case as well.

Electroless Deposition in the Presence of Interfering Reactions. According to the mixed-potential theory, the total current density, i_{total}, is a result of simple addition of current densities of the two partial reactions, i_a and i_c. However, in the presence of interfering (or side) reactions, i_a and/or i_c may be composed of two or more components themselves, and verification of the mixed-potential theory in this case would involve superposition of current–potential curves for the electroless process investigated with those of the interfering reactions in order to correctly interpret the total i–E curve. Two important examples are discussed here.

The first example is the electroless deposition of gold from solutions where the concentration of $K[Au(CN)_2]$ is greater than 2×10^{-4}M. Okinaka (21) has shown that there are two types of interference in the anodic partial reaction: (1) interference involving oxidation of the substrate—in this case the anodic partial current density i_a is the sum of the two components $i_a = i_a(BH_3OH^-) + i_a(Au)$, where $i_a(BH_3OH^-)$ is the anodic partial current density for the oxidation of BH_3OH^- and $i_a(Au)$ is that for oxidation of Au, measured in the base electrolyte alone; and (2) interference involving adsorption of complexed Au ions. Adsorption of $[Au(CN)_2]^-$ interferes with the anodic oxidation of BH_3OH^-. A lower i_a value is observed for a complete solution due to the adsorption of cathodic electroactive species; a part of the surface is blocked by $[Au(CN)_2]^-$ and thus is not available for oxidation of the reducing agent, which results in a lower i_a value. As a result of this interference, the gravimetrically determined deposition rates are in agreement with i_c rather than with i_a.

The second example is electroless deposition of copper from solutions containing dissolved oxygen (49,53). In this case the interfering reaction is the reduction of the oxygen, and the cathodic partial current density i_c is the sum of two components:

$$i_c = i_c(M^{z+}) + i_c(O_2)$$

where $i_c(M^{z+})$ is the cathodic partial current density for reduction of metal ions M^{z+} and $i_c(O_2)$ is that for reduction of the oxygen.

Interaction Between Partial Reactions. The original mixed-potential theory assumes that the two partial reactions are independent of each other (1). In some cases this is a valid assumption, as was shown earlier in this chapter. However, it was shown later that the partial reactions are not always independent of each other. For example, Schoenberg (13) has shown that the methylene glycol anion (the formaldehyde in an alkaline solution), the reducing agent in electroless copper deposition, enters the first coordination sphere of the copper tartrate complex and thus influences the rate of the cathodic partial reaction. Ohno and Haruyama (37) showed the presence of interference in partial reactions for electroless deposition of Cu, Co, and Ni in terms of current–potential curves.

Conclusions. The discussion in this section shows the validity of the mixed-potential theory for electroless deposition of Cu, Ni, and Au. The discussions in the sections "Electroless Deposition in the Presence of Interfering Reactions" and "Interaction Between Partial Reactions" illustrate the complexities of electroless processes and the presence of a variety of factors that should be taken into account when applying the mixed-potential theory to the electroless processes.

8.4. CATHODIC PARTIAL REACTION

Kinetic Scheme. Generally, metal ions in a solution for electroless metal deposition have to be complexed with a ligand. Complexing is necessary to prevent formation of metal hydroxide, such as $Cu(OH)_2$, in electroless copper deposition. One of the fundamental problems in electrochemical deposition of metals from complexed ions is the presence of electroactive (charged) species. The electroactive species may be complexed or noncomplexed metal ion. In the first case, the kinetic scheme for the process of metal deposition is one of simple charge transfer. In the second case the kinetic scheme is that of charge transfer preceded by dissociation of the complex. The mechanism of the second case involves a sequence of at least two basic elementary steps:

1. Formation of electroactive species
2. Charge transfer from the catalytic surface to the electroactive species

Electroactive species M^{z+} are formed in the first step by dissociation of the complex $[ML_x]^{z+xp}$:

$$[ML_x]^{z+xp} \rightarrow M^{z+} + xL^p \tag{8.13}$$

where p is the charge state of the ligand L, z is the charge of the noncomplexed metal ion, and $z + xp$ is the charge of the complexed metal ion.

The charge transfer

$$M^{z+} + ze \rightarrow M \tag{8.14}$$

proceeds in steps, usually with the first charge transfer (one-electron transfer) serving as the rate-determining step (RDS):

$$M^{z+} + e \xrightarrow{\ \ \text{RDS}\ \ } M_{lattice} \tag{8.15}$$

Thus, from the kinetic aspects, the cathodic partial reaction is an electrochemical reaction [Eq. (8.14)] that is preceded by a chemical reaction [Eq. (8.13)]. Paunovic (31)

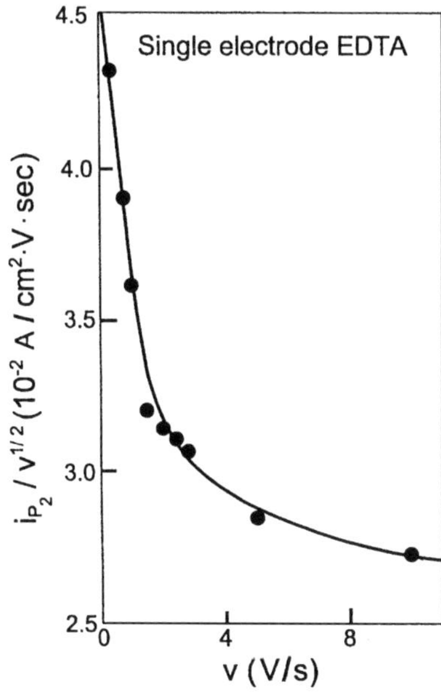

Figure 8.7. Potential sweep function for the partial cathodic process in electroless copper deposition; dissociation and reduction of Cu(II)EDTA complex. (From Ref. 31, with permission from the Electrochemical Society.)

studied the first step in the cathodic partial reaction of electroless copper deposition by chronopotentiometry and potential sweep (potentiodynamic) methods. The potentiodynamic current function related to this problem is the $i_p/v^{1/2}$ function, where i_p is the peak current and v is the rate of the potential scan (V/s). In the case of a single-step electrochemical reaction, $i_p/v^{1/2}$ is a constant value for constant concentration and is independent of the rate of the potential scan. However, when an electrochemical reaction is preceded by a chemical reaction, i_p and $i_p/v^{1/2}$ vary with the scan rate v. This variation depends on the equilibrium constant for the complex and the rate constants of the preceding chemical reactions (rate of dissociation and complexation). Figure 8.7 shows the change of $i_p/v^{1/2}$ versus the scan rate for the reduction of Cu(II)EDTA complex. A decrease of $i_p/v^{1/2}$ with an increase in scan rate is a diagnostic criterion for the kinetic scheme where the charge transfer is preceded by a chemical reaction. Chronopotentiometric results lead to the same conclusions. Thus, both methods show that reduction of the Cu(II)EDTA complex is preceded by dissociation of the complex.

Kinetics. The major factors determining the rate of the partial cathodic reaction are concentrations of metal ions and ligands, pH of the solution, and type and concentration of additives. These factors determine the kinetics of partial cathodic reaction in a general way, as given by the fundamental electrochemical kinetic equations discussed in Chapter 6.

Effect of Additives. Schoenberg (13,16) and Paunovic and Arndt (44) have shown that additives may have two opposing effects: acceleration and inhibition. For example, the accelerating effect of guanine and adenine on the cathodic reduction of Cu^{2+}

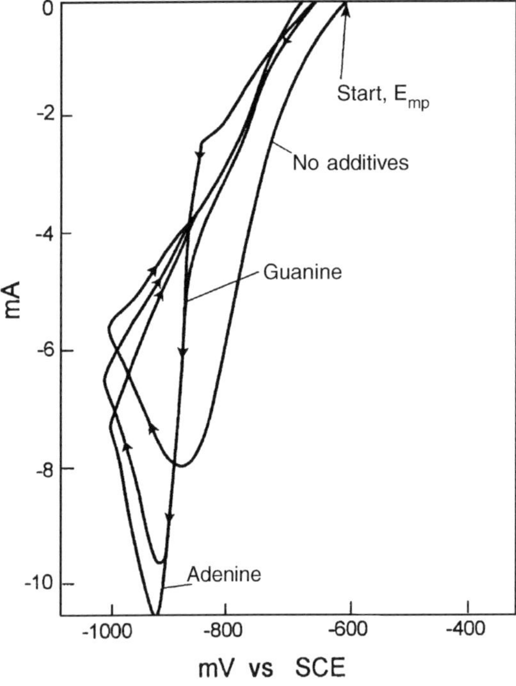

Figure 8.8. Cyclic voltammograms of a Cu electrode in electroless copper solution in the absence and the presence of additive; effect of additives on the reduction of Cu^{2+}. The scan rate is 100 mV/s. (From Ref. 44, with permission from the Electrochemical Society.)

ions in electroless copper solution is shown in Figure 8.8. It can be seen that adenine and guanine show an increase in peak current compared with peak current in the absence of additives. The same additives show an increase in the rate of electroless copper deposition. The inhibiting effect of NaCN for the same reaction was studied by potentiostatic and potentiodynamic techniques (45,50). Figure 8.9 shows that the

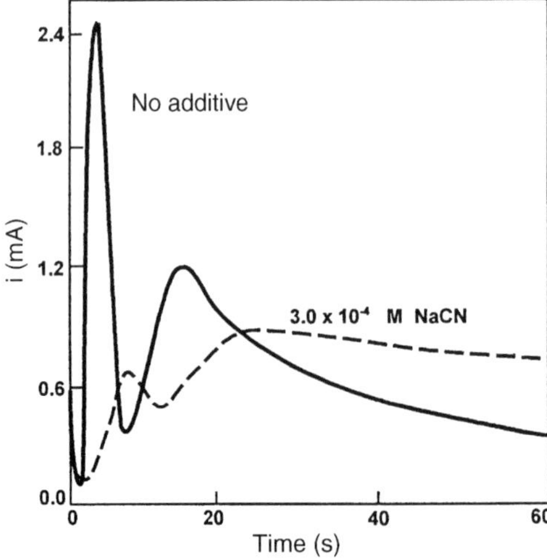

Figure 8.9. Potentiostatic current–time transients of a Pt electrode in electroless copper solution showing the effect of NaCN; $E = -900$ mV. (From Ref. 50, with permission from the Electrochemical Society.)

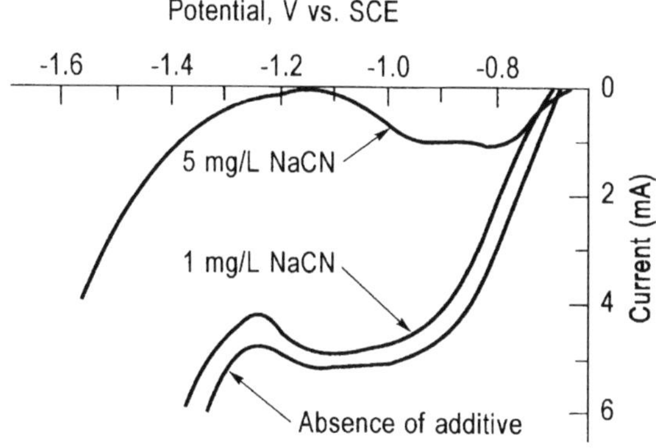

Figure 8.10. Effect of NaCN on the current–potential curves for reduction of Cu^{2+} in an electroless solution at 25°C containing 0.05 M $CuSO_4$, 0.075 M EDTP [1,1,1′,1′-(ethylenedinitrilo)tetra-2-propanol], and 5.8 mL/L of HCHO. The Pt cathode (0.442 cm^2) was rotated at 100 rpm; the scan rate is 10 mV/s. (From Ref. 45, with permission from the American Electroplaters and Surface Finishers Society.)

height and position of the maximum on the time coordinate in the potentiostatic experiment depend on the additive present in solution.

Figure 8.9 clearly illustrates the inhibiting effect of NaCN. A potentiodynamic current–potential curve in the presence and absence of NaCN is shown in Figure 8.10, where it is seen that the addition of NaCN changes considerably the shape and magnitude of the current–potential relationship. The addition of NaCN to an electroless copper solution results in a decrease in current density at a given potential. This inhibition increases with an increasing amount of NaCN in solution. *Cathodic passivation* may be seen in the range -1100 to -1200 mV versus SCE when 5.0 mg/L $(1.0 \times 10^{-3}$ M) of NaCN is present in solution.

The major factors probably responsible for the acceleration effect of additives are (1) the charge density of the electron system of the additive and (2) the exchange of electrons between electrode, π-bonded additive molecule, and the complexed metal ions in the solution. Inhibition effect and cathodic passivation are explained in terms of blocking of the catalytic surface, which results in a decrease in the available surface area (45).

8.5. ANODIC PARTIAL REACTION

Mechanism. The overall anodic partial reaction, Eq. (8.5), usually proceeds in at least two elementary steps (like the cathodic partial reaction): formation of an electroactive species, and charge transfer. The formation of electroactive species (R) usually proceeds in two steps through an intermediate (Red_{interm}):

$$Red \rightarrow Red_{interm}$$
$$Red_{interm} \rightarrow R$$

Van den Meerakker (38) proposed the following general mechanism for formation of electroactive species R from the intermediate R_{interm}, now represented by R—H:

$$Red_{interm} = R—H \rightarrow R_{ads} + H_{ads} \qquad (8.16)$$

where R_{ads} is the electroactive species R adsorbed at the catalytic surface. According to this mechanism [Eq. (8.16)], the electroactive species R_{ads} is formed in the process of dissociative adsorption (dehydrogenation) of the intermediate R—H, which involves breaking of the R—H bond.

The adsorbed hydrogen, H_{ads}, may be desorbed in the chemical reaction

$$H_{ads} \rightarrow \tfrac{1}{2} H_2 \tag{8.17}$$

or in the electrochemical reaction

$$H_{ads} \rightarrow H^+ + e \tag{8.18}$$

For example, in electroless deposition of copper, when the reducing agent is formaldehyde and the substrate is Cu, H_{ads} desorbs in the chemical reaction (8.17). If the substrate is Pd or Pt, hydrogen desorbs in the electrochemical reaction (8.18).

The most studied anodic partial reaction is the oxidation of formaldehyde, Red $= H_2CO$. The overall reaction of the electrochemical oxidation of formaldehyde at the copper electrode in an alkaline solution proceeds as

$$H_2CO + 2OH^- \rightarrow HCOO^- + H_2O + \tfrac{1}{2} H_2 + e \tag{8.19}$$

The mechanism of this reaction involves the following sequence of elementary steps (6,38):

1. Formation of electroactive species R in three steps:
   a. Hydrolysis of H_2CO:

$$H_2CO + H_2O = H_2C(OH)_2 \qquad \text{(methylene glycol)} \tag{8.20}$$

   b. Dissociation of methylene glycol:

$$H_2C(OH)_2 + OH^- = H_2C(OH)O^- + H_2O \tag{8.21}$$

   c. Dissociative adsorption of the intermediate, $H_2C(OH)O^-$ (R—H), involving breaking the C—H bond in the R—H molecule:

$$H_2C(OH)O^- = [HC(OH)O^-]_{ads} + H_{ads} \tag{8.22}$$

   where $[HC(OH)O^-]_{ads}$ is R_{ads}.
2. Charge transfer, the electrochemical oxidation (desorption) of electroactive species R_{ads}:

$$[HC(OH)O^-]_{ads} + OH^- \rightarrow HCOO^- + H_2O + e \tag{8.23}$$

where $HCOO^-$ is the oxidation product of R_{ads} (Ox).

A similar kinetic scheme can be applied to other reducing agents, such as borohydride (Red $= BH_4^-$), hypophosphite ($H_2PO_2^-$), and hydrazine (NH_2NH_2) where the electroactive species RH are $[BH_2OH^-]_{ads}$, $[HPO_2^-]_{ads}$, and $[N_2H_3]_{ads}$, respectively (21,38).

Parallel Reactions.   There are parallel reactions in some cases of oxidation of reducing agents. For example, in the case of oxidation of BH_4^- and $H_2PO_2^-$, the parallel reactions are probably cathodic reactions, resulting in incorporation of B and P into the metal deposit, respectively. Thus, when electroless Ni is deposited from solutions containing BH_4^- as the reducing agent, we designate this deposit as Ni(B), and when the reducing agent is $H_2PO_2^-$ the deposit is designated as Ni(P).

Kinetics.   The major factors determining the rate of the anodic partial reaction are pH and additives. Since OH^- ions are reactants in the charge-transfer step [e.g., Eq. (8.23)], the effect of pH is direct and significant (see, e.g., Ref. 32). Additives may have an inhibiting or an accelerating effect.

8.6.   ACTIVATION OF NONCATALYTIC SURFACES

Noncatalytic surfaces (e.g., nonconductors, noncatalytic metals, noncatalytic semiconductors) have to be activated (i.e., made catalytic) prior to electroless deposition. This activation is performed by generating catalytic nuclei on the surface of a noncatalytic material. Two major types of processes have been used to produce catalytic nuclei: electrochemical and photochemical.

Electrochemical Activation.   In the electrochemical method, catalytic nuclei of metal M on a noncatalytic surface S may be generated in an electrochemical oxidation–reduction reaction,

$$M^{z+} + Red \rightarrow M + Ox \qquad (8.24)$$

where M^{z+} is the metallic ion and M is the metal catalyst. The preferred catalyst is Pd, and thus the preferred nucleating agent M^{z+} is Pd^{2+} (from $PdCl_2$). The preferred reducing agent Red in this case is Sn^{2+} ion (from $SnCl_2$). In this example the overall reaction of activation, according to a simplified model, is

$$Pd^{2+} + Sn^{2+} \rightarrow Pd + Sn^{4+} \qquad (8.25)$$

Sn^{2+} can reduce Pd^{2+} ions since the standard oxidation–reduction potential of Sn^{4+}/Sn^{2+} is 0.15 V and that of Pd^{2+}/Pd is 0.987 V. As shown in Section 5.7 and Figure 5.10, the flow of electrons is from a more electronegative couple (here Sn^{4+}/Sn^{2+}) toward a less electronegative (more positive) couple (here Pd^{2+}/Pd).

Since the standard potential of Au^+/Au is 1.692 V, Sn^{2+} ions can reduce Au^+ to produce Au catalytic nuclei. Electrochemical activation using $PdCl_2$ and $SnCl_2$ may be performed in either two steps or one step.

Some nonconductors, such as the polymers polycarbonates and polystyrenes, must be subjected to a surface treatment prior to activation to ensure good adhesion of palladium nuclei. Surface treatment can include the use of chemical etchants for plastics or reactive gas plasma treatments (66).

Two-Step Activation Process. A simplified model of two-step activation is as follows. In the first step, sensitization, Sn^{2+} ions are adsorbed on the nonconducting substrate S from the solution of Sn^{2+} ions:

$$S + Sn^{2+}_{solution} \rightarrow S\ Sn^{2+}_{ads} \tag{8.26}$$

where $S \cdot Sn^{2+}_{ads}$ represents the adsorbed Sn^{2+} at the surface S. The amount of tin on the surface of the sensitized substrate is about $10\,\mu g/cm^2$ (14), and the surface coverage is less than 25% (12). The product of sensitization on Kapton, du Pont polyamide (14), graphite (12), glass, quartz, mica, and Formvar (11) is in the form of particulate matter with particles on the order of $10\,\text{Å}$ in diameter (11,12). These particles tend to agglomerate into dense clumps that are about 100 to $250\,\text{Å}$ in size. The clumps are composed of particles about $25\,\text{Å}$ in size. A typical sensitizing solution formula is (7)

- $SnCl_2$, $10\,g/L$
- HCl (37%), $40\,mL/L$

In the fabrication of printed circuit boards the sensitizer is applied to the substrate S by immersion of the substrate into the solution for 1 to 3 min. Alternatively, the surface of a nonconductor may be sprayed with sensitizer. Addition of aged stannic chloride ($SnCl_4$) solution to the tin sensitizer solution results in an improved sensitizer (17). The improved sensitizer yields a greater number of active centers per unit surface area (greater density) and a more uniform distribution. The density of adsorbed centers, using the conventional and improved sensitizers, is 10^{11} and 10^{12} particles per square centimeter, respectively. The diameter of adsorbed particles for both types of sensitizers is about 10 to $15\,\text{Å}$.

The second step in the two-step process is nucleation. A typical nucleating solution used in industry is

- $PdCl_2$, 0.1 to $1.0\,g/L$
- HCl (37%), 5 to $10\,mL/L$

Nucleation is performed by immersion of a sensitized nonconductor into the nucleating solution for 0.5 to 2 min. The surface reaction between the stannous ions, Sn^{2+}, adsorbed on the surface of the substrate and the palladium ions, Pd^{2+}, in the nucleator solution is

$$S \cdot Sn_{ads}^{2+} + Pd_{solution}^{2+} \rightarrow S \cdot Pd_{ads} + Sn_{solution}^{4+} \qquad (8.27)$$

This reaction has been studied on Kapton (14), graphite (28), Teflon (18), and glass (20). Cohen et al. (14), using Mössbauer spectroscopy, found that when the sensitized tin on Kapton is placed in the palladium, essentially all the divalent tin on the surface is eliminated and the amount of palladium picked up on the surface corresponds to the amount of tin oxidized. The total amount of tin on the substrate remains the same, about $10 \, \mu g/cm^2$, before and after nucleation. The Mössbauer spectra argue strongly in favor of the reaction according to Eq. (8.27). Qualitatively, the same results have been obtained on graphite (28): After sensitization the graphite surface contains Sn^{2+} and Sn^{4+}, but after activation, only Sn^{4+}.

The nucleation process produces small Pd catalytic sites dispersed on the surface of a substrate in an island network. Marton and Schlesinger (11) estimated that these islands are less than $10 \, \text{Å}$ in diameter. The height of these islands is approximately $40 \, \text{Å}$ (12).

The catallytic metallic Pd covers only a small fraction of the surface. The amount of Pd on the glass substrate (11) is 0.04 to $0.05 \, \mu g/cm^2$. Assuming uniform distribution, this amount corresponds roughly to 0.3 of a monolayer of Pd on a glass substrate. The surface density of catalytic sites σ depends on substrate material. For glass the maximum value of σ was found to be 10^{14} sites/cm^2.

We have presented here a simple redox model of the electrochemical activation using $SnCl_2$ and $PdCl_2$ solutions. An advanced model takes into account the presence of mono-, di-, tri-, and tetrachlorostannate(II) species in a solution of $SnCl_2$: for example, $[SnCl]^+$, $SnCl_2$, $[SnCl_3]^-$, and $[SnCl_4]^{2-}$, respectively. Chloride ion acts as a ligand (donor) in these complex ions.

One-Step Activation Process. In a one-step activation process, the sensitizing and nucleating solutions are combined into one solution. It is assumed that when this solution is made up, it contains various Sn–Pd chloride complexes (24). These complexes may subsequently transform into colloidal particles of metallic Pd or a metallic alloy (Sn/Pd) to form a colloidal dispersion (19,28). This dispersion is unstable. It may be stabilized by addition of an excess of Sn^{4+} ions. In this case, Pd particles adsorbed on the nonconductor surface are surrounded by Sn^{4+} ions. The latter must be removed by solubilizing before electroless plating so that the catalytic Pd on the surface will become exposed, freely available, to subsequent plating. An example of such a solubilizing solution is a mixture of fluoroboric and oxalic acids in a dilute solution, or just plain NaOH or HCl.

Photochemical Activation. Photochemical processes have been used to produce catalytic metallic nuclei in three ways, each characterized by a different kinetic scheme: photoelectrochemical, photoelectron, or intramolecular kinetic. These processes have been reviewed by Paunovic (35) and Zhang et al. (67) with numerous references. Here we present only the basic ideas.

The photoelectrochemical kinetic scheme involves a photochemical reaction that is followed by an electrochemical reaction. The photochemical reaction is used to produce or deactivate the reducing agent. Catalytic metallic nuclei are formed in the subsequent electrochemical reaction. For example, the Fe^{2+} reducing agent (Red) needed for reaction (8.24) is generated in the photochemical reduction of complexed Fe^{3+} ions. This redox photolysis is the ligand-to-metal charge transfer with the overall reaction of oxidation of $[C_2O_4]^{2-}$:

$$C_2O_4{}^{2-} \xrightarrow{\ h\nu\ } 2CO_2 + 2e \qquad (8.28)$$

where $h\nu$ is the energy of one photon (ν is the radiation frequency and h is Planck's constant). Electrons generated in reaction (8.28) are used for reduction of Fe^{3+} to Fe^{2+}. The Fe^{2+} ions are then used in the subsequent electrochemical reaction to produce catalytic Pd nuclei according to Eq. (8.24).

In the photoelectron method, electrons necessary for the formation of catalytic metallic nuclei M from metallic cations M^{z+} are generated in a direct absorption process of a photon by a semiconductor crystal, which results in the generation of an electron and a hole. For example, absorption of photon by a solid, TiO_2, promotes an electron e from the filled $(2p)$ valence band to the vacant $(3d)$ conduction band when the energy of the photon is equal to or greater than the bandgap energy ($\lambda < 4100\,\text{Å}$, where λ is the wavelength; $h\nu \geqslant 3\,\text{eV}$). The result of this absorption is the generation of a free electron in the conduction band and a free hole p^+ in the valence band. This process can be represented by a simplified kinetic scheme:

$$TiO_2 \xrightarrow{\ h\nu\ } e + p^+(TiO_2) \qquad (8.29)$$

resulting in the formation of a hole–electron pair. The mobile electron e, generated in the process according to Eq. (8.29), may be captured by a species adsorbed on TiO_2, for instance, a metal ion M^{z+}, which is thus reduced to metal M:

$$ze + M_{ads}^{z+} = M_s \qquad (8.30)$$

where z is the number of electrons and the subscripts s and ads designate surface and adsorbed species, respectively. Metallic nuclei M formed in the reduction step (8.30) can catalyze electroless metal deposition. The metal M may be any of Pd, Pt, Au, Ag, or Cu. Photoholes p^+, generated in process (8.29), may be trapped by the surface O^- ions and form O_2, recombined with electrons, or participate in destruction of the metallic phase (M) to form metal ions (M^{z+}).

In the intramolecular photoreduction kinetics scheme, catalytic metallic nuclei are formed in the intramolecular ligand-to-metal electron-transfer process. For example, catalytic metallic Cu nuclei can be formed in the photochemical reaction ($\lambda < 3500$ Å) of cupric acetate (CuA):

$$CuA \xrightarrow{h\nu} Cu + Ox_A \qquad (8.31)$$

where Ox_A stands for the oxidation products of the acetate ion A. Pd, Pt, Au, or Ag catalytic metallic nuclei may be produced using this method.

Activation by Displacement Deposition. Silicon can be made catalytic for electroless deposition of Ni by replacing the surface Si atoms with Ni atoms (58,62):

$$2Ni^{2+} + Si \rightarrow 2Ni + Si^{4+} \qquad (8.32)$$

This reaction is called *displacement deposition,* because the nickel ions in solution simply displace the silicon at the surface. The substrate, Si, acts here as a reducing agent, as discussed in Chapter 9. Copper may be deposited on Si from HF acid solutions (69). In the presence of HF, Si is oxidized into $[SiF_6]^{2-}$.

Similarly, aluminum substrate can be activated by a displacement reaction (56,59):

$$3Pd^{2+} + 2Al \rightarrow 3Pd + 2Al^{3+} \qquad (8.33)$$

Activation by Thermal Decomposition of Metallic Oxides. The surface of alumina, Al_2O_3, may be activated by employing laser or ultraviolet irradiation to decompose Al_2O_3 (68). Decomposition of Al_2O_3 results in the generation of aluminum particles that are catalytic for electroless deposition of Cu (the first reaction probably is displacement deposition).

8.7. KINETICS OF ELECTROLESS DEPOSITION

Steady-state electroless metal deposition at mixed potential E_{mp} is preceded by a non-steady-state period, called the *induction period.*

Induction Period. The induction period is defined as the time necessary to reach the mixed potential E_{mp} at which steady-state metal deposition occurs. It is determined in a simple experiment in which a piece of metal is immersed in a solution for electroless deposition of a metal and the potential of the metal is recorded from the time of immersion (or the time of addition of the reducing agent, i.e., time zero) until the steady-state mixed potential is established. A typical recorded curve for the electroless deposition of copper on copper substrate is shown in Figure 8.11.

Paunovic (31) studied the induction period for the overall process, dividing it into dependence of the open-circuit potential (OCP) on the oxidation and reduction

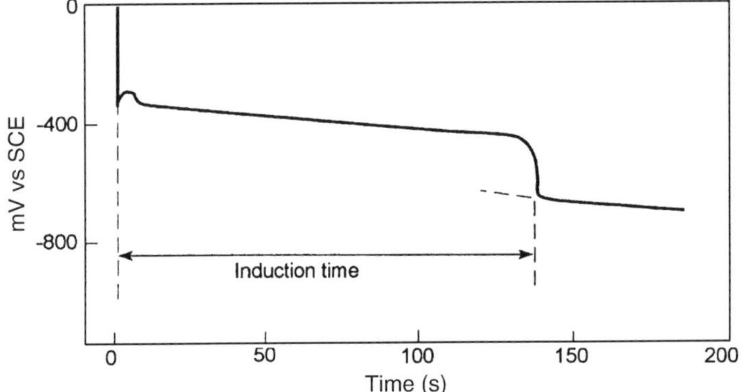

Figure 8.11. Induction period for the solution 0.3 M EDTA, 0.05 M CuSO$_4$, pH 12.50, 2.5 g/L paraformaldehye, Cu electrode, 2.2 cm^2, 25°C, SCE reference electrode, argon atmosphere. (From Ref. 31, with permission from the Electrochemical Society.)

partial reactions, that is, into individual induction periods for each partial process. A typical curve representing the change in the OCP with time, for the reducing agent, is given in Figure 8.12. The OCP of the Cu/Cu^{2+} system is reached instantaneously. From a comparison of these OCP curves, one can conclude that the rate of setting of the OCP of the reducing agent, CH$_2$O, is the rate-determining partial reaction in the setting of the steady-state mixed potential.

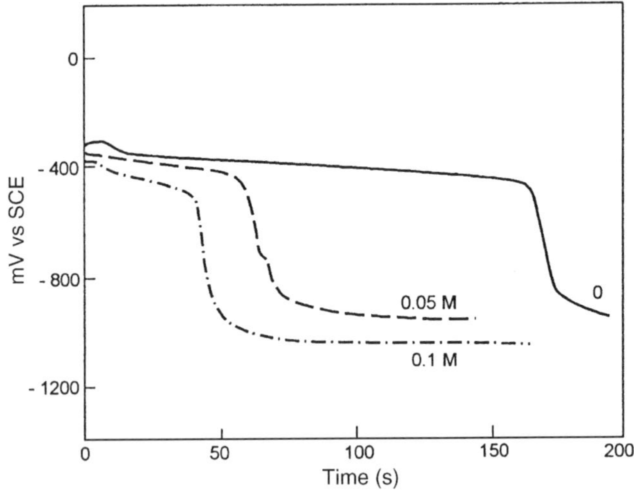

Figure 8.12. Open-circuit potential for the solution 1 g/L paraformaldehyde, pH 12.50, 25°C, 1 M KCl, Cu electrode, SCE reference electrode, EDTA variable. (From Ref. 31, with permission from the Electrochemical Society.)

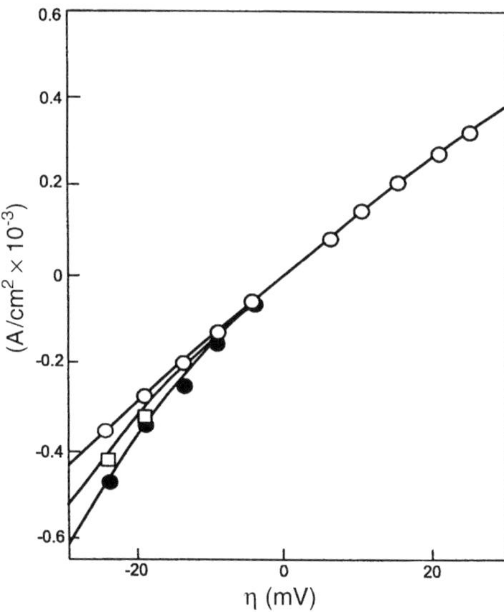

Figure 8.13. Polarization data in the vicinity of the mixed potential for electroless copper deposition; 0.05 M $CuSO_4$, 0.15 M EDTA, 0.072 M HCHO, NaOH to pH 12.50, nitrogen atmosphere, 25°C. (From Ref. 34, with permission from the Electrochemical Society.)

The major factors that determine the time necessary to reach the rest potential of the reducing agent are the type and concentration of the ligand present. The influence of the type of ligand present may be illustrated by the following example. In a supporting electrolyte that is 1 M KCl, pH 12.50, 25°C, with 1 g/L of paraformaldehyde, the time to reach the OCP, t_{OCP}, in the absence of a ligand is 210 s. Addition of 0.1 M EDTA to the electrolyte gives a t_{OCP} of 45 s. The influence of the concentration of the ligand is illustrated in Figure 8.12. For a concentration higher than 0.1, there is no further change.

Steady-State Kinetics. There are two electrochemical methods for determination of the steady-state rate of an electrochemical reaction at the mixed potential. In the first method (the *intercept method*) the rate is determined as the current coordinate of the intersection of the high overpotential polarization curves for the partial cathodic and anodic processes, measured from the rest potential. In the second method (the *low-overpotential method*) the rate is determined from the low-overpotential polarization data for partial cathodic and anodic processes, measured from the mixed potential. The first method was illustrated in Figures 8.3 and 8.4. The second method is discussed briefly here. Typical current—potential curves in the vicinity of the mixed potential for the electroless copper deposition (average of six trials) are shown in Figure 8.13. The rate of deposition may be calculated from these curves using the Le Roy equation (29,30):

$$i_{dep} = \frac{\sum_{j=1}^{n} i_j E_j}{\sum_{j=1}^{n} E_j^2} \tag{8.34}$$

$$E_j = 10^{\eta_j/b_a} - 10^{-\eta_j/b_c}$$

(8.35)

where i_j and η_j are current density and overpotential, respectively, at the jth point on the i–E curve; and b_a and b_c are the anodic and cathodic Tafel slopes, respectively.

A comparison of the results using this method and the rate of electroless copper deposition determined gravimetrically shows that the best results are obtained with the Le Roy equation applied to the polarization data in the anodic range. It is interesting to note that here, in the metal deposition as in the corrosion (9), the partial reaction, which does not involve destruction or building of a crystal lattice of metal substrate, gives better results (this is hardly surprising, of course).

Ohno (54) used ac polarization data and Ricco and Martin (55) used an acoustic wave device for in situ determination and monitoring of the rate of deposition. Various empirical rate equations were derived for electroless deposition of copper (15,33).

Empirical Modeling. The effect of process variables on the rate of deposition and properties of electrolessly deposited metals is usually studied by one-factor-at-a-time experiments (one-factor experiments are discussed further later in the book). In these experiments the effect of a single variable (factor), such as x_1, in the multivariable process with the response y, $y = f(x_1, x_2, x_3, \ldots, x_n)$, is studied by varying the value (level) of this variable while holding the values of the other independent variable fixed, $y = f(x_1)_{x_2, x_3, \ldots, x_n}$. Any prediction (extrapolation) of the effect of a single variable on the response y at a different level of the other variables, made on the basis of these one-factor experiments, implies the assumption that the effect of one variable is independent of the levels of the other variables. However, in almost all cases, the effect of one variable depends on the level of other variables. This dependence is due to interaction between variables, which occurs frequently in chemical and electrochemical processes. Interaction between variables can be detected easily by the use of factorial design of experiments. In a full factorial design, a fixed number of levels for each variable is selected, and then experiments are run with all possible combinations and levels. Thus, in factorial experiments the effect of a factor is estimated at several levels of the other factors, and the results are valid over a range of experimental conditions.

The factorial design of experiments was used by Paunovic et al. (65) in the study of electroless deposition of cobalt. In this study five-factor (variable) five-level experiments were performed to obtain a second-order interpolation polynomial. The five independent variables in that work were $CoSO_4$, NaH_2PO_2, $C_6H_5Na_3O_7$ (3Na—citrate), pH, and temperature. One finding in that work was the interaction between pH and citrate concentration in the electroless Co deposition process. Figure 8.14 shows that the pH effect depends on the concentration of citrate. The rate of deposition as a function of increasing pH increases at first, reaches a maximum, and then decreases. The location of the maximum depends on the concentration of the citrate. Thus, in general, the positive and negative signs of $\partial r/\partial \text{pH}$ (where r is the rate of deposition) may exist at any concentration of citrate. Figure 8.15 shows the rate of deposition as a function of pH at various concentrations of citrate obtained from the

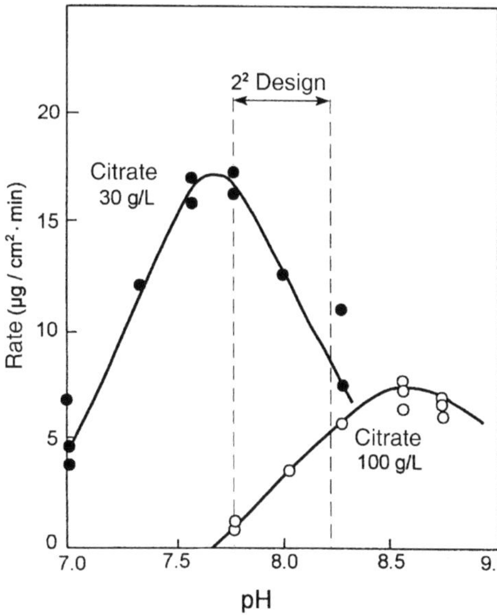

Figure 8.14. Rate of electroless cobalt deposition as a function of pH at 30 and 100 g/L citrate; one-factor experiments. (From Ref. 65, with permission from the Electrochemical Society.)

second-order polynomial. The polynomial curve (Fig. 8.15) and the one-factor experimental curve (Fig. 8.14) show that the magnitude and sign of the slope of $r = f(\text{pH})$ are functions of the citrate concentration.

The approximating polynomial was also used to obtain response surfaces. Figure 8.16 shows a 3D response surface and a 2D contour plot for the rate of deposition as a function of the concentration of cobalt sulfate and pH. The response surface in Figure 8.16 shows that the rate of deposition first increases, reaches a maximum, and then decreases with increase in pH. The value of this maximum increases with an increase in the concentration of cobalt sulfate.

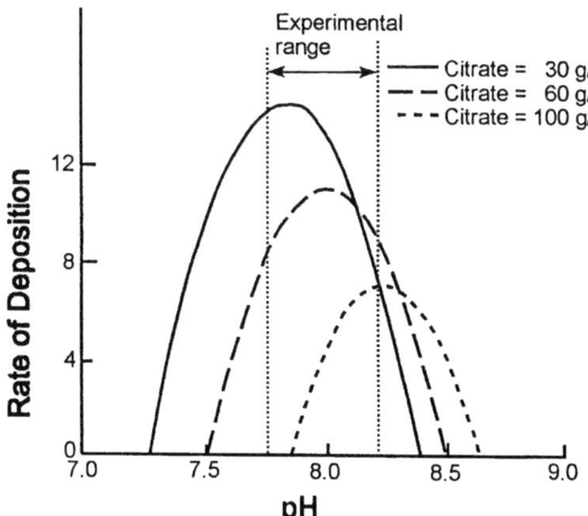

Figure 8.15. Extended approximating polynomial for the rate of electroless cobalt deposition as a function of pH at different concentrations of citrate. (From Ref. 65, with permission from the Electrochemical Society.)

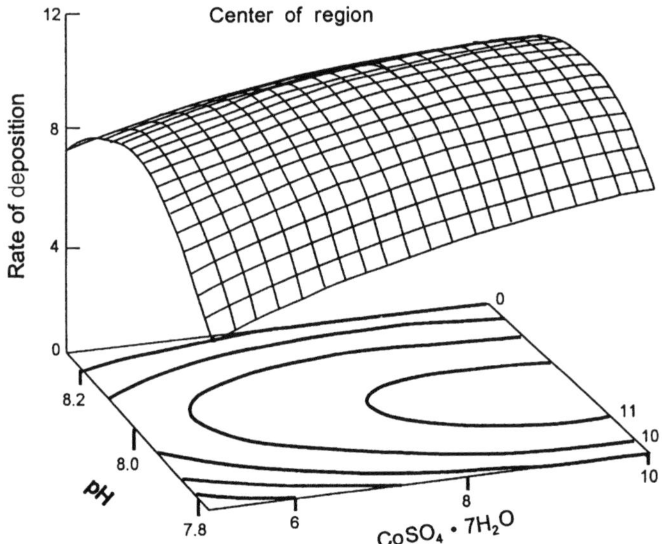

Figure 8.16. Response surface and contour plot for the rate of electroless cobalt deposition (in $\mu g/cm^2 \cdot min$) as a function of the concentration of cobalt sulfate (in g/L) and pH. (From Ref. 65, with permission from the Electrochemical Society.)

8.8. MECHANISM OF ELECTROLESS CRYSTALLIZATION

Electroless crystallization proceeds in two basic stages: (1) the thin-film stage (up to 3 μm) and (2) the bulk stage.

Thin-Film Stage. The mechanism of thin-film formation is characterized by three simultaneous crystal-building processes: nucleation (formation), growth, and coalescence of three-dimensional crystallites (TDCs). In the initial stages of electroless deposition, the average TDC density increases with the time of deposition; in this time interval, nucleation is the predominant process. Later, the average TDC density reaches a maximum and then decreases with time. In the time interval of decreasing TDC density, coalescence is the predominant crystal-building process. A continuous electroless film is formed by lateral growth and coalescence of TDC (40).

Bulk Stage. After formation of a continuous thin film, the deposition of thick (3- to 25-μm) copper or nickel film proceeds, in most cases, by the following process: (1) preferential growth of favorably oriented grains, (2) restriction (inhibition) of vertical growth of nonfavorably oriented grains, (3) lateral joining of preferentially growing grains, (4) cessation of growth of initial grains, and (5) nucleation of new layers of grains.

The TDCs formed in the thin-film stage grow vertically and laterally. In this process of vertical and lateral growth, a preferentially growing, favorably oriented grain (TDC) increases in width and subsequently joins laterally with other preferentially growing grains. After this lateral joining of growing grains, the width of preferentially oriented grains becomes constant. The result of these processes of electroless crystallization is a columnar structure of the deposit.

There is no adequate theory for lamellar growth of Ni(P). Periodic fluctuations in the content of phosphorus in electroless Ni(P) are possible causes of the lamellar structure.

8.9. UNIQUE PROPERTIES OF SOME DEPOSITS

In this section we show that some electroless deposits have unique properties compared to electrodeposited, evaporated, or sputtered metal deposits. Our discussion is limited to mechanical and diffusion barrier properties.

Mechanical Properties. One interesting example is electroless Ni(P). Electroless Ni(P) is harder and has better corrosion resistance than that of electrodeposited Ni(P). Nonmagnetic electroless Ni(P), or NiCu(P), is used as an underlayer in high-density metallic memory disk fabrication to improve the mechanical finish of the surface. Thus, hardness, wear resistance, and corrosion resistance have been major properties determining technological applications of electroless Ni(P) in the electronic, aerospace (stators for jet engines), automotive, machinery, oil and gas production, power generation, printing, and textile industries.

It is interesting to note that Brenner and Riddell (2–4) accidentally encountered electroless deposition of nickel and cobalt during electrodeposition of nickel–tungsten and cobalt–tungsten alloys (in the presence of sodium hypophosphite) on steel tubes in order to produce material with better hardness than that of steel. They found deposition efficiency higher than 100%, which was explained by an electroless deposition contribution to the electrodeposition process.

Diffusion Barriers. Diffusion barriers are used in the production of various components in the electronic industry. For example, electrochemically deposited nickel is used as a barrier layer between gold and copper in electronic connectors and solder interconnections. In these applications the product is a trilayer of composition Cu/Ni/Au. In another example, Ni and Co are considered as diffusion barriers and cladding materials in the production of integrated circuits and multichip electronic packaging. In this case the barrier metal (BM), Co or Ni, is the diffusion barrier between conductor and insulator (i.e., Cu and insulator), and the product trilayer is of composition Cu/BM/insulator. The common couple in these applications is the Cu/BM bilayer (BM, the diffusion barrier metal; Co, Ni, or Ni–Co alloy).

A comparison was made between Ni and Co diffusion barriers produced by electroless, electro-, and evaporation deposition (64). This comparison shows that only electrolessly deposited metals and alloys, at a thickness of $1000\,\mu m$, have barrier properties for Cu diffusion. For Co(P) 1000-μm-thick barriers, annealed for $14\,h$, the amount of Cu interdiffused into Co(P) is less than 1 at %. Thicker barriers of Ni(P), Ni(B), and Co(B) are required for the same degree of Cu interdiffusion. The same metals, if electrodeposited, both do and do not have inferior barrier properties. This difference between electrolessly and electrodeposited metals may be ascribed to the presence of metal phosphides (e.g., Ni_3P, Co_3P) in the electroless deposits (probably in grain boundaries). The best electroless barriers for Cu diffusion are Co(P), Ni–Co(P) alloys, and Ni(P) deposited from sulfamate solutions. Evaporated Ni and Co, 1000 μm thick, do not have barrier properties.

Thin-film interdiffusion is discussed further in Chapter 18.

A comparative study of great practical value has been carried out between several Ni-based diffusion barrier properties. Those were produced by means of electroless deposition from nickel sulfate and nickel sulfamate deposition solutions (73). It was concluded on the basis of Auger depth profiling (see Section 13.3) that Ni(P) sulfamate has much better diffusion barrier properties than Ni(P) sulfate. This conclusion is a telling example of the influence of anions on the physical properties of electrochemically deposited metals.

Magnetic Properties. Cobalt magnetic thin films with predesigned coercivity values ranging from soft magnetic to hard magnetic material can be produced using programmed changes of down to only one component of the electroless cobalt deposition solution (74,77). For example, electroless Co(P) deposited at 70°C from a solution containing 0.07 to 0.10 M citrate ions as complexing agent for the Co^{2+} ions, 0.3 to 0.4 M boric acid as buffer, sodium hydroxide to pH 8.00, 0.05 to 0.07 M sodium hypophosphite as the reducing agent, and 0.02 to 0.03 M cobalt sulfate as the source of Co^{2+} ions yields a hard magnetic material with an in-plane coercivity of 514 Oe and no in-plane anisotropy. The Co(P) film thickness in this example was 3152 Å (74). Magnetic materials are classified as hard or soft depending on the value of coercivity (75–77). Materials with coercivity exceeding 200 Oe are considered by most authors as magnetically hard.

In general, the effect of additives (supporting electrolyte) to the electroless Co(P) solution results in changes of the coercivity values of the product films. In line with this, coercivity of up to 1000 Oe in electrolessly produced Co(P) films was shown to be possible in Ref. 77.

The dependence of the coercivity of Co(P) deposit on the concentration of the sulfamic acid (H_3NO_3S) and sodium sulfate (Na_2SO_4) in solution is shown in Table 8.1, from which it can be seen (1) that the coercivity of Co(P) deposits is a function of the concentration of sulfamic acid and sodium sulfate in a basic solution, and (2) that an increase in the concentration of additives to the basic solution results in lowering of

TABLE 8.1. In-Plane Coercivity Data for Five Electrolessly Deposited Co(P) Films

Addition to the Solution	Co(P) Deposit Thickness (Å)	In-Plane Coercivity (Oe)
No[a]	3152	514
0.05 M H_3NO_3S	4107	279
0.20 M H_3NO_3S	3655	2.61
0.20 M Na_2SO_4	3395	5.21
0.20 M Na_2SO_4	6108	2.77

[a]Basic solution: 0.07 to 0.10 M citrate, 0.3 to 0.4 M boric acid, sodium hydroxide to pH 8.00, 0.05 to 0.07 M sodium hypophosphite, 0.02 to 0.03 M cobalt sulfate; temperature 70°C.

the coercivity. The variation of the coercivity with the thickness of the Co(P) deposit is discussed in Ref. 78.

One possible, although speculative explanation of the effect of the addition of sulfamic acid or sodium sulfate may be based on Eq. (4.9). According to this equation, the variation in the concentration c of a nonreacting electrolyte changes the thickness of the metal–solution interphase, the double-layer thickness d_{dl}. It appears that as the thickness of the double layer, d_{dl}, decreases, the coercivity of the Co(P) deposited decreases as well.

REFERENCES AND FURTHER READING

1. C. Wagner and W. Traud, *Z. Electrochem.* **44**, 391 (1938).
2. A. Brenner and G. Riddell, *J. Res. Natl. Bur. Stand.* **37**, 31 (1946).
3. A. Brenner and G. Riddell, *Am. Electroplat, Soc. Annu. Proc.* **33**, 23 (1946).
4. A. Brenner and G. Riddell, *Proc. Am. Electroplat. Soc.* **34**, 56 (1947).
5. M. Stern and A. L. Geary, *J. Electrochem. Soc.* **104**, 56 (1957).
6. R. P. Buck and L. R. Griffith, *J. Electrochem. Soc.* **109**, 1005 (1962).
7. E. B. Saubestre, *Met. Finish.* **60**(6), 67 (1962).
8. M. Saito, *J. Met Finish. Soc. Jpn.* **17**, 14 (1966).
9. E. McCaffery and A. C. Zettlemoyer, *J. Phys. Chem.* **71**, 2444 (1967).
10. M. Paunovic, *Plating* **55,** 1161 (1968).
11. J. P. Marton and M. Schlesinger, *J. Electrochem. Soc.* **115**, 16 (1968).
12. R. Sard, *J. Electrochem. Soc.* **117**, 864 (1970).
13. L. N. Schoenberg, *J. Electrochem. Soc.* **118**, 1571 (1971).
14. R. L. Cohen, J. F. D'Amico, and K. W. West, *J. Electrochem. Soc.* **118**, 2042 (1971).
15. F. M. Donahue, *Oberflaeche-Surf.* **13**(12), 301 (1972).
16. L. N. Schoenberg, *J. Electrochem. Soc.* **119**, 1491 (1972).
17. N. Feldstein, S. L. Chow, and M. Schlesinger, *J. Electrochem. Soc.* **120**, 875 (1973).
18. N. Feldstein and J. A. Weiner, *J. Electrochem. Soc.* **120**, 475 (1973).
19. R. L. Cohen and K. W. West, *J. Electrochem. Soc.* **120**, 502 (1973).
20. C. H. deMinjer and P. F. J. V. D. Boom, *J. Electrochem. Soc.* **120**, 1644 (1973).
21. Y. Okinaka, *J. Electrochem. Soc.* **120**, 739 (1973).
22. F. Mansfeld, *J. Electrochem. Soc.* **120**, 515(1973).
23. F. Mansfeld, *Corrosion* **29**, 397 (1973).
24. N. Feldstein, M. Schlesinger, N. E. Hedgecock, and S. L. Chow, *J. Electrochem. Soc.* **121**, 738 (1974).
25. A. Molenaar, M. F. Holdrinet, and L. K. H. van Beek, *Plating* **61**, 238 (1974).
26. L. K. H. van Beek, *Plating* **61**, 238 (1974).
27. G. O. Mallory, *Plating* **61**(11), 1005 (1974).
28. R. L. Meek, *J. Electrochem. Soc.* **122**, 1177, 1478 (1975).
29. R. L. LeRoy, *Corrosion* **31**, 173 (1975).
30. R. L. LeRoy, *J. Electrochem. Soc.* **124**, 1006 (1977).

31. M. Paunovic, *J. Electrochem. Soc.* **124**, 349 (1977).

32. M. Paunovic, *J. Electrochem. Soc.* **125**, 173 (1978).

33. S. M. El-Raghy and A. A. Abo-Salama, *J. Electrochem. Soc.* **126**, 171 (1979).

34. M. Paunovic and D. Vitkavage, *J. Electrochem. Soc.* **126**, 2282 (1979).

35. M. Paunovic, *J. Electrochem. Soc.* **127**, 441C (1980).

36. H. J. Choi and R. Weil, *Plat. Surf. Finish.* **68**(5), 110 (1981).

37. I. Ohno and S. Haruyama, *Surf. Technol.* **13**, 1 (1981).

38. J. E. A. M. Van den Meerakker, *J. Appl. Electrochem.* **11**, 395 (1981).

39. D. W. Baudrand, *Plat. Surf. Finish.* **68**(12), 57 (1981).

40. M. Paunovic and C. Stack, in *Electrocrystallization*, R. Weil and R. G. Baradas, eds., *Proceedings*, Vol. 81-6, Electrochemical Society, Pennington, NJ, 1981, p. 205.

41. D. C. Montgomery and E. A. Peck, *Introduction to Linear Regression Analysis*, Wiley, New York, 1982.

42. S. Nakahara and Y. Okinaka, *Acta Metall.* **31**, 713 (1983).

43. M. Paunovic, *Plat. Surf. Finish.* **70**, 62 (1983).

44. M. Paunovic and R. Arndt, *J. Electrochem. Soc.* **130**, 794 (1983).

45. D. Vitkavage and M. Paunovic, *Plat. Surf. Finish.* **70**(4), 48 (1983).

46. D. C. Montgomery, *Design and Analysis of Experiments*, Wiley, New York, 1984.

47. A. Brenner, *Plat. Surf. Finish.* **71**(7), 24 (1984).

48. J. Kim, S. H. Wess, D. Y. Jung, and R. W. Johnson, *IBM J. Res. Dev.* **8**, 697 (1984).

49. T. Hayashi, *Met. Finish.* **85**(6), 85 (1985).

50. M. Paunovic, *J. Electrochem. Soc.* **132**, 1155 (1985).

51. M. Paunovic and R. Zeblisky, *Plat. Surf. Finish.* **71**(2), 52 (1985).

52. G. E. Box and N. R. Draper, *Empirical Model Building and Response Surfaces*, Wiley, New York, 1987.

53. J. W. M. Jacobs and J. M. G. Rikken, in *Electroless Deposition of Metals and Alloys*, M. Paunovic and I. Ohno, eds., *Proceedings*, Vol. 88-12, Electrochemical Society, Pennington, NJ, 1988, p. 75.

54. I. Ohno, in *Electroless Deposition of Metals and Alloys*, M. Paunovic and I. Ohno, eds., *Proceedings*, Vol. 88-12, Electrochemical Society, Pennington, NJ, 1988, p. 129.

55. A. J. Ricco and S. J. Martin, in *Electroless Deposition of Metals and Alloys*, M. Paunovic and I. Ohno, eds., *Proceedings*, Vol. 88-12, Electrochemical Society, Pennington, NJ, 1988, p. 142.

56. M. Paunovic and C. Ting, in *Electroless Deposition of Metals and Alloys*, M. Paunovic and I. Ohno, eds., *Proceedings*, Vol. 88-12, Electrochemical Society, Pennington, NJ, 1988, p. 170.

57. R. Weil, J. H. Lee, and K. Parker, *Plat. Surf. Finish.* **76**(2), 62 (1989).

58. C. H. Ting and M. Paunovic, *J. Electrochem. Soc.* **136**, 456 (1989).

59. C. H. Ting, M. Paunovic, P. L. Pai, and J. Chiu, *J. Electrochem. Soc.* **137**, 462 (1989).

60. E. R. Ott and E. G. Schilling, *Process Quality Control*, McGraw-Hill, New York, 1990.

61. M. Paunovic, in *Electrochemistry in Transition*, O. J. Murphy, S. Srinivasan, and B. E. Conway, eds., Plenum Press, New York, 1992, p. 479.

62. C. H. Ting and M. Paunovic, U.S. patent 5,169,680, Dec. 8, 1992.

63. M. Paunovic, L. A. Clevenger, J. Gupta, C. Cabral, Jr., and J. M. E. Harper, *J. Electrochem. Soc.* **140**, 2690 (1993).

64. M. Paunovic, P. J. Bailey, R. G. Schad, and D. A. Smith, *J. Electrochem. Soc.* **141**, 1843 (1994).

65. M. Paunovic, T. Nguyen, R. Mukherjee, C. Sambucetti, and L. Romankiw, *J. Electrochem. Soc.* **142**, 1495 (1995).

66. M. Charbonnier, M. Alami, and M. Romand, *J. Electrochem. Soc.* **143**, 472 (1996).

67. J.-Y. Zhang, H. Esrom, and I. W. Boyd, *Appl. Surf. Sci.* **96–98**, 399 (1996).

68. J.-Y. Zhang, I. W. Boyd, and H. Esrom, *Appl. Surf. Sci.* **109–110**, 253 (1997).

69. G. J. Norga, M. Platero, K. A. Black, A. J. Reddy, J. Michel, and L. C. Kimerling, *J. Electrochem. Soc.* **144**, 2801 (1997).

70. J. Duffy, L. Pearson, and M. Paunovic, *J. Electrochem. Soc.* **130**, 876 (1983).

71. J. O'M Bockris and S. U. M. Khan, *Surface Electrochemistry*, Plenum Press, New York, 1993.

72. A. Hung and I. Ohno, *J. Electrochem. Soc.* **137**, 918 (1990).

73. M. Paunovic, *IBM Tech. Discl. Bull.* **35**, 183 (1992).

74. M. Paunovic and C. Jahnes, unpublished data (see also Ref. 77).

75. R. Weil, in *Modern Electroplating*, 4th ed., M. Schlesinger and M. Paunovic, eds., Wiley, New York, 2000, p. 55.

76. C. Kittel, *Introduction to Solid State Physics*, 7th ed., Wiley, New York, 1996.

77. M. Schlesinger, X. Meng, W. T. Evans, J. A. Saunders, and W. P. Kampert, *J. Electrochem. Soc.* **137**, 1706 (1990).

78. J. S. Judge, J. R. Morrison, and D. E. Speliotis, *J. Appl. Phys.* **53**, 948 (1965).

PROBLEMS

8.1. In electroless deposition of a metal M, the current–potential relationships for the partial reaction may be written as linear functions:

$$
\begin{aligned}
\text{Partial cathodic reaction:} &\qquad E_c = -800 - 110 \log i \\
\text{Partial anodic reaction:} &\qquad E_a = -920 - 280 \log i
\end{aligned}
$$

The electrode potentials (E_c, E_a) are given in mV and the current densities i in mA/cm^2. Determine (a) E_{mp}, the mixed potential; (b) i_{dep}, the rate of deposition for this process; and (c) the transfer coefficients α for the cathodic and anodic partial reactions. Solve this problem algebraically by finding an intersection of two strait lines; do not plot any $E = f(i)$ functions.

8.2. The electroless deposition of copper is usually done in solutions containing EDTA as a complexing agent. The stability constant for the CuEDTA complex is

$$
K = \frac{Cu \cdot EDTA}{(Cu^{2+})(EDTA)} = 10^{19}
$$

and the relative standard electrode potential E^0 of the Cu/Cu^{2+} electrode is 0.337 V.

(a) Calculate the concentration of Cu^{2+} ions in a solution of 0.1 M CuSO$_4$ and 0.175 M EDTA.

(b) Calculate the equilibrium potential E_{eq} of a Cu electrode in this solution

(c) Compare your results with the experimental results presented in Fig. 8.4.

8.3. The value of the transfer coefficient α is usually 0.50. In the electrochemical oxidation of an organic molecule the transfer coefficient α may be considerably less than 0.50. One example is the oxidation of formaldehyde in electroless deposition of copper.

(a) Calculate the transfer coefficient α for the oxidation of formaldehyde from the anodic polarization curve shown in Fig. 8.4.

(b) Calculate the transfer coefficient α for the oxidation of formaldehyde from the following experimental curve-fitting equation (32):

$$\alpha_{Red} = -10.59 + 1.78 \, pH - 7.40 \times 10^{-2} \, (pH)^2$$

(c) Compare the results obtained in parts (a) and (b).

(d) Explain why in some cases α values are significantly less than 0.5. (See Ref. 70 and references therein, and Ref. 71.)

9

Displacement Deposition

9.1. INTRODUCTION

In Chapter 8 we have shown that in the case of electroless deposition the reducing agent Red in the solution is the electron source, the electron-donating species that give electrons to the catalytic surface and to the metal ions M^{z+} at the interface. In this chapter we show that the substrate itself can also be the electron-donating species. We also show that, in general, Ox/Red (M^{z+}/M) couples with high standard electrode potentials are reduced by Ox/Red (M^{z+}/M) couples with low standard electrode potentials. In other words, low-potential couples reduce high-potential couples (see Table 5.1 and Fig. 5.10). The thickness of the metal deposited in this case is self-limiting since the displacement deposition process needs exposed (free) substrate surface in order to proceed. This technique is referred to by a variety of terms, depending on the application: immersion deposition, galvanic deposition (galvanic corrosion), conversion, cementation (in the metal recovery industry), and so on.

9.2. ELECTROCHEMICAL MODEL

The overall displacement deposition reaction in general Ox/Red (M^{z+}/M) terms is given by

$$M_l + M_r^{z+} \rightarrow M_r + M_l^{z+} \tag{9.1}$$

for the cell

$$M_l \mid M_l^{z+}(aq) \parallel M_r^{z+}(aq) \mid M_r \tag{9.2}$$

Fundamentals of Electrochemical Deposition, Second Edition.
By Milan Paunovic and Mordechay Schlesinger
Copyright © 2006 John Wiley & Sons, Inc.

The partial (half-cell) reactions are:

$$\text{oxidation:} \qquad M_l \rightarrow M_l^{z+} + ze \tag{9.3}$$

$$\text{reduction:} \qquad M_r^{z+} + ze \rightarrow M_r \tag{9.4}$$

Metal substrate M_l is dissolving into the solution [Eq. (9.3)] and thus supplying electrons necessary for the reduction deposition reaction [Eq. (9.4)]. The relationship between partial reactions (9.3) and (9.4) is shown in Figure 9.1.

We described one example of this type of electrochemical deposition in Section 5.7 when we considered processes on a strip of Zn placed in a solution of $CuSO_4$ (Fig. 5.11). In Chapter 5 we stated that there are two partial reactions in that system, as in an electroless system. In displacement deposition of Cu on Zn, electrons are supplied in the oxidation reaction of Zn:

$$Zn \rightarrow Zn^{2+} + 2e \tag{9.5}$$

where Zn from the substrate dissolves into the solution and thus supplies electrons necessary for the reduction deposition reaction:

$$Cu^{2+} + 2e \rightarrow Cu \tag{9.6}$$

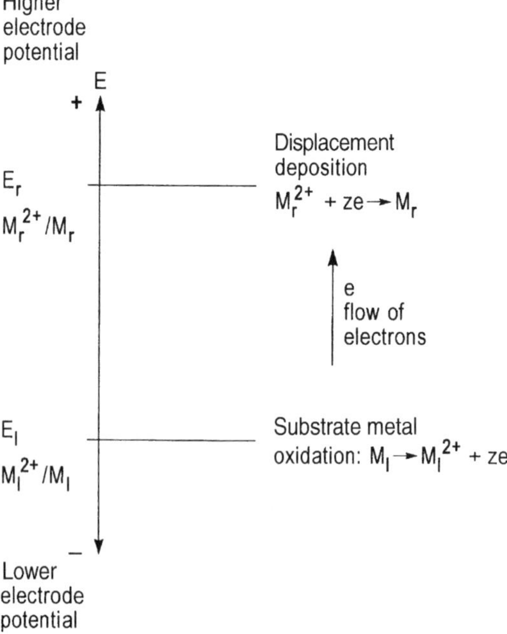

Figure 9.1. Relationship between partial reactions in displacement deposition.

The overall displacement deposition reaction is

$$Zn + Cu^{2+} \rightarrow Zn^{2+} + Cu \tag{9.7}$$

This is obtained via combination of the two partial electrode reactions, oxidation and reduction, reaction (9.5) and (9.6), respectively. Thus, in the displacement deposition of Cu on a Zn substrate, a layer of metallic Cu is deposited on the zinc while Zn dissolves into solution (Fig. 5.11). We stated that this reaction is possible since the Zn/Zn^{2+} system has an electrode potential lower than that of the Cu/Cu^{2+} system (Table 5.1 and Fig. 5.10). The overall displacement deposition reaction according to Eq. (9.7) can be considered as the reaction of the electrochemical cell

$$Zn \,|\, ZnSO_4\,(aq) \,\|\, CuSO_4\,(aq) \,|\, Cu$$

9.3. PREDICTIONS OF THERMODYNAMIC FEASIBILITY OF REACTION

The thermodynamic criterion for spontaneity (feasibility) of a chemical and electrochemical reaction is that the change in free energy, ΔG have a negative value. Free-energy change in an oxidation–reduction reaction can be calculated from knowledge of the cell voltage $\mathcal{E}$:

$$\Delta G = -nF\mathcal{E} \tag{9.8}$$

where n is the number of electrons and F is the Faraday constant. If the concentrations of all species are at unit activity values, ΔG is the standard free-energy change for the cell reaction, ΔG^0:

$$\Delta G^0 = -nF\mathcal{E}^0 \tag{9.9}$$

where $\mathcal{E}^0$ is the standard cell voltage. From Eqs. (9.8) and (9.9) it follows that a spontaneous reaction must have a positive $\mathcal{E}^0$ value in order to have a negative value for ΔG^0.

The standard cell voltage $\mathcal{E}^0$ can in turn be calculated from the standard electrode potentials E^0 for the partial reactions using the expression

$$\mathcal{E}^0 = E_r^0 - E_l^0 \tag{9.10}$$

where the labels l (left) and r (right) refer to the electrodes as they are written in the cell description. In Eq. (9.9) or (9.8), both partial reactions, (9.3) and (9.4), are written as reductions.

From Eq. (9.10) it follows that when $\mathcal{E}$ is positive,

$$\mathcal{E} > 0, \quad E_r > E_l \tag{9.11}$$

When E_r is larger than E_l, reduction occurs at the right-hand electrode [Eqs. (9.4) and (9.6)]. Thus, when $\mathcal{E} > 0$, the overall displacement deposition reaction [Eqs. (9.1) and (9.7)] will occur from left to right. The reaction is spontaneous (feasible) in a direction from left to right since ΔG is negative for positive values of $\mathcal{E}$ [Eqs. (9.8) and (9.9)]. This is in agreement with earlier discussions in Chapter 5 and Figure 5.10.

For example, let us use Eq. (9.10) to evaluate $\mathcal{E}^0$ for the displacement deposition of Cu on a Zn substrate [Eq. (9.7)]. We have

$$E^0(\text{Cu}^{2+}/\text{Cu}) = +0.34 \text{ V}$$

$$E^0(\text{Zn}^{2+}/\text{Zn}) = -0.76 \text{ V}$$

and from Eq. (9.10) it follows that

$$\mathcal{E}^0 = E_r^0 - E_l^0 = 0.34 - (-0.76) = 1.10 \text{ V}$$

The same value of $\mathcal{E}^0$ is obtained experimentally. Since $\mathcal{E}^0$ is positive, $+1.10$ V, and ΔG is negative [Eq. (9.9)], the overall displacement deposition reaction (9.7) proceeds spontaneously from left to right.

9.4. COMPLEXED METAL IONS IN DISPLACEMENT DEPOSITION

Let us determine whether we can use the displacement deposition technique to deposit Sn on a Cu substrate. The simplest way to determine this is to use the principle presented in Figures 5.10 and 9.1. According to this principle, Sn cannot be deposited by displacement on a Cu substrate since the standard electrode potential of a Cu^{2+}/Cu couple is more positive than that of an Sn^{2+}/Sn couple:

$$E^0(\text{Sn}^{2+}/\text{Sn}) = -0.136 \text{ V}$$

$$E^0(\text{Cu}^{2+}/\text{Cu}) = +0.34 \text{ V}$$

Evaluation of $\mathcal{E}^0$ leads to the same conclusion. For the evaluation we consider the cell

$$\text{Cu} \,|\, \text{Cu}^{2+}(\text{aq})a = 1 \,\|\, \text{Sn}^{2+} (\text{aq}) \, a = 1 \,|\, \text{Sn}$$

where the Sn^{2+}/Sn electrode (the electrode with the ion we would like to deposit by displacement) is on the right-hand side. To deposit Sn on a Cu substrate, the following two partial reactions—the reduction (deposition) reaction and the oxidation partial reaction (the reaction that supplies electrons)—must proceed:

$$\text{Sn}^{2+} + 2\text{e} \rightarrow \text{Sn} \tag{9.12}$$

$$\text{Cu} \rightarrow \text{Cu}^{2+} + 2\text{e} \tag{9.13}$$

The overall displacement deposition reaction, obtained from addition of reactions (9.12) and (9.13), is given by

$$Sn^{2+} + Cu \rightarrow Sn + Cu^{2+} \tag{9.14}$$

From Eq. (9.10) the standard cell voltage $\mathcal{E}^0$ for the preceding [Eq. (9.14)] Cu/Sn cell reaction is

$$\mathcal{E}^0 = E_r^0 - E_l^0 = E^0(Sn^{2+}/Sn) - E^0(Cu^{2+}/Cu) = -0.136 - 0.34 = -0.476 \, V$$

Since $\mathcal{E}^0$ is negative, ΔG is positive and the conclusion is that reaction (9.14) is not feasible (cannot proceed spontaneously from left to right).

However, reaction (9.14) can be made to proceed from left to right spontaneously if the potential of the Cu^{2+}/Cu electrode is made more negative than that of Sn^{2+}/Sn electrode. This can be achieved by complexing the Cu^{2+} ions in solution. The preferred complexing agent is CN^- ion. When CN^- ions are added to the solution of $CuSO_4$, the concentration of Cu^{2+} ions is reduced and the electrode potential of the Cu^{2+}/Cu electrode is moved to the negative value. We illustrate this in the following example.

If sufficient NaCN (or KCN) is added to the solution of Cu(II) ions to form the complexed ions $[Cu(CN)_3]^{2-}$, and if the excess of CN^- ions is such that the concentration of free CN^- is 1×10^{-4}M, the concentration of free Cu^+ ions can be calculated from the dissociation constant of the complex. The dissociation constant of the complexed ion $[Cu(CN)_3]^{2-}$ is 5.6×10^{-28}:

$$\frac{(Cu^+)(CN^-)^3}{[Cu(CN)_3]^{2-}} = 5.6 \times 10^{-28} \tag{9.15}$$

For 0.05 M $[Cu(CN)_3]^{2-}$ and 1×10^{-4}M CN^- from this equation, one concludes that the concentration of Cu^+ is 2.8×10^{-17}M. The reversible electrode potential of the Cu^+/Cu electrode can be calculated for this concentration of Cu^+ from the Nernst equation:

$$E = 0.55 + 0.059(\log 2.8 \times 10^{-17}) = -0.43 \, V \tag{9.16}$$

where 0.55 is the standard electrode potential for Cu^+/Cu. Thus, in this case Cu^+/Cu is more negative than Sn^{2+}/Sn and reaction (9.14) will go spontaneously from left to right, resulting in displacement deposition of Sn on Cu. In this approximate calculation we considered that only $[Cu(CN)_3]^{2-}$ complexed ions are present in the solution and that the CN^- ions do not affect the electrode potential of the Sn^{2+}/Sn electrode. However, in reality there is a mixture of different CN^- complexes of copper in the solution.

Reaction (9.14) is a very important displacement deposition reaction in the printed circuit industry. It is used to help the soldering capability of copper.

9.5. KINETICS AND MECHANISM

The kinetics and mechanisms of the displacement deposition of Cu on a Zn substrate in alkaline media was studied by Massee and Piron (5). They determined that at the beginning of the deposition process, the rate is controlled by activation. The activation control mechanism changes to diffusion control when the copper covers enough of the Zn surface to facilitate further deposition of copper. This double mechanism can explain the kinetic behavior of the deposition process.

The mechanisms of the crystal-building process of Cu on Fe and Al substrates were studied employing transmission and scanning electron microscopy (1). These studies showed that a nucleation–coalescence growth mechanism (Section 7.10) holds for the Cu/Fe system and that a displacement deposition of Cu on Fe results in a continuous deposit. A different nucleation–growth model was observed for the Cu/Al system. Displacement deposition of Cu on Al substrate starts with formation of isolated nuclei and clusters of Cu. This mechanism results in the development of dendritic structures.

The properties of deposits may be controlled by changing the kinetics of the deposition and the mechanism of crystallization. One way to achieve this is by complexing the depositing ions, as stated above.

REFERENCES AND FURTHER READING

1. L. E. Murr and V. Annamalai, *Metall. Trans.* **B9**, 515 (1978).

2. D. S. Lashmore, *Plat. Surf. Finish.* **65**(4), 44 (1978).

3. M. Paunovic and C.H. Ting, in *Electroless Deposition of Metals and Alloys,* M. Paunovic and I. Ohno, eds., *Proceedings*, Vol. 88–12, Electrochemical Society, Pennington, NJ, 1988.

4. C. H. Ting, in *Electroless Deposition of Metals and Alloys,* M. Paunovic and I. Ohno, eds., *Proceedings*, Vol. 88–12, Electrochemical Society, Pennington, NJ, 1988.

5. N. Massee and D. L. Piron, *J. Electrochem. Soc.* **140**, 2818 (1993).

6. G. J. Norga, M. Platero, K.A. Black, A. J. Reddy, J. Michel, and L. C. Kimerling, *J. Electrochem. Soc.* **144**, 2801 (1997).

PROBLEMS

9.1. Determine the spontaneity (feasibility) of the following reactions by determining the standard cell voltage $\mathcal{E}^0$ and the standard free-energy change ΔG^0 for the following cell reactions:

(a) Zincate processing (displacement deposition of Zn) on Al and Cu substrates:

$$3Zn^{2+} + 2Al \rightarrow 3Zn + 2Al^{3+}$$
$$Zn^{2+} + Cu \rightarrow Zn + Cu^{2+}$$

(b) Tinning processing (displacement deposition of Sn) on Al:

$$3Sn^{2+} + 2Al \rightarrow 3Sn + 2Al^{3+}$$

The standard electrode potential E^0 are:

Metal/Ion Couple	Standard Electrode Potential, E^0 (V), 25^0C
Zn/Zn^{2+}	-0.763
Al/Al^{3+}	-1.66
Cu/Cu^{2+}	0.337
Sn/Sn^{2+}	-0.136

9.2. Some industrial formulas for displacement deposition of Sn on Cu (tinning process) contain a $1.0\,M$ solution of NaCN (or KCN). Assume that only $[Cu(CN)_3]^{2-}$ complexed ions are present in the solution. The dissociation (instability) constant of the complexed ion $[Cu(CN)_3]^{2-}$ is 5.6×10^{-28}. The standard electrode potential for Cu/Cu^+ is $E^0 = 0.55$ V.

(a) Determine the concentration of free Cu^+ ions in a solution of 0.1 M $CuSO_4$ and $1.0\,M$ NaCN.

(b) Determine the reversible electrode potential of the Cu/Cu^+ electrode in the solution in part (a).

9.3. (a) The spontaneous displacement deposition reaction

$$Cu^{2+} + Zn \rightarrow Cu + Zn^{2+}$$

began under standard conditions ($Cu^{2+} = Zn^{2+} = 1$ M; assume that activity $a = 1$). Calculate the cell voltage $\mathcal{E}$ when the cell reactions are (1) 90% completed; (2) 95% completed.

The standard cell voltage for the cell reaction is $\mathcal{E}^0 = 1.10$ V at 25°C.

(b) An excess of Zn metal is placed into a 1 M (assume unit activity) solution of $CuSO_4$, where the above-mentioned spontaneous displacement deposition of Cu occurs. Eventually, an equilibrium will be established. Evaluate (1) the equilibrium constant $K = \Pi$ (products)/Π (reactants), where Π represents the product of the concentrations (activities, a) raised to the power of their stoichiometric numbers; (2) the equilibrium concentration of Cu^{2+} ions in terms of K (write an expression for the equilibrium concentration of Cu^{2+} ions).

[*Hints*: For part (a), $\mathcal{E} = f$(concentrations) is given by Eq. (5.9) in Section 5.3; for part (b), $\Delta G^0 = -RT \ln K$ (thermodynamics); $\Delta G^0 = -nF\mathcal{E}^0$, Eq. (9.9).]

10

Effect of Additives

10.1. INTRODUCTION

Most solutions used in electrodeposition of metals and alloys contain one or more inorganic or organic additives that have specific functions in the deposition process. These additives affect deposition and crystal-building processes as adsorbates at the surface of the cathode. Thus, in this chapter we first describe adsorption and the factors that determine adsorbate–surface interaction. There are two sets of factors that determine adsorption: substrate and adsorbate factors. *Substrate factors* include electron density, *d*-band location, and the shape of substrate electronic orbitals. *Adsorbate factors* include electronegativity and the shape of adsorbate orbitals.

After discussing adsorption, we discuss the effects of additives on the kinetic parameters of the deposition process and on the elementary processes of crystal growth. The general effect of additives on electroless deposition is discussed in Section 8.4.

10.2. ADSORPTION

Chemisorption and Physisorption. One classification of adsorption phenomena is based on the *adsorption energy:* the energy of the adsorbate–surface interaction. In this classification there are two basic types of adsorption: chemisorption (an abbreviation of *chemical adsorption*) and physisorption (an abbreviation of *physical adsorption*). In *chemisorption* the chemical attractive forces of adsorption are acting between surface and adsorbate (usually, covalent bonds). Thus, there is a chemical combination between the substrate and the adsorbate where electrons are shared and/or transferred. New electronic configurations are formed by this sharing of electrons. In *physisorption* the physical forces of adsorption, van der Waals or pure electrostatic forces, operate between the surface and the adsorbate; there is no electron transfer and no electron sharing. Adsorption energy for chemisorbed species is greater than

Fundamentals of Electrochemical Deposition, Second Edition.
By Milan Paunovic and Mordechay Schlesinger
Copyright © 2006 John Wiley & Sons, Inc.

that for physisorbed species. Typical values for chemisorption are in the range 20 to 100 kcal/mol and for physisorption, around 5 kcal/mol.

The difference between physisorption and chemisorption can be explained using a potential-energy diagram. The potential-energy diagram for physisorption and chemisorption of an A–A molecule (e.g., H_2) is shown in Figure 10.1. Curve P in Figure 10.1 gives the potential energy of molecule A_2 for cases in which only physical forces of attraction are operating. It is seen that as the molecule approaches the surface, its energy falls as it becomes physisorbed. The minimum of this curve represents the equilibrium state for the physisorbed molecule. The potential well q_p (ΔH_p), the heat of adsorption for the physisorption, is relatively shallow and is due to long-range forces (e.g., van der Waals forces), so the equilibrium distance from the surface (d_p) is relatively large. It is located at the sum of the van der Waals radii for the surface atom and adsorbate molecule A_2. An attempt to decrease the distance of separation below the equilibrium value leads to gradually increasing repulsion. Curve C corresponds to the interaction of two A atoms with the surface. The minimum of this curve represents the equilibrium state for chemisorbed atoms A. The minimum is deeper and at a smaller distance (d_c) than the minimum for physisorption (d_p). The two curves (P and C, Fig. 10.1) cross, and the adsorbate can pass from the first to the second. The transition from physisorption to chemisorption occurs at the crossing point of curves P and C. The energy at this point is the activation energy E_a, which is the excess of energy of that for the separated metal and molecule A_2. It is the activation energy for the transition from physical to chemical adsorption. Figure 10.1 shows that chemisorption in this case involves dissociation of physisorbed molecules.

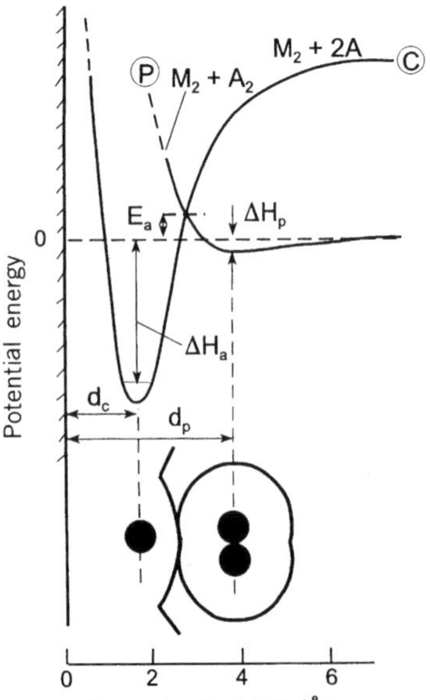

Figure 10.1. Potential-energy diagram for physisorption and chemisorption of an A–A molecule.

Adsorption Equilibrium. Since the additive is not used up in many cases of electrodeposition in the presence of an additive (the additive is not incorporated in the deposit), one can conclude that the adsorption equilibrium is dynamic. In a dynamic adsorption equilibrium state the adsorbed molecules are continually desorbing at a rate equal to the rate at which dissolved molecules from the solution become adsorbed. If the rates of the adsorption and desorption processes are high and of the same order of magnitude as that of the cathodic deposition process, no incorporation or entrapment of additives in the deposit will occur. However, if they are much smaller, additive molecules will be entrapped in the deposit via propagating steps (growing crystallites). Thus, at a current density higher than the optimum value, additives (brighteners or levelers) will be incorporated into the deposit. This incorporation can result in poor quality of the resulting deposit.

Adsorption Isotherms. Adsorption isotherms describe the relationship between the coverage θ of the surface by the adsorbate and the concentration of the adsorbate in the bulk solution, c^b, at a given temperature. The surface coverage θ is defined as

$$\theta = \frac{N_{occ}}{N} \tag{10.1}$$

where N_{occ} and N are the number of adsorption sites occupied and the total number of adsorption sites available, respectively. From this equation it follows that $N\theta = N_{occ} =$ number of adsorption sites occupied, and $N - N_{occ} = N - N\theta = N(1 - \theta) =$ number of vacant sites.

The relationship $\theta = f(c^b)$ can be derived from a kinetic model assuming that the rate of adsorption r_a is proportional to the number of vacant sites $N(1 - \theta)$ and also to the bulk solution concentration c^b:

$$r_a = k_a N(1 - \theta)c^b \tag{10.2}$$

and that the rate of desorption r_d is proportional to the number of adsorption sites occupied, $N\theta$:

$$r_d = k_d N\theta \tag{10.3}$$

where k_a and k_d are adsorption and desorption rate constraints, respectively. At dynamic equilibrium $r_a = r_d$ and

$$k_a N(1 - \theta)c^b = k_d N\theta \tag{10.4}$$

Solving for θ, one obtains the Langmuir isotherm (1):

$$\theta = \frac{Kc^b}{1 + Kc^b} \tag{10.5}$$

where $K = k_a/k_d$ is the adsorption equilibrium constant.

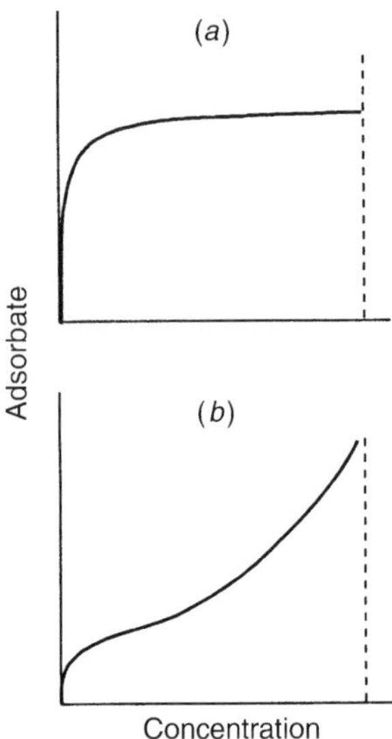

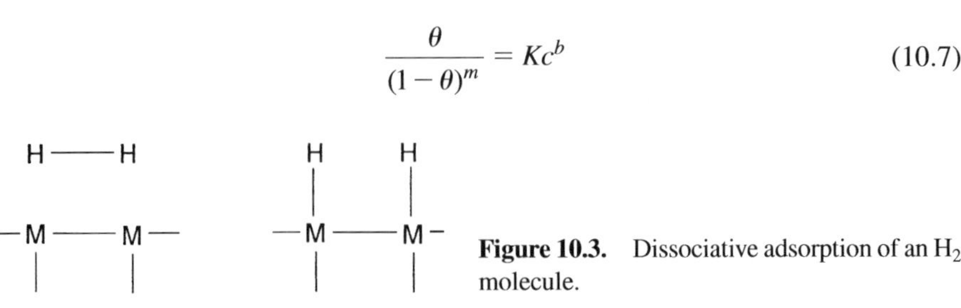

Figure 10.2. Types of adsorption isotherms: (*a*) Langmuir type; (*b*) physisorption, multilayer.

The Langmuir isotherm is based on the simplest model that involves the following assumptions: (1) the adsorption energy of all sites is the same and is unaffected by adsorption on neighboring sites; (2) the adsorption is immobile; (3) each site accommodates only one adsorbed particle; and (4) adsorbed atoms (molecules) do not interact with each other. Figure 10.2*a* shows that the Langmuir-type isotherm for chemisorption has a limiting adsorption that corresponds to a monolayer coverage. In contrast, the isotherm for physisorption (Fig. 10.2*b*) does not show a saturation plateau but indicates a multilayer formation.

Equation (10.5) is valid for cases where there is no dissociation on adsorption. However, in many cases chemisorption is dissociative, involving, for example, adsorption of hydrogen (H_2). In these cases the chemisorption process can be formulated as shown in Figure 10.3. If the adsorbed molecule dissociates into n fragments, the Langmuir isotherm has the form

$$\theta = \frac{K(c^b)^{1/n}}{1 + K(c^b)^{1/n}} \tag{10.6}$$

If a molecule is being adsorbed on m sites on the surface, without dissociation, the following equation holds:

$$\frac{\theta}{(1-\theta)^m} = Kc^b \tag{10.7}$$

Figure 10.3. Dissociative adsorption of an H_2 molecule.

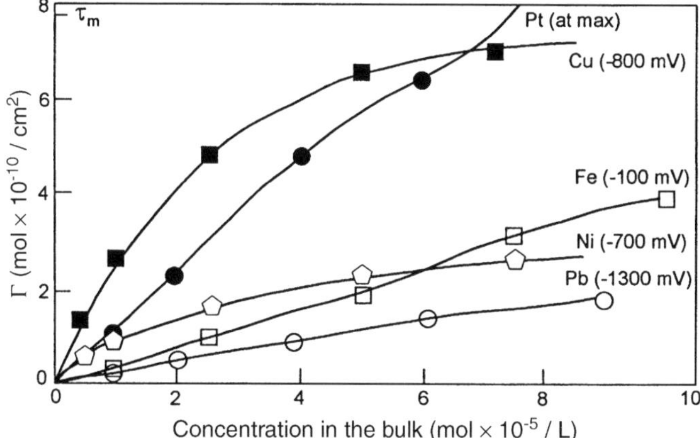

Figure 10.4. Adsorption isotherms for *n*-decylamine on Ni, Fe, Cu, Pb, and Pt at the potential of maximum adsorption. (From Ref. 5, with permission from the Electrochemical Society.)

Adsorption isotherms for *n*-decylamine on Ni, Fe, Cu, Pb, and Pt at the potential of maximum adsorption are shown in Figure 10.4. It is seen that a limiting coverage is approached in each case except on Pt, where multilayer formation occurs. The coverage θ in this case is defined as

$$\theta = \frac{\Gamma}{\Gamma_s} \tag{10.8}$$

where Γ is the surface concentration of adsorbate (mol/cm^2) and Γ_s is the saturation coverage of electrode by adsorbate (or Γ_{max}). In this case Γ_s is 7.9×10^{-10} mol/cm^2 (Γ_m in Fig. 10.4). Thus, a very small amount of material is involved in adsorption. This case is of interest since this small amount of adsorbent can influence the type of deposit, as shown in Section 10.5.

Simultaneous Adsorption of Two or More Species. If there are N different additives adsorbed at the electrode, the total surface coverage θ_T is given by

$$\theta_T = \theta_1 + \theta_2 + \theta_3 + \cdots + \theta_N \tag{10.9}$$

where θ_1, θ_2, θ_3, ..., θ_N is the surface coverage of additives A_1, A_2, A_3, ..., A_N, respectively.

Adsorbate Molecular Orientation at Electrode Surface. Adsorption of some molecules from solution produces an oriented adsorbed layer. For example, nicotinic acid (NA, or 3-pyridinecarboxylic acid, niacin, or vitamin B_3) is attached to a Pt(111) surface primarily or even exclusively through the N atom with the ring in a (nearly) vertical orientation (12) (Fig. 10.5*a*).

In another example, benzoic acid (BA) is an aromatic compound that orients horizontally at a Pt(111) surface (Fig. 10.5*b*). This horizontal orientation of adsorbed

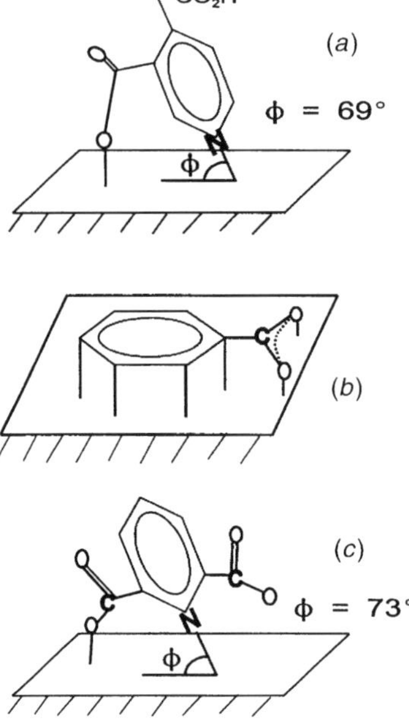

Figure 10.5. Adsorbate molecular orientation at the electrode surface: (*a*) nicotinic acid; (*b*) benzoic acid; (*c*) 2,6-pyridinedicarboxylic acid. (From Ref. 12, with permission from the American Chemical Society.)

BA involves coordination of the carboxylic acid to the Pt surface. The coordination depends on surface potential (12). BA adsorbed at negative potentials is coordinated to the Pt through the aromatic ring and (primarily) one carboxylate oxygen; when adsorbed at positive potentials, it is coordinated to the Pt surface through two equivalent oxygens. 2,6-Pyridinedicarboxylic acid (2,6-PDA) adsorbs in a tilted fashion with $\theta = 73°$, as shown in Figure 10.5*c*. When adsorbed at negative potentials, it is coordinated to the Pt surface by one, not both, carboxylates; in contrast, when adsorbed at positive electrode potentials, it is coordinated to Pt by both carboxylates, each through one oxygen.

Change in adsorbate concentration in solution can also result in orientational changes of molecules on the surface. For example, Soriaga et al. (10) have shown that diphenols and quinones are adsorbed on Pt electrodes with the diphenol or quinonoid ring parallel to the substrate at low concentration and that they reorient irreversibly to edgewise orientations as the concentration is increased. Figure 10.6 illustrates the way in which the average area occupied by a single molecule in the adsorbed layer depends on the orientation of the adsorbate on the surface, and how the molecular packing density, expressed in mol/cm^2, depends on adsorbate molecular orientation. Thus, the adsorbate orientation and orientational transitions are of fundamental and practical interest because of their influence on the kinetics and mechanism of electrochemical deposition.

Adsorption of Polymers. The three major characteristics of polymers in the metal–solution interphase of interest in metal deposition are the polydispersity, large number

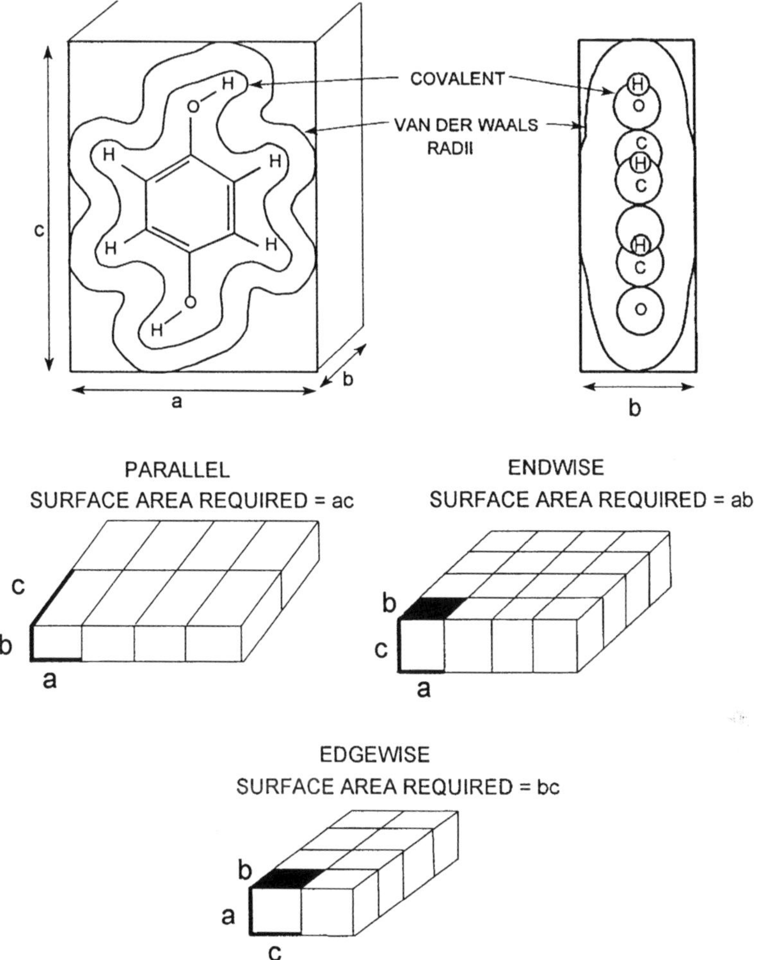

Figure 10.6. Orientation of adsorbate on the surface and the molecular packing density. (From Ref. 10, with permission from Elsevier.)

of configurations, and number of points of attachments. Polymers used as additives (e.g., wetting agents) are as prepared, generally polydisperse. Their adsorption has to be treated as a multicomponent system. For good reproducibility in metal deposition, it is important to use fraction of relatively limited molecular weight. A large number of configurations and the number of points of attachment are factors that determine the rate of attainment of adsorption equilibrium.

The shape of a flexible polymer molecule in the vicinity of the surface is greatly distorted from its average shape in solution. Adsorbed polymer molecules are attached to the surface by stretches of segments at the surface alternating with loops out of the surface (Fig. 10.7).

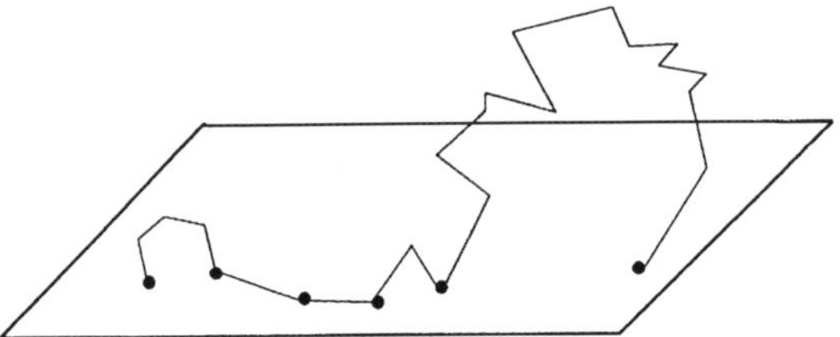

Figure 10.7. Schematic representation of adsorbed polymer molecules.

10.3. EXPERIMENTAL METHODS FOR STUDY OF ADSORPTION

Potentiodynamic Technique. Adsorption of methanol on Pt in acid solution was studied by Breiter and Gilman (3) using a potentiostatic technique. The anodic sweep, with a sweep rate of 800 V/s, was started at rest potential and extended to 2.0 V with respect to a hydrogen reference electrode in the same solution. As shown in Figure 10.8, the current was recorded as a function of potential (time) in the absence (curve A) and in the presence (curve B) of methanol. The increase in current in curve B is due to oxidation of the adsorbed methanol on the platinum electrode. Thus, shaded area 2 minus shaded area 1 (Fig. 10.8) yields the change Q_M (C/cm^2) required for oxidation of the adsorbed methanol:

$$Q_M = \int_0^t i_B \, dt - \int_0^t i_A \, dt \tag{10.10}$$

The applicability of this technique is based on four assumptions: (1) the number of coulombs per square centimeter used for the anodic formation of an oxygen layer ($2OH^- = O_{ads} + H_2O + 2e$) and the oxygen evolution during the sweep are the

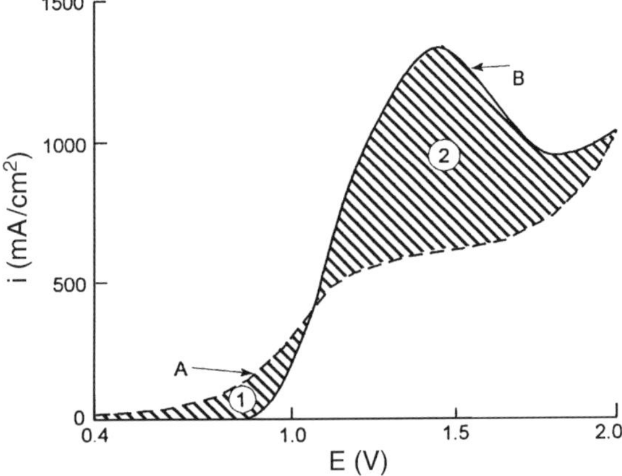

Figure 10.8. Potentiostatic *i–E* curves with $V = 800$ V/s in $HClO_4$ + 1 M CH_3OH (curve B) and in 1 N $HClO_4$ (curve A) starting from the open-circuit potential. (From Ref. 3, with permission from the Electrochemical Society.)

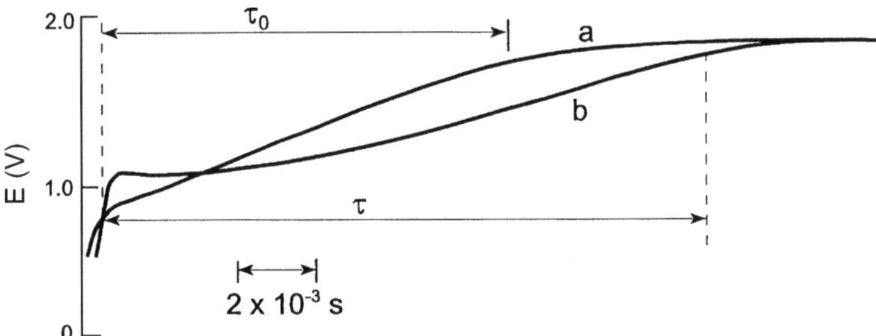

Figure 10.9. Anodic charging curves from 0.4 V during the galvanostatic transients' anodic potential sweep at 91 mA/cm² in 1 N HClO₄ (curve a) and in 1 N HClO₄ + HCOOH (curve b). (From Ref. 3, with permission from the Electrochemical Society.)

same in either the absence or presence of methanol; (2) the same number of electrons per molecule, independent of the amount of the adsorbed methanol, are used up in the oxidation; (3) the double-layer charging current is the same in the presence or absence of methanol; and (4) the anodic sweep is sufficiently fast (here, 800 V/s) that oxidation of methanol, which diffuses from the bulk solution to the electrode during the sweep, is negligible (i.e., only the adsorbed methanol is oxidized).

Galvanostatic Transient Technique. Breiter (4) measured the adsorption of formic acid (HCOOH) on platinum in a solution of perchloric acid (HClO₄) using a galvanostatic transient technique. Figure 10.9 shows two anodic galvanostatic transients at 91 mA/cm² on Pt in HClO₄ (curve a) and in HClO₄ + HCOOH (curve b). Curve a corresponds to the anodic formation of an oxygen layer (Pt–O) on Pt. The completion of about a monolayer of Pt–O is reached at the transition time τ_0 (after τ_0 is the region of oxygen evolution). Curve b corresponds to the simultaneous oxidation of adsorbed HCOOH molecules and formation of the oxygen layer. A plateau at about 1.0 V results from oxidation of the adsorbed HCOOH. The transition time τ_F for the oxidation of the adsorbed HCOOH (formic acid) molecules alone is

$$\tau_F = \tau - \tau_0 \tag{10.11}$$

This equation is written assuming that (1) charging of the double layer requires approximately the same charge Q (C/cm²) for curves a and b, (2) the number of coulombs per square centimeter for the anodic formation of an oxygen layer on Pt is the same for curves a and b; and (3) oxidation of HCOOH molecules that diffuse to the Pt electrode is negligible since the transition times are short enough (order of 10^{-3} s). The coverage θ_F of formic acid molecules is given as

$$\theta_F = \frac{\tau_F}{\tau_{F,m}} = \frac{\tau - \tau_0}{\tau_m - \tau_0} \tag{10.12}$$

where $\tau_{F,m}$ refers to a monolayer of HCOOH molecules (or, τ_m is the limiting value of τ_F).

Chronopotentiometry. Paunovic and Oechslin (8) measured the adsorption of peptone on lead–tin alloy electrodes using chronopotentiometric and double-layer measurements. This case is different from the adsorption of HCOOH because peptone is not an electroactive species in the conditions studied but only blocks the surface used for the electrodeposition of lead–tin alloys from solutions containing Sn^{2+} and Pb^{2+} ions. Chronopotentiometric analysis is based on the following principles (7). In the absence of adsorption, the relationship between the transition time τ (for reduction of Sn^{2+} and Pb^{2+} in this case), the bulk concentration c^0 of the substance reacting at the electrode, and the current I is given by the equation

$$\sqrt{\tau} = \frac{nFA\sqrt{\pi D}}{2I} c^b \tag{10.13}$$

where n is the number of electrons involved in the reaction, F is the Faraday constant, A is the surface area of the electrode, and D is the diffusion coefficient (diffusivity). For a given system and for a constant current, Eq. (10.13) reduces to

$$\sqrt{\tau} = kAc^0 \tag{10.14}$$

where the constant $k = (nF\sqrt{\pi}\, D)/2I$. In the presence of adsorption, the adsorbed substances that are neither reduced nor oxidized at the electrode block a part of the electrode surface, the result of which is a decrease in available surface area for the electrode reaction (here the deposition of Pb and Sn).

Thus, the surface area A in Eqs. (10.13) and (10.14) must be replaced by the available (free, unoccupied) surface area. This surface area, which is free for the electrode reaction, can be expressed in terms of the surface coverage θ. The fractional surface coverage θ of the electrode surface by an adsorbed substance is defined as

$$\theta = \frac{A_1}{A} \tag{10.15}$$

where A is the total surface area and A_1 the occupied surface area of the electrode. From Eq. (10.15) it follows that the unoccupied surface area A_2 is

$$A_2 = A - A_1 = A(1 - \theta) \tag{10.16}$$

since, from Eq. (10.15), $A_1 = \theta A$.

Finally, the chronopotentiometric equation for a given system and for a constant current in the presence of adsorption is obtained by substituting A in Eq. (10.14) with A_2 from Eq. (10.16); thus,

$$\sqrt{\tau} = kA_2 c^0 = kA(1 - \theta)c^0 \tag{10.17}$$

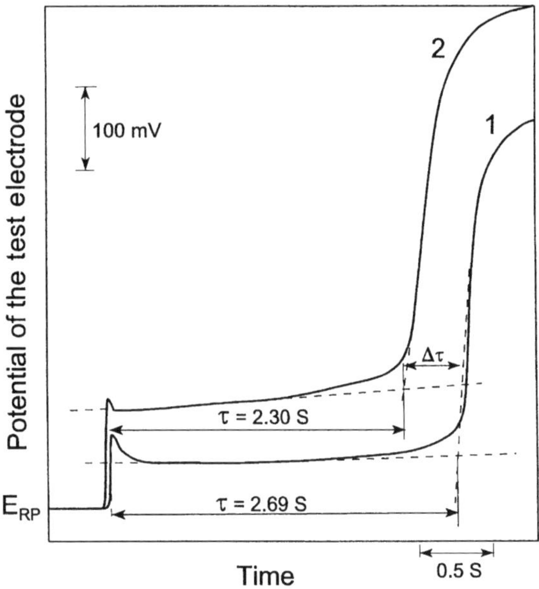

Figure 10.10. Change of transition time for reduction of Pb^{2+} and Sn^{2+} due to adsorption of peptone: 1, chronopotentiogram in the absence of peptone; 2, chronopotentiogram in the presence of 4 g/L peptone; E_{RP}, the rest potential. (From Ref. 8, with permission from the American Electroplaters and Surface Finishers Society.)

It is seen from Eq. (10.17) that an increase in θ causes a decrease in the transition time τ. If τ_0 is the transition time in the absence and τ that in the presence of adsorption of the electrochemically inactive additives, the difference,

$$\Delta\tau = \tau_0 - \tau \tag{10.18}$$

is a function of the amount of additive adsorbed. Typical chronopotentiograms for the reduction of Sn^{2+} and Pb^{2+} in both the absence and presence of peptone are shown in Figure 10.10. At the transition time τ, all the Sn^{2+} and Pb^{2+} ions available at the electrode surface are reduced and a new process starts (evolution of hydrogen). The difference $\Delta\tau$ is a function of the bulk concentration of the additive and increases with increasing concentration of the additive according to an adsorption isotherm. Thus, measurements of $\Delta\tau$ can be used to determine concentration or adsorption properties of additives in a solution.

The fractional surface coverage θ can be calculated from the experimental values of τ_0, τ, and τ_m using the equation

$$\theta = \frac{\tau_0 - \tau}{\tau_0 - \tau_m} \tag{10.19}$$

where τ_m is the transition time for the limiting (saturation) value of adsorption.

Galvanostatic Transient Technique: Double-Layer Capacitance Measurements.
The value of the fractional surface coverage θ may be inferred by means of double-layer capacitance data. As discussed in Section 6.9, the double-layer capacitance C may, in turn, be determined by means of a transient technique. In the galvanostatic transient technique (as in Fig. 6.18), the duration of the constant-current (density) pulse is on the order of microseconds. In the microsecond time range the only process taking place at the electrode is charging of the double layer. Hence, in this case, Eq. (6.96) reduces to

$$i = i_{dl} \tag{10.20}$$

Since the definition of current (density) in terms of moving charges is

$$i = \frac{dq}{dt} \tag{10.21}$$

and the definition of capacitance C is

$$q = CV \tag{10.22}$$

we obtain for Eq (10.20) the form

$$i = C \frac{dV}{dt} \tag{10.23}$$

Introducing the overpotential η instead of V in Eq. (10.23) and using Eq. (10.20), we obtain

$$C = i_{dl} \frac{1}{d\eta/dt} \tag{10.24}$$

The derivative $d\eta/dt$ can be determined experimentally from the slope of the galvanostatic transient $\eta = f(t)$. Using Eq. (10.24) and experimental values of i_{dl} and $d\eta/dt$, one is able to calculate the double-layer capacitance C.

The relationship between the fractional surface coverage θ and the double-layer capacitance C may be better understood in terms of the following model. The double-layer capacitance at the electrode in the presence of adsorption can be viewed as consisting of two capacitors connected in parallel. One capacitor corresponds to the electrode areas that are unoccupied (free) and the other to the electrode areas that are occupied (covered) with adsorbate (13–15). These two condenser–capacitors have different dielectrics and thus different capacitances. The capacitance of a parallel combination of capacitors is equal to the sum of the individual capacitances.

On the basis of this model, the double-layer capacitance is given by (8, 13–17)

$$C = C_0(1 - \theta) + C_m\theta \tag{10.25}$$

where C, C_0, and C_m are double-layer capacitances for the coverage θ, $\theta = 0$, and $\theta = 1$, respectively. From Eq. (10.25) it follows that

$$\theta = \frac{C_0 - C}{C_0 - C_m} = \frac{\Delta C}{\Delta C_m} \tag{10.26}$$

This equation is analogous to Eq. (10.19). A simplified empirical equation proposed by Breiter (18) reads

$$\theta = \frac{C}{C_m} \tag{10.27}$$

One example of the application of Eq. (10.26) is given in Ref. 8 and in Problem 10.1.

10.4. EFFECT OF ADDITIVES ON KINETICS AND MECHANISM OF ELECTRODEPOSITION

In the discussion of atomistic aspects of electrodeposition of metals in Section 6.8 it was shown that in electrodeposition the transfer of a metal ion M^{n+} from the solution into the ionic metal lattice in the electrodeposition process may proceed via one of two mechanisms: (1) a direct mechanism in which ion transfer takes place on a kink site of a step edge or on any site on the step edge (any growth site) or (2) the terrace-site ion mechanism. In the terrace-site transfer mechanism a metal ion is transferred from the solution (OHP) to the flat face of the terrace region. At this position the metal ion is in an adion state and is weakly bound to the crystal lattice. From this position it diffuses onto the surface, seeking a position with lower potential energy. The final position is a kink site.

Adsorbed additives affect both of these mechanisms by changing the concentration of growth sites c_{gs} on the surface [n_{gs}/cm^2 (where n_{gs} is the number of growth sites)], concentration of adions, c_{adi} on the surface, diffusion coefficient D_{adi}, and activation energy E_{adi} of the surface diffusion of adions.

10.5. EFFECT OF ADDITIVES ON NUCLEATION AND GROWTH

In Chapter 7 various growth models were described: layer growth (Section 7.9), nucleation–coalescence growth (Section 7.10), development of texture (Section 7.11), columnar microstructure (Section 7.12), and other structural forms (Section 7.13). In this section we discuss the effects of additives on these growth mechanisms.

Nucleation. In the presence of adsorbed additives, the mean free path for lateral diffusion of adions is shortened, which is equivalent to a decrease in the diffusion coefficient D (diffusivity) of adions. This decrease in D can result in an increase in adion concentration at steady state and thus an increase in the frequency of the two-dimensional nucleation between diffusing adions.

Layer Growth. In Section 7.9 we showed that many monoatomic steps can unite (bunch, coalesce) to form polyatomic steps in the presence of impurities. Additives can also influence the propagation of microsteps and cause bunching and formation of macrosteps.

Dependence of Deposit Type on Surface Coverage by an Additive. The type of deposit obtained at constant current density may depend on the value of the surface coverage θ by an additive. Damjanovic et al. (25) studied the effect of various values of the surface coverage θ of n-decylamine on the growth form of copper on the (100) plane of copper single crystals at $5\,mA/cm^2$. The surface coverage θ was varied by addition of a known amount of n-decylamine to a highly purified solution of $CuSO_4$, H_2SO_4, and H_2O. The coverage θ of n-decylamine was estimated from the adsorption isotherm for n-decylamine on copper in 1 N $NaClO_4$ (Fig. 10.4). It was found that when $\theta < 10^{-2}$ (at the bulk concentration of n-decylamine $<10^{-3}\,mol/L$), the crystal growth form was of the layer type. At $\theta \geqslant 10^{-2}$, the deposit observed was of the ridge type. Thus, variation in the surface coverage θ resulted in two entirely different types of deposit.

10.6. LEVELING

Leveling was defined initially as progressive reduction in surface roughness during deposition. Surface roughness may be the result of coarse mechanical polishing. In this case, scratches on the cathode represent the initial roughness, and the result of cathodic leveling is a smooth (flat) deposit or a deposit of reduced roughness. During this type of leveling, more metal is deposited in recessed areas than on peaks. This leveling is of great value in the metal finishing industry.

Leveling is of great importance in the electronics industry for deposition through polymeric masks and in electroforming of micromechanical devices (e.g., sensors, actuators, micromotors). In these cases the leveling process means uniform deposition in microprofiles (filling up recesses), which can be defined by masks or without masks. Typical microprofiles are shown in Figure 10.11. Leveling can be achieved in solutions both in the absence and in the presence of addition agents (leveling agents). Thus, there are two types of leveling processes: (1) geometric leveling, corresponding to leveling in the absence of specific agents, and (2) true or electrochemical leveling, corresponding to leveling in the presence of leveling agents. We discuss the theory of leveling for both processes.

Leveling in the Absence of Additives: Geometric Leveling. Geometric leveling, produced by uniform current distribution, is illustrated in Figure 10.12, Figure 10.12*a* shows uniform deposition thickness in a groove after a time t_1; Figure 10.12*b* shows

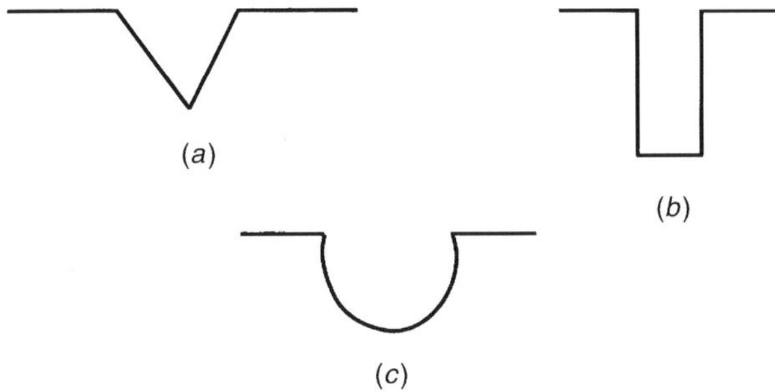

Figure 10.11. Typical microprofiles.

evolution of a groove profile during deposition, resulting in geometric leveling. DuRose et al. (28) observed leveling experimentally by uniform current deposition.

Dukovic and Tobias (37) and Madore and Landolt (39) developed theoretical models of leveling during electrodeposition in both the absence and presence of additives. Both models show that significant leveling of semicircular and triangular grooves by uniform current distribution (geometric leveling) is achieved when the deposit thickness is at least equal to the groove depth. We show in the next section that leveling in the presence of leveling agents (true leveling) is achieved much earlier.

Leveling in the Presence of Leveling Agents: True Leveling. We first consider the case of uneven current distribution that results in increased roughness of the surface.

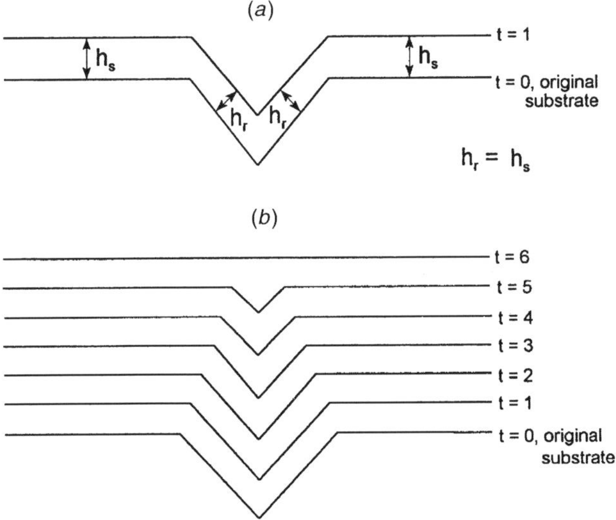

Figure 10.12. (*a*) Uniform deposition thickness in a groove after a time t_1; (*b*) evolution of a groove profile during deposition, resulting in geometric leveling.

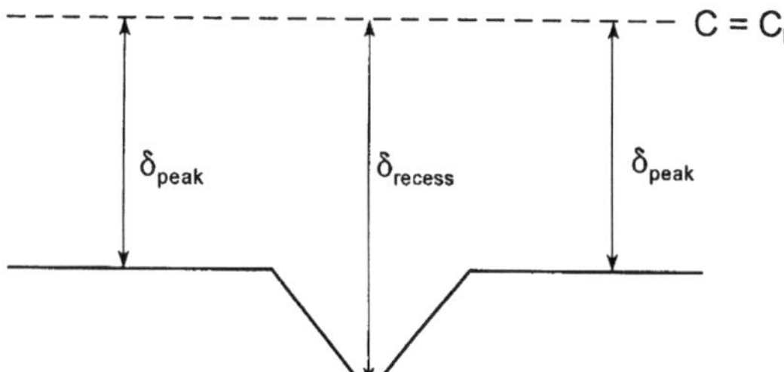

Figure 10.13. Variations in the effective thickness of diffusion layer δ over an electrode with a microprofile.

The two major causes of uneven current distribution are diffusion and ohmic resistance. Nonuniformity due to diffusion originates from variations in the effective thickness of the diffusion layer δ over the electrode surface as shown in Figure 10.13. It is seen that δ is larger at recesses than at peaks. Thus, if the mass-transport process controls the rate of deposition, the current density at peaks i_p is larger than that at recesses i_r since the rate of mass transport by convective diffusion is given by

$$j = \frac{D(c_b - c_0)}{\delta} \tag{10.28}$$

and the current density by

$$i = -DnF \frac{c_b - c_0}{\delta} \tag{10.29}$$

Mass-transport deposition control occurs when the exchange current density i^0 is high and the limiting current density i_1 is low. Ohmic resistance can be a cause of nonuniformity if there is an appreciable difference in solution resistance from the bulk of the solution to peaks or to recesses. Distribution of the current density will be such that $i_p > i_r$ and peaks will receive a larger amount of deposit than will recesses. Distribution of deposit in the triangular groove under conditions of mass transport and ohmic control nonuniform deposition, with $i_p > i_r$, is shown in Figure 10.14.

Under the conditions shown in Figure 10.14, the roughness of the surface increases. Thus, to get leveling of the surface it is necessary to change from diffusion and ohmic control to activation control. Activation control can result in uniform deposition ($i_p = i_r$) or in nonuniform deposition (with $i_r > i_p$). In nonuniform deposition with

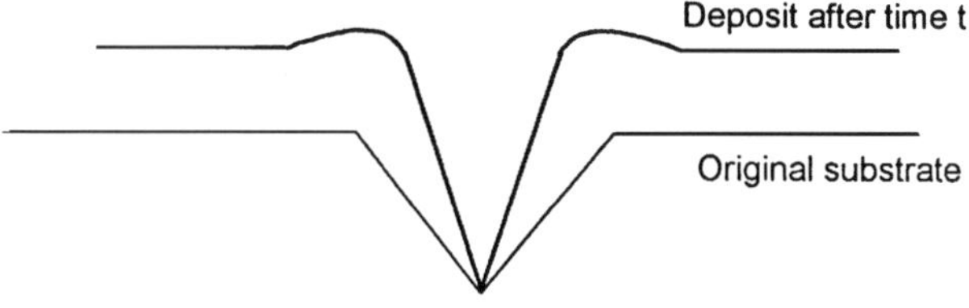

Figure 10.14. Distribution of deposit in a triangular profile when $\delta_r > \delta_p$ and $i_r < i_p$; high i^0, low i_L; diffusion and ohmic control.

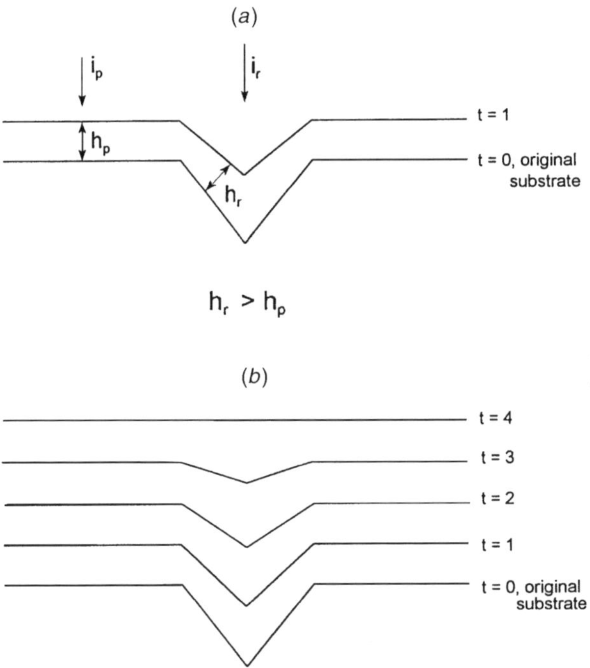

Figure 10.15. True leveling on a V-groove produced by nonuniform current distribution, $i_r > i_p$, in the presence of a leveling agent; (*a*) nonuniform current density and deposit thickness, h_s and h_g, after time $t = 1$; (*b*) evolution of a groove profile during deposition; activation control.

$i_r > i_p$, a higher rate of deposition at recesses than at peaks, and thus leveling, can be achieved by the introduction of additives, leveling agents. Leveling produced by a nonuniform current distribution, $i_r > i_p$, in the presence of leveling additives is shown in Figure 10.15.

Theories of leveling by additives are based on (1) the correlation between an increase in polarization produced by the leveling agents (29) and (2) preferential adsorption of a leveling agent on high points (peaks or flat surfaces) (30). Theories and modeling of superconformal electrodeposition are discussed in Section 19.4.

10.7. BRIGHTENING

Kardos and Foulke (32) distinguish three possible mechanisms for bright deposition: (1) diffusion-controlled leveling, (2) grain refining, and (3) randomization of crystal growth. Leveling and the increased frequency of nucleation, which produces grain refinement, were discussed earlier in the chapter. Here we discuss only the theory of random electrodeposition proposed by Hoar (19), which asserts that there are two basic mechanisms of deposition: selective and random. *Selective deposition* occurs when atoms deposit on favorable surface sites. Favorable sites for deposition are kinks, steps, and the surface ends of screw dislocations, discussed in Chapters 6 and 7.

In *random deposition* there is no such regularity of growth on favorable sites. Deposition occurs randomly on all available surface sites. Random deposition can be achieved by the use of additives that produce random distribution of isolated surface sites uncovered by the additive and available for deposition. This random distribution of uncovered (free) sites must be in a state of rapid flux from site to site. This can be achieved if the equilibrium surface coverage of the additive θ is large (on the order of 0.9) and if the rates of the adsorption and desorption processes are high. With large surface coverage many selective, favorable sites will be blocked and random deposition will be favored. Random coverage of a surface is considered to be essential for bright deposition. The rate of adsorption and desorption of additive must be of the same order of magnitude as that of the cathodic deposition process. If they are much smaller, additive molecules can be entrapped in the deposit, and this can lead to a selective deposition mechanism. Random deposition can also be promoted by an increase in current density. Some fundamental aspects of brightening and leveling have been reviewed by Onicin and Muresa (38).

10.8. CONSUMPTION OF ADDITIVES

Additives can be consumed at the cathode by incorporation into the deposit and/or by electrochemical reaction at the cathode or anode. Consumption of coumarin in the deposition of nickel from a Watts-type solution has been studied extensively. Thus, in this section we discuss the consumption of coumarin, which is used as a leveler and partial brightener. In a series of papers (33, 36), Rogers and Taylor, described the effects of coumarin on the electrodeposition of nickel. They found that the coumarin concentration decreases linearly with time at $-960\,\text{mV}$ (versus SCE and 485 to 223 rpm at a rotating-disk electrode, for plating times of 8 to 75 min. A rotating-disk electrode was used to achieve a uniform and known rate of transport of additive to the cathode. Rogers and Taylor found that the rate of coumarin consumption is a function of coumarin bulk concentration. Figure 10.16 shows that the rate of consumption

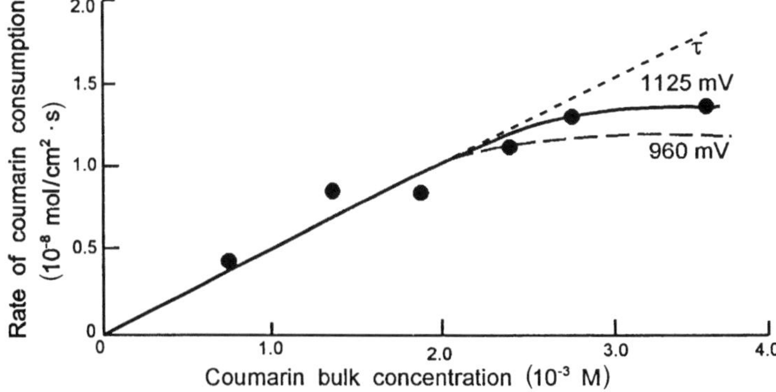

Figure 10.16. Rate of coumarin consumption as a function of coumarin concentration; constant-potential experiments: -960 and $-1125\,\text{mV}$ versus SCE, 985 rpm, pH 4; τ, theoretical rate transport. (From Ref. 33, with permission from Elsevier.)

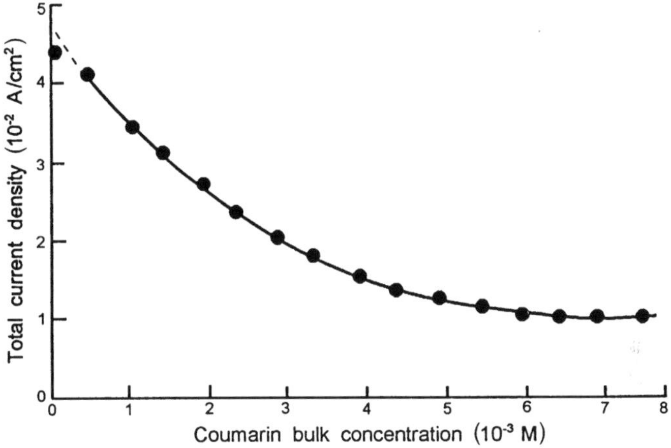

Figure 10.17. Total current density as a function of coumarin concentration; constant-potential experiments: $-960\,\text{mV}$ versus SCE, 980 rpm, pH 4. (From Ref. 33, with permission from Elsevier.)

increases with increase in the bulk concentration of coumarin. The inhibiting effect of coumarin is shown in Figure 10.17. The total current density in this figure is the sum of the current densities for nickel deposition i_{Ni}, hydrogen evolution, i_{H}, and additive reduction, i_R.

Electroreduction of Coumarin. The principal cathodic reaction of coumarin is the reduction to melilotic acid (Fig. 10.18). A second reduction product is an approximately equimolar mixture of o-hydroxyphenylpropanol and o-propylphenol (Fig. 10.18). Hence, there are three reactions involving coumarin.

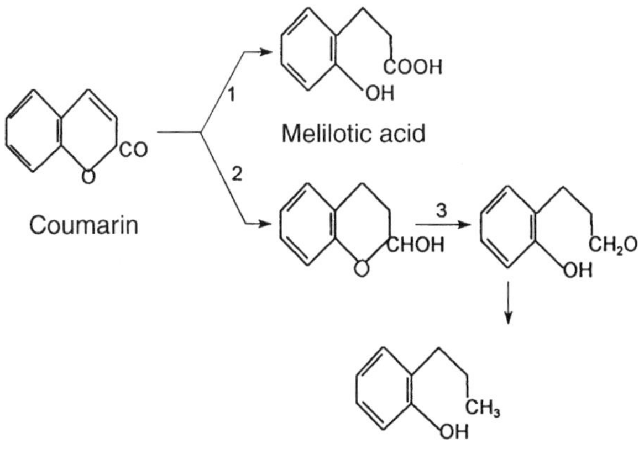

Figure 10.18. Electroreduction of coumarin.

TABLE 10.1. Effect of Additives on Tensile Strength and Elongation of Ni Deposits

Type of Solution	Additive	Tensile Strength (psi)	Elongation (%)
Fluoborate	None	74,500	16.6
	Saccharin, 1 g/L	203,000	1.0
Sulfamate	None	80,000	8.0
	Naphthalene trisulfonic acid, 8 g/L	140,000	1.0
Watts	None	56,000	28.0
	Nickel benzene sulfonate, 7.5 g/L, and triamine tolylphenyl methane chloride, 5–10 mg/L	212,000	5.0

Incorporation of Carbon from Coumarin in Deposits. Incorporation of carbon from coumarin was studied by Rogers and Taylor (36) and Edwards and LeWett (34) using a radiotracer technique with ^{14}C-labeled coumarin. They found that the consumption of coumarin by incorporation into nickel deposits is small compared with consumption by reduction reaction, by a factor of about 10.

The preceding discussion concerning the consumption of coumarin in a Watts-type solution can serve as a model in a general discussion on the consumption of additives in electrodeposition.

10.9. EFFECT OF ADDITIVES ON PROPERTIES OF DEPOSITS

We present here three specific examples of the effect of additives on the properties of electrodeposited Ni. Table 10.1 shows the effect of additives on tensile strength and elongation of Ni produced by three different Ni electrodeposition solutions (40, 41). For additional examples the reader is referred to Refs. 40 and 41. In general, it may be stated that additives which increase the tensile strength of Ni deposits will decrease the elongation (ductility) (42). The effect of additives on the properties of electroless deposits is discussed in Section 8.9.

REFERENCES AND FURTHER READING

1. I. Langmuir, *J. Am. Chem. Soc.* **40**, 1361 (1918).
2. G. C. Bond, *Catalysis by Metals*, Academic Press, New York, 1962.
3. M. W. Breiter and S. Gilman, *J. Electrochem. Soc.* **109**, 622 (1962).
4. M. W. Breiter, *Electrochim. Acta* **8**, 447 (1963).
5. J. O'M. Bockris and T. Swinkels, *J. Electrochem. Soc.* **111**, 736 (1964).
6. G. C. Bond, *Discuss. Faraday Soc.* **41**, 200 (1966).
7. M. Paunovic, *J. Electroanal. Chem.* **14**, 447 (1967).

8. M. Paunovic and R. Oechslin, *Plating* **58**, 599 (1971).

9. A. W. Adamson, *Physical Chemistry of Surfaces*, Interscience, New York, 1976.

10. M. P. Soriaga, P. H. Wilson, and A. T. Hubbard, *J. Electroanal. Chem.* **142**, 317 (1982).

11. A. T. Hubbard, *Chem. Rev.* **88**, 633 (1988).

12. D. A. Stern, L. L. Davidson, D. G. Frank, J. Y. Gui, C. H. Lin, F. Lu, G. N. Salaita, N. Walton, D. Z. Zapien, and A. T. Hubbard, *J. Am. Chem. Soc.* **111**, 877 (1989).

13. B. E. Conway and J. O'M. Bockris, *Plating* **46**, 371 (1959); Sec. VII.

14. A. N. Frumkin and B. B. Damaskin, in *Modern Aspects of Electrochemistry*, No. 3, J. O'M. Bockris and B. E. Conway, eds., Butterworth, Washington, DC, 1964, pp. 156, 181.

15. P. Delahay, *Double Layer and Electrode Kinetics,* Interscience, New York, 1965.

16. R. Parson, *Trans. Faraday Soc.* **55**, 999 (1959).

17. M. W. Breiter and P. Delahay, *J. Am. Chem. Soc.* **81**, 2938 (1959).

18. M. W. Breiter, *Electrochim. Acta* **7**, 533 (1962).

19. T. P. Hoar, *Trans. Inst. Met. Finish.* **29**, 302 (1953).

20. H. Fischer, *Electrochim. Acta* **2**, 50 (1960).

21. G. C. Bond, *Catalysis by Metals*, Academic Press, New York, 1962.

22. D. O. Hayward and B. M. W. Trapnell, *Chemisorption*, Butterworth, London, 1964.

23. G. Fabricius, K. Kontturi, and G. Sundholm, *Electrochim. Acta* **39**, 2353 (1994).

24. H. Fischer, *Electrochim. Acta* **2**, 50 (1960).

25. A. Damjanovic, M. Paunovic, and J. O'M. Bockris, *J. Electroanal. Chem.* **9**, 93 (1965).

26. J. O'M. Bockris and G. A. Razumney, *Fundamental Aspects of Electrocrystallization*, Plenum Press, New York, 1967.

27. R. Weil, *Annu. Rev. Mater. Sci.* **19**, 165 (1989).

28. A. H. DuRose, W. P. Karash, and K. S. Wilson, *Proc. Am. Electroplat. Soc.* **37,** 193 (1950).

29. H. Leidheiser, Jr., *Z. Elektrochem.* **59**, 756 (1955).

30. O. Kardos, *Am. Electroplat. Soc.* **43**, 181 (1956).

31. G. T. Rogers, M. J. Ware, and R. V. Fellows, *J. Electrochem. Soc.* **107**, 677 (1960).

32. O. Kardos and D. G. Foulke, in *Advances in Electrochemistry and Electrochemical Engineering*, Vol. 2, C. W. Tobias, ed., Interscience, New York, 1962.

33. G. T. Rogers and K. J. Taylor, *Electrochim. Acta* **8**, 887 (1963); **11**, 1685 (1966); **13**, 109 (1968).

34. J. Edwards and M. J. LeWett, *Trans. Inst. Met. Finish.* **41**, 157 (1964).

35. J. Edwards, *Trans. Inst. Met. Finish.* **41**, 169 (1964).

36. G. T. Rogers and K. J. Taylor, *Trans. Inst. Met. Finish.* **43**, 75 (1965).

37. J. O. Dukovic and C. Tobias, *J. Electrochem. Soc.* **137**, 3748 (1990).

38. L. Onicin and L. Muresan, *J. Appl. Electrochem.* **21**, 565 (1991).

39. C. Madore and D. Landolt, *J. Electrochem. Soc.* **143**, 3936 (1996).

40. W. H. Safranek, ed., *The Properties of Electrodeposited Metals and Alloys*, 2nd ed., American Electroplaters and Surface Finishers Society, Orlando, FL, 1986.

41. J. W. Dini, *Electrodeposition: The Materials Science of Coating and Substrates*, Noyes, Park Ridge, NJ, 1993, Chap. 7.

42. G. A. DiBari, in *Modern Electroplating*, 4th ed., M. Schlesinger and M. Paunovic, eds., Wiley, New York, 2000, Chap. 3, Sec. 4.1.

PROBLEMS

10.1. Electrodeposition of lead–tin alloy films is usually performed in the presence of peptone as an additive. Peptone is adsorbed on the metal surface during the electrodeposition process. The fractional surface coverage θ of the lead–tin electrode may be determined from the double-layer capacitance C measurements, and/or chronopotentiometric measurements. For a solution containing 9.0 g/L of tin and 13.0 g/L of lead, the following relationship between the concentration of peptone, the double-layer capacitance C, and the transition time $\Delta\tau$ is observed (8).

Concentration of Peptone, c(g/L)	C (μF/cm^2)	$\Delta\tau$ (s)
0	211	0
5	132	0.64
10	85.5	0.91
15	76.0	1.13
20	76.6	1.24
25	76.6	1.31

(a) Calculate the fractional surface coverage θ for the concentrations of peptone given in the table using the C and $\Delta\tau$ values, listed assuming that C_m and $\Delta\tau_m$ are reached at 25 g/L of peptone.

(b) Plot θ as a function of peptone concentration in the solution.

10.2. Calculate and plot the fractional surface coverage θ as a function of bulk solution concentrations: 0.2, 0.4, 0.6, 0.8, 1.0, 1.5, 2.0, 2.5, and 3.0 for the equilibrium constant $K = 1, 5,$ and 10 (assume that K is independent of the potential).

10.3. Coumarin is used as an additive in the electrodeposition of nickel from a Watts-type solution. It is reduced at the cathode, and it may be incorporated in the deposit (Section 10.8). The rate of consumption of coumarin as a function of coumarin concentration is shown in Figure 10.16.

(a) Calculate the consumption of coumarin (mol/cm^2) after 1, 30, and 60 min of electrodeposition of Ni in a solution containing 1.0, 2.0, and 3.0 $\times$ 10^{-3} M of coumarin.

(b) After what length of time will 25% of coumarin be consumed in a solution containing 1.0, 2.0, and 3.0 $\times$ 10^{-3} M coumarin? These results are of practical importance in the determination of replenishment times for the solution. Present your results in the form of a table.

11

Electrodeposition of Alloys

11.1. INTRODUCTION

Alloy deposition is almost as old an art and/or science as is the electrodeposition of individual metals. (Brass deposition, for instance, was invented circa 1840!) In the last analysis, as can well be expected, alloy deposition is subject to the same scientific principles as individual metal plating. Indeed, progress in either of the two has almost always depended on similar advances in electrodeposition science and/or technology.

The subject of alloy electroplating is being dealt with by an ever-increasing number of scientific publications (close to 200 in the last five years in the *Journal of the Electrochemical Society* alone!). The reason for this seems to be the vastness of the number of possible alloy combinations and the concomitant possible practical applications.

Similar to the present chapter, the contribution by C. L. Faust in the 1974 edition of *Modern Electroplating* (1) draws attention to a number of fundamentally important points regarding alloys and their properties. Many of these are as relevant today as they were nearly three decades ago. Thus, for instance, it is stated that properties of alloy deposits superior to those of single-metal electroplates are commonplace and are widely described in the literature. In other words, it is recognized that, more often than not, alloy deposition, provides properties not obtained by employing electrodeposition of single metals. It is further asserted that relative to the single-component metals involved, alloy deposits can have different properties in certain composition ranges. They can be denser, harder, more corrosion resistant, more protective of the underlying basis metal, tougher and stronger, more wear resistant, superior with respect to magnetic properties, more suitable for subsequent electroplate overlays and conversion chemical treatment, and superior in antifriction applications.

Electrodeposited binary alloys may or may not be the same in phase structure as those formed metallurgically. By way of illustration, we note that in the case of brass (Cu–Zn alloy), x-ray examination reveals that apart from the superstructure of

β-brass, virtually the same phases occur in alloys deposited electrolytically as in those formed in the melt. Phase limits agree closely with those in the bulk. Debye–Scherrer interference rings indicate the presence of a strong distortion of the lattice, particularly in the α-phase brass. Electrodeposited α-brass, for instance, is harder than the cast form, but hardness of the β, γ, and ε phases is essentially unaffected. Somewhat opposite is the case of Ag–Pb alloy. In the cast alloy form, it contains "large" silver crystals, with lead present in the grain boundaries as dendrites. In the electrodeposited form, the alloy contains exceedingly fine grains exhibiting no segregation of lead.

Using specific metal combinations, electrodeposited alloys can be made to exhibit hardening as a result of heat treatment subsequent to deposition. This, it should be noted, causes solid precipitation. When alloys such as Cu–Ag, Cu–Pb, and Cu–Ni are coelectrodeposited within the limits of diffusion currents, equilibrium solutions or supersaturated solid solutions are in evidence, as observed by x-rays. The actual type of deposit can, for instance, be determined by the work value of nucleus formation under the overpotential conditions of the more electronegative metal. When the metals are codeposited at low polarization values, formation of solid solutions or of supersaturated solid solutions results. This is so even when the metals are not mutually soluble in the solid state according to the phase diagram. Codeposition at high polarization values, on the other hand, results, as a rule, in two-phase alloys even with systems capable of forming a continuous series of solid solutions.

11.2. PRINCIPLES

The electrodeposition of an alloy requires, by definition, the codeposition of two or more metals. In other words, their ions must be present in an electrolyte that provides a *cathode film*, where the individual deposition potentials can be made to be close or even the same. Figure 11.1 depicts typical polarization curves, that is, deposition

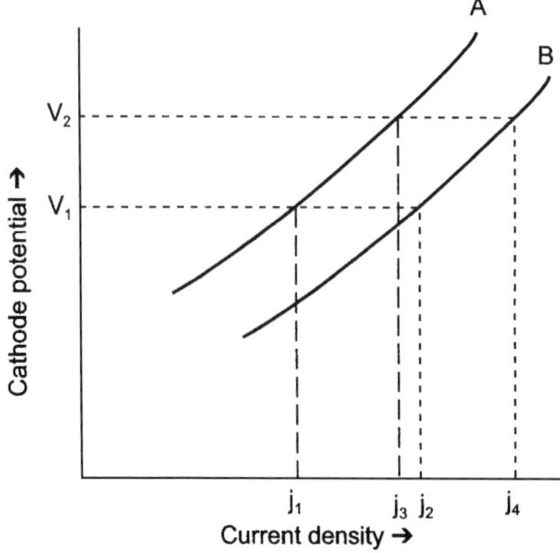

Figure 11.1. Polarization curves for the deposition of alloys. (From *Science and Technology of Surface Coating*, a NATO Advanced Study Institute publication, 1974, with permission from Academic Press.)

potentials as a function of current density for two metals (A and B; corresponding to curves A and B in Fig. 11.1) separately. From such curves it is possible to infer that from a deposition bath that contains both metal ions it would be, at potential V_2, possible to codeposit the two metals A and B in the ratio j_3/j_4.

One recognizes three main steps in the cathodic deposition of alloys or single metals:

1. *Ionic migration,* meaning that the hydrated ion(s) in the electrolyte migrate(s) toward the cathode under the influence of the applied potential as well as through diffusion and/or convection.

2. *Electron transfer,* meaning that at the cathode surface the hydrated metal ion(s) enter(s) the diffusion double layer where the water molecules of the hydrated ion are aligned by the weak field present in this layer. Subsequently, the metal ion(s) enter(s) the fixed double layer, where, due to the higher field present, the hydrated shell is lost. Then on the cathode surface, the individual ion is neutralized and is adsorbed.

3. *Incorporation,* meaning that the adsorbed atom wanders to a growth point on the cathode and is incorporated in the growing lattice.

In general, when a metal is immersed in a solution of (i.e., containing) its own ions, some surface atoms in the metal lattice do become hydrated and dissolve into the solution. At the same time, ions from the solution are deposited on the electrode. The rate of these two opposing processes is controlled by the potential differences at the metal–solution interface. The specific potentials at which these two reaction rates are equal, called *standard potentials*, are usually given in the literature for solutions at 25°C (room temperature) and at an activity value of unity.

The activity of an electrolyte or ion is defined for use in determining true (actual as opposed to theoretical) equilibrium constants. By definition, the activities are equal to the concentrations in very dilute solutions, and the difference between activities and concentration in more concentrated solutions depends on interaction between *all* the components of the solution, causing the ions to behave differently than they would at a high degree of dilution.

The reversible/equilibrium (as understood above) potential at a metallic electrode that is in equilibrium with its ions (no net reaction: e.g., Ag, Ag$^+$) is given by the well-known Nernst expression:

$$E = E^0 + \frac{RT}{vF} \ln a_{\text{ion}} \qquad (11.1)$$

where E is the reversible electrode potential, E^0 is the standard oxidation electrode potential, v is the valence charge, F is the Faraday, and a_{ion} is the actual ionic activity.

The *standard electrode potential* (E^{ion}, sometimes referred to as E_0) is defined as the potential that exists when the electrode is immersed in a solution of ions at unit activity. The factor a_{ion}, the actual ionic activity, is the activity of the depositing cation in the film of the plating bath at the cathode face.

By definition, $E_0^{Pt,H_2,H^+} = 0$. Also, hydrogen-ion activities are determined down to very low values. [To obviate the need to write small numbers with many zeros, the term *pH* was proposed; thus, pH $= -\log a_{H^+} \doteq \log (1/a_{H^+})$.] If there is a greater tendency toward oxidation, that is, toward loss of electrons in an electrode reaction rather than in the reference reaction $\frac{1}{2} H_2 \rightarrow H^+ + e$, the corresponding electrode is given a value $E_0 > 0$, and vice versa (by convention). It is important to understand that Eq. (11.1) refers to reversible (from the point of view of thermodynamics) equilibrium.

Deposition, as a rule, occurs as an irreversible process. The expression suitable for this state of affairs has to be modified from (11.1) to read, instead,

$$E^d = E^0 + \frac{RT}{vF} \ln a^{v+} + \eta \qquad (11.2)$$

In other words, it now includes the term $\eta = E^d - E^s$, which represents the difference between Eqs. (11.1) and (11.2). The factor a^{v+} represents, again, the activity value of the cation being deposited (i.e., cation in the film or layer of the bath at the cathode face). Thus, η is the overpotential (deposition factor). It is the extra potential needed to maintain deposition at a given desired rate suitable to the nature and properties of the cathode film. In practice, then, calculating the metal deposition potential as above means that the practitioner must know the values of a^{v+} and η for a fixed plating condition, including bath parameters such as current density and temperature, as well as ionic parameters such as concentration, valence, and mobility.

The E^0 values for Cu and Zn (-0.345 and 0.762 V, respectively) are far apart, so alloy deposition may seem to be virtually impossible. The difference can be eliminated, however, or even reversed by modifying the activity values, as discussed below. The difference as it stands, however, should not be considered to be without merit or without exploitable features. Indeed, the difference may be exploited to set up a cell using these two elements, as in the following example.

Example 11.1. Given the following rather standard cell reaction and the electromotive force in the cell as

$$\text{Zn}; \text{Zn}^{2+}(a = 1) \| \text{Cu}^{2+}(a = 1); \text{Cu} \qquad (11.3)$$

The reaction is as follows:

$$\text{left (written as oxidation)}: \tfrac{1}{2}\,\text{Zn} = \tfrac{1}{2}\,\text{Zn}^{2+} + e$$

$$\underline{\text{right (written as reduction)}: \tfrac{1}{2}\,\text{Cu}^{2+} + e = \tfrac{1}{2}\,\text{Cu}}$$

$$\tfrac{1}{2}\,\text{Zn} + \tfrac{1}{2}\,\text{Cu}^{2+} = \tfrac{1}{2}\,\text{Zn}^{2+} + \tfrac{1}{2}\,\text{Cu}$$

The sum of these two half-cell potentials is the electromotive force (EMF) of the cell:

$$E_0 = E_0^{Zn;Zn^{2+}} + E_0^{Cu^{2+};Cu} = 0.762 - (-0.345) = 1.107 \text{ V} \qquad (11.4)$$

Since E of the cell is positive (by convention, as indicated above), it follows that the electrode on the left is an anode; thus, metallic zinc will reduce copper ions to copper.

More generally stated, if we have a reaction of the type

$$aA + bB \rightarrow gG + hH \tag{11.5}$$

then we have (in case $a \neq 1$)

$$E = E_0 - \frac{RT}{vF} \ln \frac{a_G'^g a_H'^h}{a_A'^a a_B'^b} = E_0 - \frac{RT}{nF} \ln Q \tag{11.6}$$

If $T = 298.1$ and $v = 1$, then $E = E_0 - 0.0591 \log Q$.   □

From this example it may be concluded that by measurement of the electromotive force in a cell, the activities for reactions can be determined.

11.3. DEPOSITION

As is evident from Eq. (11.4), copper and zinc are very far apart in the standard EMF series, so alloy codeposition seems next to impossible. Fortunately, the difference can be eliminated (even reversed) by changing the values of the activities. This can be achieved by inducing a considerable change in ionic concentrations via complex ion formation, as discussed in detail below.

Example 11.1 illustrates the point that in a solution containing "simple" salts of zinc and copper, the ion concentration and activity values are so close together that the large difference between the E_0 values for copper and for zinc will make alloy deposition virtually impossible.

In a mixed copper–zinc solution of complex cyanide, however, the Cu^+ ion concentration can be reduced to the order of 10^{-18} mol/L and the concentration ratio (zinc ion)/(copper ion) will be made very large. A detailed calculation for this case is given by Faust in the 1974 edition of *Modern Electroplating* (1). It is shown there, and in detail below, that the copper cyanide complex is $Cu(CN)_3^{2-}$, for which the dissociation value is known. The dissociation constant for the zinc cyanide complex, $Zn(CN)_4^{2-}$, is also well known. Using those values that determine the fraction concentration of the "free" metal ion in solution and assuming an initial specific molar concentration, it is shown below that their respective reversible electrode potentials [see also Eq. (11.1)] can be brought together.

To express the preceding in a different, more specific way, we state that codeposition of two or more metals is possible under suitable conditions of potential and polarization. The necessary condition for simultaneous deposition of two or more metals is that the cathode potential–current density curves (polarization curves) be similar and close together.

By way of illustration, once again, let us suppose that curves A and B in Figure 11.1 are the polarization curves for metals A and B, respectively. At potential V_1, the deposition rates of metals A and B are given by current densities j_1 and j_2, respectively. Similarly, at V_2 the deposition rates for metals A and B are given by current densities j_3 and j_4, respectively. The amount of the metals deposited during a given interval of time may be determined from the current densities, using the relationship

$$Q = FE_q \tag{11.7}$$

where Q is the quantity of electricity (in coulombs) passed, F is the Faraday constant (96,490 C per gram equivalent), and E_q is the number of gram-equivalent weights of metal deposited. The value of Q can be determined from the expression

$$Q = \int_0^t i \, dt \tag{11.8}$$

where i is the current in amperes and t is the time in seconds.

It must be understood that in a case such as that illustrated in Figure 11.1, the plating bath is being depleted of metal B ions more quickly than of metal A ions. To keep matters under control (i.e., maintain uniform deposition conditions), metal ions must be replenished in direct proportion to their rates of deposition dictated by the specific alloy. It is clear, therefore, that ideally, the polarization curves of the competent metals being codeposited should be identical. It is next to impossible to realize this condition in practice.

We now return to the case of codeposition of metals whose standard electrode potentials are wide apart. As stated, the deposition potentials [Eq. (11.2)] are brought together by complexing the more noble metal ions, as illustrated below for the case of the codeposition of copper and zinc as *brass*.

The reversible potential [Eq. (11.1)] of copper in conditions of unit activity for copper ions is -0.34 V, and for zinc ions, 0.77 V. It is clear that if a solution contains unit activities of Cu and Zn ions, zinc will not codeposit with copper unless the overpotential for copper deposition is high enough to compensate for this large difference in deposition potentials.

Fortunately, the deposition potentials of these metals can be brought together by adjustment of their ionic concentrations. Thus, if KCN is added to the solution of the salts of these metals, it binds the Cu and Zn ions as rather stable $\{Cu(CN)_4\}^{3-}$ and $\{Zn(CN)_4\}^{2-}$ complexes, respectively. In solution, the copper complex dissociates to cuprous and cyanide ions according to

$$\{Cu(CN)_3\}^{2-} = Cu^+ + 3(CN)^- \tag{11.9}$$

The value of the *instability constant* (i.e., the concentration, indicated in brackets) *ratios* for the preceding is

$$\frac{[(Cu^+)][(CN^-)]^3}{[Cu(CN)_3{}^{2-}]} = 10^{-28} \qquad \text{(at 25°C)} \tag{11.10}$$

so at unit concentrations of the other components, that of $[Cu^+]$ is 10^{-28} g-molecule/L.

The instability constant for Zn is

$$\frac{(Zn^{2+})(CN^-)^4}{[Zn(CN)_4{}^{2-}]} = 10^{-17} \tag{11.11}$$

so that the concentration of $[Zn^{2+}]$ under the same conditions will be 10^{-17} g-molecule/L. With these ionic concentrations, the deposition potentials of copper and zinc in the absence of any polarization can each be calculated from Eq. (11.1) to be about $-1.30\,V$. It should be mentioned here again that in practice, Eq. (11.1) refers to *reversible equilibrium,* a condition in which no net reaction takes place. In practice, electrode reactions are irreversible to an extent. This makes the potential of the anode more noble and the cathode potential less noble than their static potentials calculated from (11.1). The overvoltage is a measure of the degree of the irreversibility, and the electrode is said to be *polarized* or to exhibit *overpotential*: hence, Eq. (11.2).

The individual polarization curves for the metals are often modified as a result of interactions resulting from codeposition. If the alloy deposition occurs at low polarization, the nobler metal will be deposited preferentially (Cu in Example 11.1). All factors, however, that increase polarization during electrodeposition, such as high current density, low temperature, and quiescent solution—factors that increase concentration polarization—will favor the deposition of the less noble metal (Zn in Example 11.1).

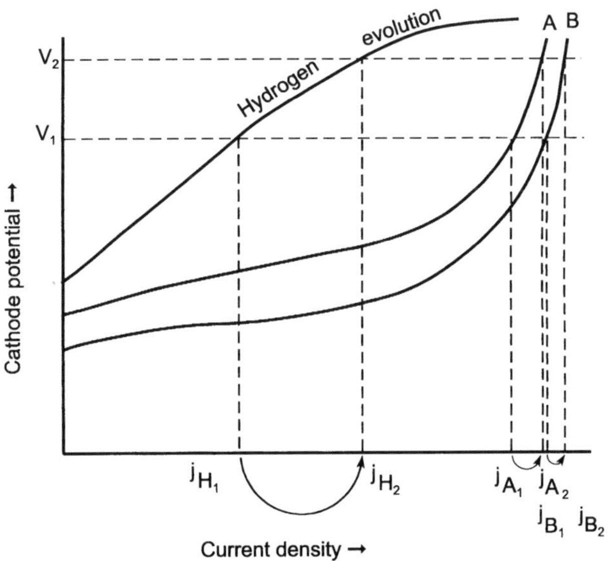

Figure 11.2. Alloy deposition with hydrogen evolution. (From *Science and Technology of Surface Coating,* a NATO Advanced Study Institute publication, 1974, with permission from Academic Press.)

Viewed electrochemically when metal deposition is accompanied by hydrogen evolution, it may be said that one deals with alloy plating in which hydrogen is the codepositing element. This is so even when hydrogen is discharged as gas since the conditions for codeposition are met. Alloy plating of metals makes it into a process of production of hydrogen and two or more metals.

The evolution of hydrogen during electrodeposition of an alloy has a significant effect on the polarization and composition of the alloy deposited. If a significant amount of hydrogen is evolved, the potential of the cathode during alloy deposition may be determined almost totally by the hydrogen evolution reaction. If, as is usually the case, the overpotential for hydrogen evolution is high in the preceding case, the currents corresponding to the individual metals will be close to limiting values. Under these conditions an increase in current will increase the amount of hydrogen evolved, resulting in a poor efficiency for alloy deposition with a minor change in the composition of the alloy deposited. This is shown in Figure 11.2.

11.4. ADVANCED CONSIDERATIONS

The codeposition behavior of zinc iron group alloys exhibits an anomaly. In the case of Zn–Co alloy deposition, for example, preferential deposition of Zn occurs even though Co is more noble, and it should deposit preferentially. A hydroxide suppression model proposed by Dahms and Croll (2) explains anomalous codeposition behavior of zinc–iron group alloys. This explanation was later supported by a number of workers (3) who measured a rise in pH near the cathode surface during the deposition of Zn–Co alloy. In this model it was assumed that the $Zn(OH)_2$ was formed during deposition as a consequence of hydrogen evolution, thus raising the pH in the vicinity of the cathode. Zinc would deposit via the $Zn(OH)_2$ layer, whereas cobalt deposition took place by discharge of Co^{2+} ions through the $Zn(OH)_2$ layer. Recent transmission electron microscopic observations of Zn–Co deposits cast doubt on this relatively simple model. In the new hydroxide suppression model, the concept of deposition of Zn–Co from an intermediate zinc hydroxide layer is not replaced, but it stresses that the existence of the layer is transient and not constant. Thus, the model considers a two-step deposition process. This is shown schematically in Figures 11.3 and 11.4. This discussion illustrates the fact that alloy deposition is still an evolving discipline.

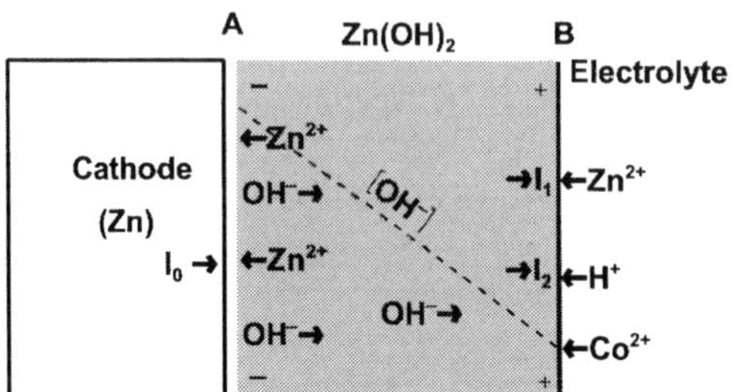

Figure 11.3. Schematic diagram showing the reactions that take place during Zn deposition via a $Zn(OH)_2$ layer: (A) metal–hydroxide interface; (B) hydroxide–electrolyte interface. (From *Electrochemically Deposited Thin Films II,* M. Paunovic, ed., Electrochemical Society, Pennington, NJ, 1995, with permission from the Electrochemical Society.)

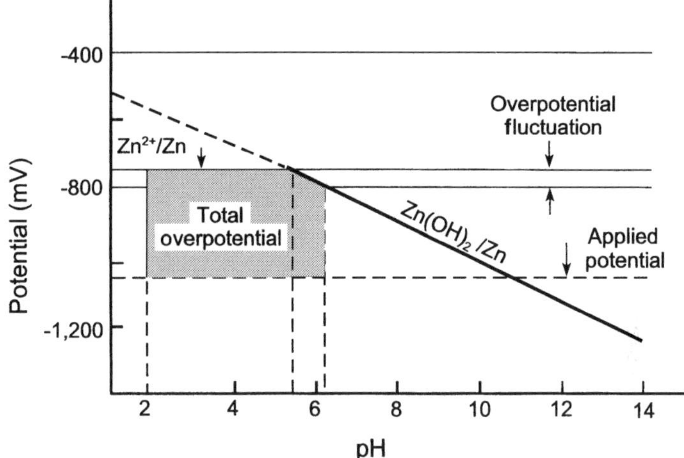

Figure 11.4. Diagrammatic representation of the effect of pH on the electrochemical potential for Zn and how overpotential fluctuations arise. (From *Electrochemically Deposited Thin Films II,* M. Paunovic, ed., Electrochemical Society, Pennington, NJ, 1995, with permission from the Electrochemical Society.)

Another path to alloy deposition is via diffusion. In this case different coatings are deposited alternately, and then heat treatment is applied to promote mutual diffusion, thus ending up with an alloy. As a specific example, an alloy of 80% Ni and 20% Cr can be produced by the deposition of alternating layers of 19-μm-thick Ni and 6-μm-thick Cr. Subsequent heating to 1000°C for 4 to 5 h produces completely diffused alloys of rather high quality as far as corrosion is concerned. Brass can also be produced by interdiffusion of Cu and Zn under suitable conditions.

REFERENCES AND FURTHER READING

1. C. L. Faust, in *Modern Electroplating,* F. A. Lowenheim, ed., Wiley, New York, 1974 (an excellent chapter on alloys).
2. H. Dahms and I. M. Croll, *J. Electrochem. Soc.* **112**, 771 (1965).
3. K. Higashi, H. Fukushina, T. Urakawa, T. Adaiya, and K. Marsudo, *J. Electrochem. Soc.* **128**, 2081 (1981).
4. A number of articles on alloys and their properties are included in M. Paunovic and I. Ohno, eds., *Electrochemically Deposited Thin Films,* Electrochemical Society, Pennington, NJ, 1993.

PROBLEMS

11.1. The Nernst equation (11.1) indicates that one can create a battery by using the same material for both half-cells but different concentrations. This is called a *concentration cell*. Calculate the EMF of the cell with concentrations as indicated:

$$Zn(s) \,|\, Zn^{2+}(0.024 \text{ M}) \,\|\, Zn^{2+}(2.4 \text{ M}) \,|\, Zn(s)$$

Comment: The Zn^{2+} molecules try to move from the concentrated half-cell to the dilute solution. That driving force gives rise to the voltage above. Naturally, at equilibrium, concentrations in the two half-cells will have to be equal, in which event the voltage will be zero.

11.2. Show that the voltage of an electrochemical cell is unaffected by multiplying the reaction equation by a positive number.

11.3. Calculate the potential of a silver electrode in a solution originally made up with $[Ag^+] = 0.01$ M and $[NH_3] = 0.1$ M. In this solution we have the equilibrium $Ag^+ + 2NH_3 \leftrightarrow Ag(NH_3)_2^+$, governed by the stability constant

$$K_{stab} = \frac{[Ag(NH_3)_2^+]}{[Ag^+][NH_3]^2} = 1.67 \times 10^7$$

The standard electrode potential of Ag/Ag^+ is -0.7991 V.

11.4. Assume that you have two half-cells, containing 0.1 M in both Fe(II) and Fe(III), with a platinum wire dipping into each. The cyanide ion reacts with both Fe^{3+} and Fe^{2+} to give complexes. For $Fe(CN)_6^{4-}$, $K_{stab} = 4.21 \times 10^{45}$, and for $Fe(CN)_6^{3-}$, $K_{stab} = 4.08 \times 10^{52}$. (Note that K_{stab} is the equilibrium constant for the reaction written as $Fe^{2+} + 6CN^- \rightarrow Fe(CN)_6^{4-}$, so $K_{stab} = [Fe(CN)_6^{4-}]/[Fe^{2+}][CN^-]^6$).

(a) What happens when enough KCN is added to only one compartment to make a solution 1.5 M (or more) in cyanide ion, CN^-?

(b) Calculate the potential of the unaffected (no CN added) half-cell.

(c) Calculate the potential of the affected (CN added) half-cell.

12

Metal Deposit and Current Distribution

12.1. INTRODUCTION

In this chapter we consider the flow of electric current in the electrochemical system and the effect it has on the process of electrodeposition.

It should be stressed that fundamentally, electricity is not a substance. Instead, an attribute called *electric charge* can be assigned to certain types of particles under certain circumstances. The smallest particles to which this attribute is assigned are often referred to as *charge carriers*. A stream of electrically charged particles is referred to as an *electric current*. The region of space where charged particles are subject to a force proportional to their electric charge is referred to as an *electric field*. In such a region, an electric *field vector* is assigned to every point such that its magnitude is the force experienced by a unit charge and the vector points in the direction in which a positive charge would move under the influence of the field. Incidentally, the assignment of positive and negative charges is quite arbitrary, and there is nothing "negative" about negative charges.

It is well known that under static conditions the *electric field vector* $\mathbf{E}$ is zero inside a perfect conductor. However, if an electric field is maintained by an external source (e.g., when one connects a conducting metal wire between the terminals of a battery), charge carriers drift in the field and there is an electric current. Within the conductor, the *current density vector* $\mathbf{J}$ points in the direction of flow of positive charges; for negative carriers, it points in the direction opposite to the flow. The *magnitude* of $\mathbf{J}$ is defined as the quantity of charge flowing through an infinitesimal surface perpendicular to the direction of flow, per unit area and unit time. Thus, current density is expressed in coulombs per square meter per second or in *amperes* per square meter, as follows:

$$\mathbf{J} = \sigma \mathbf{E} \tag{12.1}$$

Fundamentals of Electrochemical Deposition, Second Edition.
By Milan Paunovic and Mordechay Schlesinger
Copyright © 2006 John Wiley & Sons, Inc.

where σ is the *conductivity* of the material. Conductivity is expressed in ampere-meters per volt or in *siemens* per meter. It is evident from this equation that for ordinary conductors, the current density is proportional to the electric field intensity.

For the purposes of discussion, we distinguish between two types of electric conductance: *metallic* and *electrolytic*, the first being a stream of electrons, as in a copper wire, the second being a stream of ions, as in the case of a salt solution in water. In this case, positive ions will drift in the direction of the cathode, whereas negative ions will drift in the direction of the anode.

The magnitude of the *drift velocity vector* **v** of conduction electrons can be calculated rather easily. It is surprisingly low. If we denote the number of conduction electrons per cubic meter n and the electronic charge as e, then

$$\mathbf{J} = ne\mathbf{v} \tag{12.2}$$

In copper there are two conduction electrons per atom and $n = 8.5 \times 10^{28}$ electrons per cubic meter. For a wire with a cross section of 1 mm carrying a current of 1 A, a value of $\mathbf{v} = 25 \times 10^{-2}\,\text{m/h}$ is obtained. For the sake of comparison, it is interesting to note that in a molar copper sulfate solution, the *absolute mobility* (mobility in a potential gradient of 1 V/cm) of copper ions is $2.5 \times 10^{-2}\,\text{m/h}$.

12.2. CURRENT EFFICIENCY

In cases of practical applications such as electrorefining, electrowinning, and plating, the practitioner is interested only in the weight of metal deposited on the cathode or dissolved from the anode. Any currents causing other charges are considered "wasted." Of course, according to Faraday's law, stated in 1833, the overall amount of chemical change produced by any given quantity of electricity can be accounted for exactly. Thus, we can define the *current efficiency* (CE) as the ratio between the actual amount of metal deposited (or dissolved), M_a, to that calculated theoretically from Faraday's laws, M_t, in percent:

$$\text{CE} = 100 \times \frac{M_a}{M_t} \tag{12.3}$$

Again, the cathode efficiency is CE as applied to the cathode reaction, and the anode efficiency is CE as applied to the anode reaction. In other words, the ratio of the weight of metal actually deposited to the weight that would have resulted if all the current had been used for depositing it is called the *cathode efficiency*. By way of illustration, in nickel plating, CE values near 100% are not uncommon, and in some copper plating, the value is actually 100%. In contrast, in chrome plating the typical CE is about 20%. CE for other plating cases tends to be between these values.

Cathode efficiency in plating, in general, depends on a number of key parameters of the electrolyte or bath, such as chemical component concentrations, pH, agitation, and (last but not least) current density.

In an ideal situation (and only in that!) the cathode and anode efficiencies should be equal; that is, as much metal should dissolve from the anode as is being deposited at the cathode, leaving the bath in perfect or constant equilibrium.

12.3. CURRENT DENSITY AND DISTRIBUTION

The sheer amount of material deposited on a cathode is of less practical significance than the distribution of the deposit over the cathode and its thickness. Indeed, it ought to be understood that in practice metal ions cannot be expected to and do not deposit as continuous sheets from one edge of the cathode to the other. Rather, metal ions become attached to the cathode at certain favored sites. The result of this is the possible presence of discontinuities in the form of pores, cracks, or other irregularities. Thus, in electroplating, current density and its distribution play a centrally important role in determining the quality of the final deposit. Defined in terms of the actual electrodeposition setup or process, the *current density* is the total current divided by the area of the electrode.

The current density can, as a rule, be controlled by the plater. It determines the CE and/or whether deposition will take place at all. The definition of current density in terms of electrodeposition as given above yields an average figure of little use in most cases. A more accurate, useful, and immediate definition is given as

$$j = \frac{di}{ds} \tag{12.4}$$

where j is the magnitude of the current density and di is the element of current impinging on ds, which is an element of cathode surface area.

Indeed, the current density over a cathode will vary from point to point. Current tends to concentrate at edges and protruding points. It tends to be low in recesses, vias, and cavities. This is so at least partially because current tends to flow more readily to points nearer the opposite electrode than to more distant points. This, in turn, is so since by definition we have a scalar (nonvector) type of relationship: electric field = voltage/distance. In other words, this relationship can be understood in terms of the local electric field being more intense in the former case than in the latter, resulting in a higher current density locally. In addition, there are geometric (e.g., edge and similar) effects to contend with. Detailed discussion of these is available in many textbooks, some of which are listed at the end of the chapter. Thus, the thickness of deposit will tend to vary over the surface of the cathode and be thicker at edges and points (bumps).

To be even more specific, one notes that the distribution of metal deposited is also influenced markedly by the variation of cathodic CE with current density. This can, however, sometimes be of help in building deposits of even (uniform) thickness. Thus, in some cyanide metal baths (e.g., Cu, Zn; see Chapter 11), especially those with a high cyanide/metal ratio, the CE value *drops* as current density *increases*; consequently, thicknesses in regions of high current density do not much exceed the

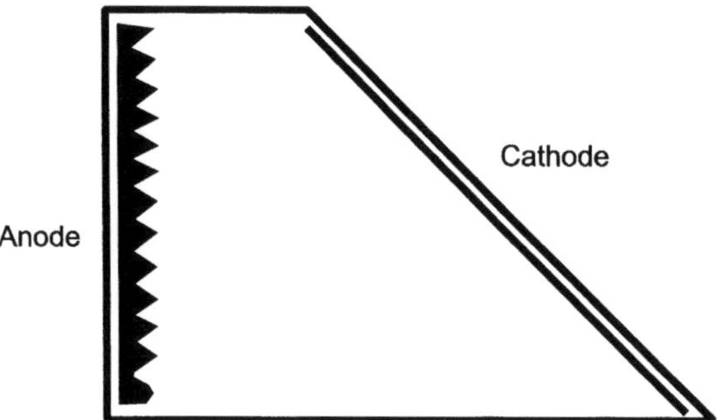

Figure 12.1. Schematic of a Hull cell. (From *Electrodeposition* by J. W. Dini © 1993 Noyes Publications, with permission.)

average. On the other hand, sometimes the opposite is the case. In chromium plating baths, for instance, the cathodic CE does increase with current density, resulting in a great degree of nonuniformity of metal thickness. On small-scale profiles (such as in the case of printed circuitry), CE varies not only with current density but also with local variations in the effective thickness of the diffusion layer. In cyanide-type baths these opposite influences tend to compensate for each other.

In brief, deposited metal distribution depends on the shape and dimensions of the object, the geometry of the plating cell, the conductivity of the bath, the shapes of the polarization curves, CE–current density (or similar) curves, and the effect of agitation.

In closing this section, we should mention a relatively simple, yet very powerful tool for studying the effects of varying current densities on the plating process. This tool is the Hull cell, made of a trapezoidal box of insulating material (Fig. 12.1). The anode is fixed against the right-angle side and the cathode is set against the slanted side. An order-of-magnitude variation in current density value is obtained in a single experiment.

12.4. THROWING POWER

Some bath compositions tend to have the property of decreasing the difference between the thinnest and thickest deposits. In other words, they give rise to uniform cathode coverage despite irregularities. Such solutions are referred to as having good *throwing power*.

12.4.1. Macro Throwing Power

A more precise way to define throwing power is as follows. The ability of a bath to produce deposits of more or less uniform thickness on cathodes having macroscopic irregularities is termed *macro throwing power*. This quantity is often measured using the Haring–Blum throwing power box (Fig. 12.2). The throwing power is then expressed as

$$T = \frac{P - M_0}{P + M_0 - 2} \times 100\%$$

$$(12.5)$$

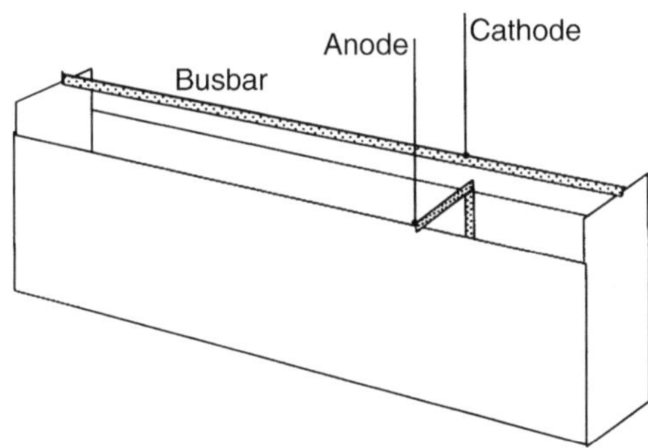

Figure 12.2. Haring–Blum throwing power box.

where T is the macro throwing power, P is the distance ratio of the cathodes from the anode, $P > 1$ (usually ≈ 5); and M_0 is the ratio of weight of metal deposited on the two cathodes.

Solutions vary a great deal in regard to their throwing power when measured this way. Although it is said that exact figures obtained this way do not matter in practice, solutions that exhibit good throwing power in the lab usually also have it on the shop floor.

Again, to be specific, Cr plating baths show $T < 0$, whereas Ni plating baths yield $T \approx 0$ and cyanide complex baths give $T \approx 40$ to 60%. Let us try to understand this. As stated, Ni plating baths (as well as other acidic baths, such as those of Cu and Zn) show poor throwing power. This is so because their CE values are $\approx 100\%$ at both low *and* high current density values, so macroscopic irregularities on a cathode will lead to nonuniform deposits. Alkaline baths, on the other hand, have better macro throwing power. This is the case since in order to remain in solution in such a bath, the metal ion to be deposited must be present in complex ions. These ions, in turn, encounter high concentration polarization. Also, in most complex baths the deposition potentials are amenable to hydrogen evolution, which "competes" with metal deposition such that CE falls as the current density is increased. That kind of behavior results in a more uniform deposit on cathodic macro irregularities. With very high concentration polarization, however, micro throwing power is rather poor.

The ability to produce a deposit over a given surface, including recesses, is called *covering power.* As a point of practical importance, it may be noted that in some cases the required potential for deposition of the metal may not be reached in recesses and vias. Other processes, such as H_2 evolution or reduction of ions such as Fe^{2+} to Fe^+, may occur instead. In such cases, a preliminary "strike" deposit may be the answer. This is made by an extremely high current density for a very short time or in a specially formulated bath.

12.4.2. Micro Throwing Power

If the depth of crevices, vias, or similar details on a cathode is small (about $10\,\mu m$ to 1 cm), the distribution of current and thus that of the deposit should be uniform. In most cases, however, one observes that deposits are thicker over micro peaks (bumps) than, say, in micro valleys. Such a state of affairs is referred to as *bad micro throw.* When the opposite is true, one talks about *true leveling.* From simple geometric considerations it follows that given a V-shaped recess, even in the case of uniform metal distribution, the deposit would still be expected to be thicker at the bottom than at the top. However, pronounced leveling is obtainable by using suitable additives, known as *levelers.*

It should be understood that even for micro surface features the potential is uniform and the ohmic resistance through the bath to peaks and valleys is about the same. Also, electrode potential against SCE will be uniform. What is different is that over micro patterns the boundary of the diffusion layer does not quite follow the pattern contour (Fig. 12.3). Rather, it thus lies farther from depth or vias than from bump peaks. The effective thickness, δN, of the diffusion layer shows greater variations. This variation of δN over a micro profile therefore produces a variation in the amount of concentration polarization locally. Since the potential is virtually uniform, differences in the local rate of metal deposition result if it is controlled by the diffusion rate of either the depositing ions or the inhibiting addition (leveling) agents.

The most convincing proof of this state of affairs is the fact that on a cathode at constant potential, variations of effective current density have been observed (Fig. 12.3). Those correspond to variations in local deposit thicknesses between micro recesses and micro peaks, where the agitation is increased, thus decreasing δN. Consequently, an increase in agitation rate may well be expected to have any of the following consequences: (1) it may increase current density if the potential remains constant; (2) it may cause the opposite, that is, decrease polarization at constant current density, resulting in poor micro throw in either of the two events; (3) it may have the opposite effect, corresponding to true leveling; or (4) it may leave both polarization and current density virtually unchanged, corresponding to good micro throw. Which of the possible

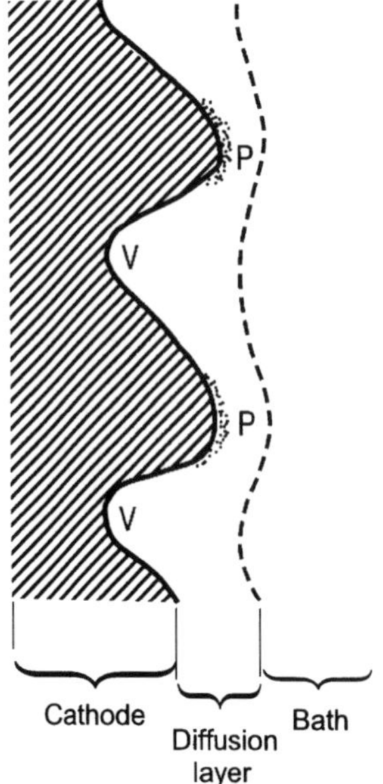

Figure 12.3. Schematic cross section showing microroughness of a cathode and the attending diffusion layer with leveling agent accumulated at peaks (P). Metal deposition is thus inhibited at peaks but not at valleys (V). Filling the latter results in smoother surfaces. (From Ref. 4, with permission from Wiley.)

effects will occur in practice depends on the nature and behavior of the addition agents, the characteristics of the metal deposition process, or possibly, both.

12.5. OPERATION CONDITIONS

12.5.1. Temperature

Effective control of the deposition process operating temperature is vital for the consistent performance of any deposition bath. Deviations of more than 5°C from optimum temperature are sufficient to harm plate quality, deposition rates, and other properties. Baths can usually be formulated, however, to operate satisfactorily at any given temperature within a relatively wide range (typically, up to 60°C). The advantages of higher temperatures for, say, decorative coatings include higher plating rates, improved anode corrosion, and the ability to operate more dilute baths without loss of performance. For protective coatings, the ability to use higher plating currents with increased deposition rates is of greatest practical importance. In either case, the complex (e.g., cyanide)/metal ratio must be regulated to yield the required throwing power and deposit quality in the current density range used.

12.5.2. Current Density

Cathode current density values must be held within the proper interval with respect to bath composition and temperature. Insufficient current for a given task will result in poor coverage of recesses/vias and a low plating rate, but the presence of excessive current does not necessarily result in increased plating rate and is liable to create other difficulties. Traces of impurities, for instance, may produce dull, burned plate at excessively high current densities. The reason for this is that over the limiting current density for good deposition, hydrogen discharge occurs. That, in turn, increases the pH level at the cathode, causing metal hydroxide to be included in the deposit. To summarize, the optimum current density range for a given plating bath depends on composition, operating conditions, and the type of plating sought. Some latitude is almost always present, so that it is possible to accommodate special requirements and/or equipment-dictated limitations.

To illustrate and clarify the foregoing, we discuss the case of cyanide–zinc plating baths. Those are operated at less than 100% cathode efficiency. In a quantitative sense, typical plating baths show functional behavior as shown in Figures 12.4 and 12.5, where we present the effects of temperature and cyanide/Zn concentration ratio on cathode efficiency versus cathode current density. As seen in the figures, if the deposition rate is most important, baths can be formulated for high current densities and operated in the range where efficiency *just* begins to decrease. It should be noted that if good throwing power is essential, steeply sloping efficiency curves are desirable.

Finally, cathode efficiencies are seldom determined directly. When questions arise in the plating shop, the experienced plater may adjust current and/or bath composition and/or temperature and other parameters to give a visibly "normal" amount of gas evolution.

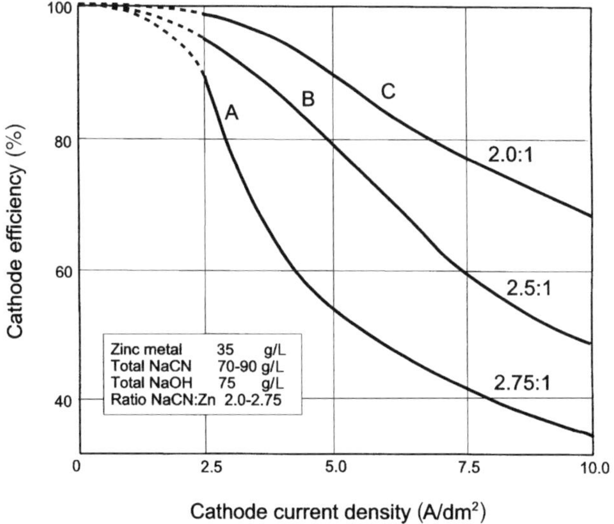

Figure 12.4. Effect of cyanide/zinc ratio on cathode efficiency. (From Ref. 4, with permission from Wiley.)

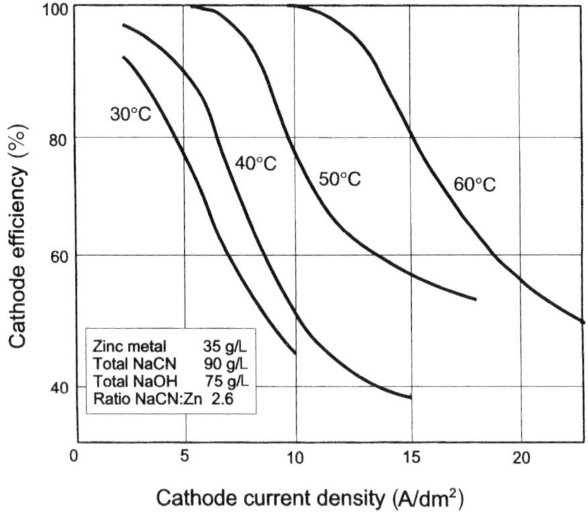

Figure 12.5. Effect of temperature on cathode efficiency. (From Ref. 4, with permission from Wiley.)

Anode current densities are also an important plating parameter, and as such should be properly controlled. This can be done through adjustment of the total anode area and the proportion of it made up of the metal being deposited. The anode, it should be remembered, plays two important roles in the plating operation: (1) by its physical location and geometry, it helps regulate the distribution of current densities on the cathode; and (2) it can also be made to replenish the metal content of the bath as it is plated out.

12.6. ALLOYS

Although the electrodeposition of alloys was the subject of Chapter 11, we mention a few relevant points here in reference to current density (5). The plating parameters—current density, agitation, operating temperature, pH, and concentration of bath constituents—together determine the ratio in which two or more metals codeposit, the physical character of the plated end-product alloy, and the rate of deposition. An appreciable variation in any one parameter may require a significant and compensating change in another parameter or combination of parameters to maintain a given plate composition. As a rule, no single variable or parameter has a well-defined independent effect on plate composition or physical properties; still, each parameter can be considered with regard to its general effect on the plating process.

An increase in current density tends to increase the proportion of the less noble metal in the alloy deposit. The extent of such change may be expected to be greater in the case of simple primary salt solutions than in complex primary salt solutions, and still more so when the codepositing metals are present in complex ions with a common anion than when the anions of the complex ions are different. In cases where the metals are associated with different complexing ions, a significant change in current density can be accommodated with relatively little change in plate composition.

The limiting current density for alloy codeposition in commercially acceptable physical form is likely to differ from the limiting current density for codeposition, without regard to physical condition. Since some additives influence the physical properties of plated deposits, limiting current densities can be altered by using the appropriate addition agents. Agitation, temperature, pH, and the like also influence the current density effects. If addition agents are used, their influence on plate composition must be considered when judging current density effects.

Since current density changes can, themselves, alter the composition of an alloy deposit, there is a question of uniform deposit composition associated with throwing power. In other words, in the case of inferior throwing power, the result may be uneven alloy composition. In other words, whenever the deposit composition is critically dependent on current density, the throwing power of alloy composition will be critical. Fortunately, however, alloy deposition baths can now be developed for which the current density range is sufficiently wide to allow for a good throwing power of composition even over a fairly irregularly shaped object. A knowledge of the current density effect on composition and current efficiency will give a good indication of expected throwing power, covering power, and deposit composition uniformity over a surface for the particular type of bath being used.

12.7. CONCLUSIONS

Low current densities tend, in general, to result in higher impurities. For example, in the case of nickel, low current densities produce a high impurity content, and this affects stress and other properties of the deposit. In Table 12.1 we show that for nickel sulfamate solution, the hydrogen and sulfur contents are considerably higher

TABLE 12.1. Influence of Current Density in Nickel Sulfamate Solution on the Impurity Content of Deposits

Current Density		Impurity Content (ppm)				
A/m^2	A/ft^2	C	H	O	N	S
54	5	70	10	44	8	30
323	30	80	3	28	8	8
538	50	60	4	32	8	6

Source: J. W. Dini and I. Johnson, *Thin Solid Films* **54**, 183 (1978), with permission from Elsevier.

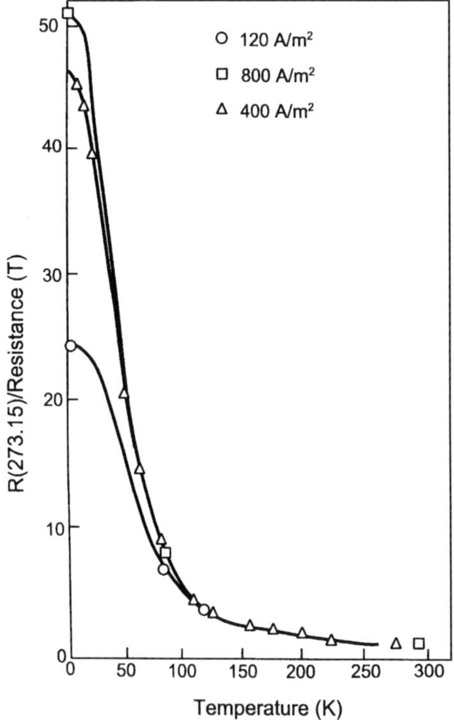

Figure 12.6. Resistance–temperature curves for electrodeposited nickel films about 20 μm thick.

for low-current-density deposits (54 A/m^2) than for the higher values. In addition, the electrical resistance of electroformed nickel films shows a strong dependence on plating current density (Fig. 12.6). Films deposited at a moderately low current density of, say, 120 A/m^2 show considerably lower residual resistance than do high-current-density films over the temperature range 4 to 40 K, as a result of codeposited impurities in the low-current-density deposits.

REFERENCES AND FURTHER READING

1. E. Hecht, *Physics–Calculus*, Brooks/Cole, Pacific Grove, CA 1996.
2. H. D. Young and R. A. Freedman, *University Physics*, Addison-Wesley, Reading, MA, 1996.

3. W. P. Crummet and A. B. Western, *University Physics: Models and Applications*, Wm. C. Brown, Dubuque, IA, 1994.

4. F. Lowenheim, ed., *Modern Electroplating*, Wiley, New York, 1974.

5. C. L. Faust, in *Modern Electroplating*, F. A. Lowenheim, ed., Wiley, New York, 1974 (an excellent chapter on alloys).

PROBLEMS

12.1. What is the current density in an extension cord used for a 100-W table lamp? Assume that the copper conducting wire in the cord is 1 mm in diameter and the voltage is 110 V. Express your result in A/cm^2 as well as in A/m^2 (compare to the values in Table 12.1).

12.2. What is the drift velocity of the conduction electrons in the cord in Problem 12.1? (*Hint*: Use the formula given in the text. Assume the electronic charge to be 1.6×10^{-19} C.)

12.3. Explain why the answer to Problem 12.2 differs from the value for the drift velocity in copper given earlier as 25×10^{-2} m/h.

12.4. Explain the basis for the operation of the Hull cell: in other words, why it is expected that different points on a slanted cathode will be electroplated to different degrees.

13

Characterization of Metallic Surfaces and Thin Films

13.1. INTRODUCTION

Clearly, the most prominent imperfection in a crystalline solid is its surface, since it represents a cutoff of the lattice periodicity. The surface can be defined as constituting one atomic–molecular layer. This definition is sometimes not particularly useful, however. In certain cases the system or property of interest requires that additional layers be considered as "the surface."

The surface structure does not conform to the three-dimensional lattice structure of the bulk. Rather, it forms a two-dimensional lattice that is described in terms of one of the five possible plane lattices (in three dimensions there are, of course, 14 Bravais or space lattices). The typical concentration of atoms or molecules at the surface of a solid can be estimated from the bulk density as follows. For the sake of simplicity, assume a given material's density to be 1 g/cm^3, the density of ice or water. In terms of the number of molecules per cubic centimeter, we have an approximate value of 5×10^{22}. The surface concentration of molecules [(number of molecules)/cm^2] is proportional to the two-thirds power of the bulk density. This is equivalent to assuming cubic-type packing. This, in turn, should be evident when one remembers that given the numerical value of the volume of a cube, the cube root of that number gives the numerical value of the cube's side height. We then have a value of 10^{15} molecules/cm^2. This, of course, is an order-of-magnitude value, due to those simplifying assumptions of the density value (of 1/cm^2) and cubic-type packing. In practice, however, density values are almost all within no more than a factor of 10 of each other, thus making this number rather useful. Often in the literature pertaining to surfaces, a parameter is defined. This parameter relates the number of surface atoms to their total number. It is important in the case of clusters or particles, where for very small particles (or thin films) that ratio may approach unity. On the other hand, in larger clusters some atoms may be surrounded by neighbors and cannot be

Fundamentals of Electrochemical Deposition, Second Edition.
By Milan Paunovic and Mordechay Schlesinger
Copyright © 2006 John Wiley & Sons, Inc.

considered as being on the surface. In that case, the parameter has a value of less than unity. A large number of chemical reactions are made possible by surface atoms of heterogeneous catalysts. Those are in small-particle form. This includes catalysts to produce high-octane fuels, for instance.

Over the last 25 years or so, an ever-expanding number of techniques have been developed as tools in the study of many surface properties. The properties most often studied are structure and composition, factors causing changes in chemical electronic and mechanical properties. In these techniques the tendency has been to concentrate on surface investigative tools that provide information on the atomic or molecular scale and have increasing sensitivity to detect fewer and fewer numbers of surface atoms. The instrumentation and methodology of surface characterization is subject to constant progress and improvement in the direction of detecting smaller and smaller details, higher spatial resolution, and better energy resolution, together with learning more about a given surface in shorter time scales. It must be stressed that no one technique is capable of providing all the information about the surface and its constituent atoms. Consequently, a combination of techniques is required. The most frequently employed techniques involve absorption, emission, or scattering of light (photons), electrons, ions, or atoms. Other methods not falling into these categories are used as well. An added complexity is the fact that many surface test methods require vacuum conditions for their application. Thus, most surface-characterizing instruments are equipped with high- and low-pressure cells; the former provide high-pressure/temperature conditions, and the latter are used for the subsequent surface analysis. Sample preparation is an integral and often the most difficult part of the characterization operation. By way of example, we remark that single crystals may be oriented using x-ray diffraction methods and then cut and polished. They may subsequently be treated chemically or otherwise to rid their surface of unwanted matter such as impurities or polycrystalline film produced by polishing of the surfaces. Thin films may be considered to constitute a surface in and of themselves, particularly when they are no thicker than two atomic layers. In the following sections we describe some of the many surface-characterizing methods and the object lessons they may provide for a scientist or technician. Detailed discussion of the various techniques is outside the scope of this book, and interested readers are referred to many excellent sources, such as the Internet. In addition, our emphasis is on metallic surfaces rather than on all surfaces in general.

13.2. SURFACE STRUCTURES

A detailed inspection of a metallic surface using an optical microscope, for instance, reveals the existence of irregularities. These irregularities exist in a variety of shapes and scales. Specifically, even in an area of no more than a square micrometer (1 μm is 10^{-4} cm), one could expect to encounter several types of surface sites. Such sites are distinguishable by their number of nearest-neighbor atoms. Figure 13.1 depicts some of the possible surface atomic sites. This depiction is somewhat idealized. It creates the impression of a perfectly rigid bulk lattice immediately under the surface so that

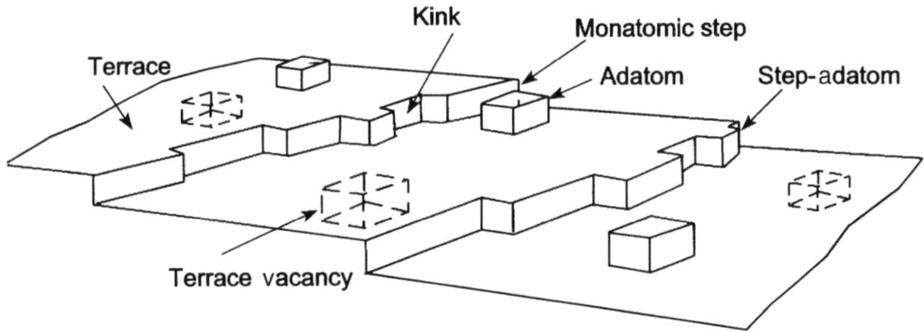

Figure 13.1. Model of a heterogeneous solid surface depicting various surface sites, which are characterized by the number of their neighbors. (From Ref. 1, p. 41, with permission from Wiley.)

lattice points can be related or projected from the bulk to the surface. This is not actually the case. Surface atoms, almost as a rule, are found at locations that are shifted significantly from locations that the bulk geometry would dictate. What one observes in practice is a contraction of the atomic layers near and parallel to the surface in the fashion depicted in Figure 13.2. It should be emphasized that even in the case of a clean surface, there is noticeable contraction between the first and second layers. If surface roughness (or open surface structure) is present, the contraction may be even more severe. If the surface adsorbs atoms or molecules, surface atoms relocate as required in terms of thermodynamic equilibrium. To formulate these phenomena, in other words, surface atoms do move on the surface, rotate off the surface, or otherwise change position as a result of imperfections and/or adsorbates, forming "new" surface structures. These characteristics, together with the many different composition types of metallic surfaces, make the subject matter of this chapter very complex, requiring sophisticated experimental methods.

When adsorption takes place on an ordered metal–crystal surface, the adsorbed material forms ordered surface structures. The cause of this lies in mutual atomic interactions, which may be categorized into adsorbate–adsorbate and adsorbate–substrate interactions. In the case of chemisorption, the former is considerably the weaker of the two. The possible long-range ordering of the overlayer formed is dominated by adsorbate–adsorbate interaction, however. Ordering of the adsorbed material is also dependent on the degree of surface coverage. Thus, for instance, at a low degree of

Figure 13.2. Schematic representation of the contraction in interlayer spacing usually observed at clean solid surfaces. (From Ref. 1, p. 41, with permission from Wiley.)

coverage, bunching of the adsorbate can be expected in a two-dimensional fashion as a result of adsorbate–adsorbate interaction and the ease of diffusion on the surface. As coverage increases to a degree such that the mean interaction distance between the adsorbate components is on the order of 1 nm, ordering in specific configurations will result.

In other words, in the case of physisorption or physical adsorption, when the adsorbate does not form chemical bonds with the substrate, adsorbate–adsorbate interactions dominate over adsorbate–substrate interactions. In such a situation the adsorbate–substrate-dicated lattice geometry will be overtaken by an adsorbate–adsorbate dictated geometry, resulting in an incommensurate structure in which the overlayer and the underlayer will generally have different lattice structures. In the case of metallic adsorbates (such as during electrodeposition), a close-packed overlayer will form. This is a result of the metal atoms attracting each other, quite strongly coalescing to covalent interatomic distances. If the atomic sizes of the overlayer and substrate metals are similar, one adsorbate atom may occupy every unit cell of the substrate, a state referred to as *epitaxial growth.* With atomic sizes farther from equal, other structures may form. In addition to the closed-packed overlayer that is present, and even before a single atomic layer is completed on the surface, metal adsorbates can form multilayers or even three-dimensional crystallites. Alloy formation via interdiffusion is also possible.

13.3. SURFACE ANALYSIS TECHNIQUES

At this point it is natural to wonder how the type of information discussed above is obtained. It is possible to group the analysis techniques for the characterization of metal surfaces and/or thin films into three major categories:

1. Mass-spectrometric techniques
2. Electron-beam techniques
3. Other techniques

The two most prominent members of the first group are the methods of secondary-ion mass spectrometry (SIMS) and Rutherford backscattering. The first method is based on the fact that ions are far more massive than electrons (the lightest of ions, H^+, is 1800 times more massive than the electron) and so are more able to transfer energy to a given surface. Ions impinging on a surface may break chemical bonds and sputter (eject) atoms, molecules, or clusters of the same. Although most sputtered species are neutral, some can be expected to be ionized by the impinging ions. The detection of these ions, called *secondary-ion mass spectrometry* (SIMS), constitutes an efficient technique for surface composition analysis. As an aside, we may point out that ion bombardment itself is an important means for cleaning surfaces of unwanted contaminants. As is true of all methods, this one has important limitations, including the destructive nature of the method, the fact that the wide variation in detection sensitivity from element to element makes quantitative observation very unlikely, and the quality of analysis, which is strongly dependent on instrument design.

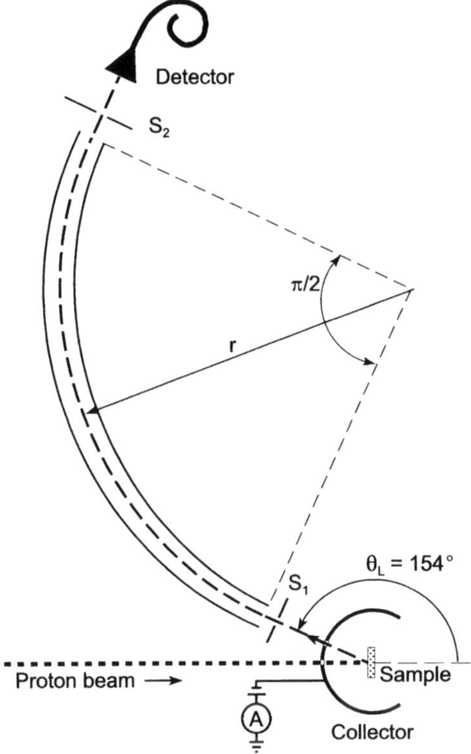

Figure 13.3. Schematic diagram of backscatter apparatus. (From Ref. 2, with permission from the Electrochemical Society.)

The other method, *proton (Rutherford) backscattering*, is best illustrated by considering Figure 13.3, where the typical experimental setup is given schematically (see also Ref. 2). The magnetically analyzed beam of protons has a fixed energy, to within one part in a thousand, of about 100 keV. An operating current density of 5×10^{-5} A/cm^2 is typical. Thin-film targets are mounted in an evacuated system. The target film is scanned by moving it across the path of the proton beam. The energy distribution spectrum of protons scattered through a 154° laboratory angle is analyzed with a cylindrical–electrostatic field sector of 90° with a radius $r = 19.05$ cm. The widths of the entrance and exit slits (Fig. 13.3) are set with a view to achieving an energy-resolving power R sufficient for a desired surface depth resolution. Information about the atomic composition of the surface region is obtained from an analysis of the energy distribution spectrum of the backscattered protons.

If the thin-film or surface target consists of two or more atomic species, the energy distribution function of the backscattered protons shows some structure. The relative heights of the components that make up this structure can be utilized to determine the concentrations of the various atomic species giving rise to them. The corresponding atomic masses can be found through calculation of the recoil energy after scattering.

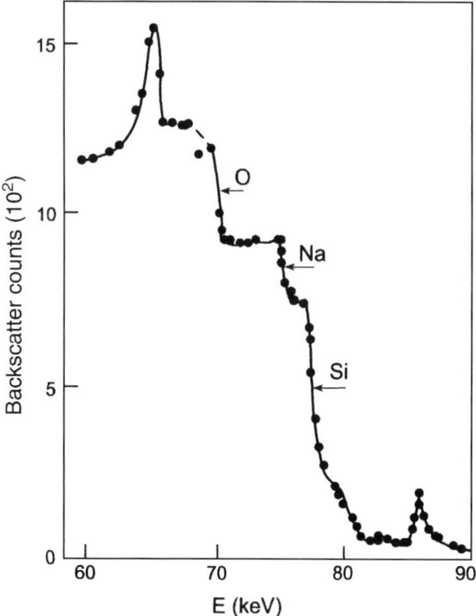

Figure 13.4. Energy distribution of backscattered protons from a glass substrate after immersion in a $PdCl_2$ bath and subsequent deposition of a thin carbon layer. Contributions due to the backscatter yield from different species are indicated. (From Ref. 2, with permission from the Electrochemical Society.)

Figure 13.4 illustrates the energy distribution of backscattered protons off a glass substrate after immersion in $PdCl_2$ bath. The various atomic species present are indicated.

As representative techniques of the second group, we discuss two methods: *x-ray photoelectron spectroscopy* (XPS), sometimes referred to as *electron spectroscopy for chemical analysis* (ESCA); and *Auger electron spectroscopy* (AES). The main principle of the first method (XPS) is the excitation of electrons in an atom or molecule by x-rays. The resulting electrons carry energy away according to the formula

$$E_{\text{kin}} = h\nu - E_B$$

where h is Planck's constant, ν is the impinging x-ray photon frequency, E_B is the binding energy, and E_{kin} is the kinetic energy of the electron emitted. Figure 13.5 illustrates typical XPS spectra obtained from nickel compared with a spectrum corresponding to NiO. As with all other techniques, the method has limitations. Data collection is slow; quantitative analysis requires one to several hours *after* an overnight pumpdown prior to analysis. The method has relatively poor lateral resolution, and sample charging effects may pose a problem with insulating samples.

The other method, *Auger electron spectroscopy*, is considered appropriate for studying the chemical makeup (composition) of surfaces, with a sensitivity down to 1% of a single atomic layer (monolayer). It is also easier to perform than many other methods of surface studies of the present group. It is based on the principle that if an

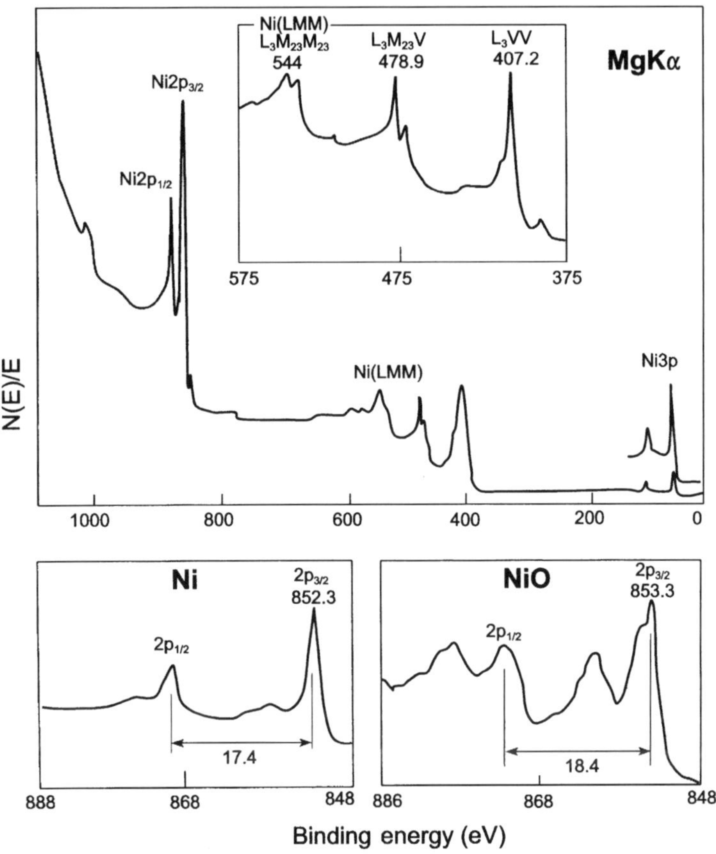

Figure 13.5. Typical XPS spectra obtained from nickel compared with a spectrum for NiO. (From Ref. 1, p. 386, with permission from the Electrochemical Society.)

energetic beam (e.g., of a few kiloelectron volts) of electrons or x-rays is directed at atoms on or near a surface, electrons with binding energies less than that of the beam will be ejected from, say, an inner atomic level. This results in a singly ionized excited atom. The electron deficiency created may subsequently be made up (filled) by the deexcitation of electron(s) from other, higher-lying energy states. The energy thus released may be transmitted to another electron in the same or in another atom (e.g., via radiation). If this quantity of transmitted energy is more than the binding energy of the receiving electron, the electron may be ejected from its parent atom, resulting in a doubly ionized atom. Electrons ejected as a consequence of the deexcitation process are known as *Auger electrons*, and their energies are characteristic of the energy-level separation in the atom. Determination of the energies of Auger electrons indicates the identity of the atoms from which they have originated.

As an example of a technique from the third group, we discuss *scanning tunneling microscopy* (STM). This method, which is capable of atomic-scale resolution, is based on the principle of quantum tunneling of electrons between a sharp metallic tip and the surface under study. The tip can be brought within distances of about 2 to 0.2 nm or less from the surface. Now, using rapid-response-time electronic feedback circuits, the tip can be held steady at these distances. It can also be moved above the surface, mounted on a piezoelectric ($BaTiO_3$) tip holder that is capable of expanding in the 0.1-nm range under applied potential. Since the tunnel current varies exponentially with distance from the surface, atom-size bumps can be detected by tracking the changes in that current. Alternatively, the tunnel current may be kept constant and the tip movements required to maintain the current value may be indicative of atom-size topographic variations. With the experimental techniques now available, ordered atomic arrays in atomic planes, the periodicity of atomic "steps," and kinks (Fig. 13.1) in the steps can all be identified. STM has a number of variants, including *atomic force microscopy* (AFM) and *magnetic force microscopy* (MFM). To put matters in context, we note that all three are variations on a method of surface imaging with near-atomic resolution collectively referred to as *scanning probe microscopy* (SPM). In these procedures, as discussed above, a small tip is passed over a surface to obtain a 3D image of the surface at the atomic level of resolution. Fine control of the tip's scanning is achieved using piezoelectrically induced motions. When the tip and surface (metal) are both conducting, the structure of the surface is detected by tunneling of electrons between the tip (usually at negative potential) and the surface (usually at a relative positive potential) (STM). Any type of surface can be examined by exploiting the molecular forces exerted by the surface against the tip (AFM). In general, the tip can be made to be in contact with the surface continuously; it can tap the surface oscillating at high frequency or be just minutely above the surface. Finally, coating the tip with magnetic material, the magnetic fields just above the surface can be imaged. Image processing software allows easy extraction of the surface parameters (MFM).

13.4. CHARACTERIZATION OF SURFACES

In the last few years, a number of factors have led practitioners to abandon the concept of *ultraclean processing* (reducing the level of contaminants to below the level detectable with state-of-the-art equipment). The approach that seems to have taken hold instead is that of "just clean enough," which requires a fundamental understanding of the specific effects of contaminants and as a consequence, the ability to define tolerable levels of contaminants. In the next section we review the characterization tools that can be used in this context.

13.4.1. Microelectronics

As indicated above, the goal in recent years in the ultralarge-scale integrated-circuit (ULSI) industry has been to reduce contamination on surfaces to below the level detectable with state-of-the-art equipment. For economic and environmental reasons, this method had to be replaced by one referred to as "just clean enough." This requires that action–reaction relationships be identified between contaminants and their influence on a given surface, to enable us to fix acceptable levels of organic and metallic contamination, surface roughness, and other surface characteristics.

A typical example of this is the procedure known among practitioners as *IMEC,* in which there is sequential removal of organic and particle contamination followed by a metal-removing step, usually by the use of diluted HF followed by drying. It should be noted that this sequence is far more environmentally friendly than the typical RCA–clean sequence, as it uses less deionized water or chemicals. The chemicals used do not need to be as clean as those in the RCA–sequence.

Since metallic contamination is considered important in the case of microelectronics, it must be reduced to low levels: at most, 10^9 to 10^{11} atoms/cm^2. The demands of 0.25-μm (and below!) technology reduce this to 10^8 atom/cm^2. A number of techniques are used routinely to try to determine the presence of unwanted metallic contaminants on surfaces. Of these, SIMS (Section 13.3) seems to offer the best way to overcome the problem of detection of light elements; the higher sensitivity of other methods is true only for some elements and not for others, particularly for the light elements. Although SIMS has different sensitivities for different elements, depending on ionization potential–electron affinities, it does have a rather good detection limit. One way to solve the problem of obtaining quantitative results is to realize that at the surface itself, ionization is not stabilized (i.e., it is subject to variations) and to apply an extra coat of polysilicon layer on the silicon wafer, for instance. This is called *polyencapsulated SIMS.* In this way the surface (now the interface) of interest is beyond the unstable ionization region. Experiments have shown the method to be quite reliable compared to other methods of surface characterization. Actually, the polysilicon coating can be dispensed with by performing SIMS with an oxygen gas beam together with oxygen flooding conditions. That leads to oxidation of the silicon surface, stabilizing the ion yields to result in light-element detection limits down to 10^9 atoms/cm^2.

Up to this point in this section we have studied the case of metallic contamination on relatively large surfaces. Here the development of instrumentation and methodology has made it possible to handle the low level of possible contamination required for the present and next generations of technologies. It is still an open question whether the efficiency of cleaning procedures may be affected adversely when applied to patterned wafers, particularly in small areas. When applied to small areas, characterization techniques, are a complicated branch of technology, outside the scope of this book. By way of illustration, we note that for an area of 1 μm^2, a detection limit of 10^8 atoms/cm^2 corresponds to single-atom detection; in other words, sample areas will determine the possible detection limits. Also, it is likely that statistical considerations will have to be brought to bear rather than average contamination levels.

13.4.2. Electroplating

Surface characterization is of vital importance in electroplating processes. First there is the preparation of the surface to receive the plating and the need to know the surface's suitability to receive and its condition after the completion of plating. Then there is the in situ characterization during plating, which is the subject of Chapter 14.

The effectiveness of a cleaning agent or a cleaning procedure can be evaluated only by determining the resulting degree of substrate cleanliness. The cleaning process involves the use of a combination of chemical, mechanical, and thermal energy, over time, in an attempt to overcome the forces holding the "soil" to the substrate. Similarly, cleanliness in an important sense is the measure of the degree of success in arriving at a completely soil- or dirt-free surface. Metals used in industry seldom, if ever, leave a production facility without some type of coating. The coating may be applied in many different ways, including anodizing, electroplating, electroless deposition, galvanizing, and painting. In all these processes, pre- as well as postcleaning is always part of the cycle. Some degree of cleaning should precede each processing step. In the case of plating, a high level of cleanliness is a necessity if one is to avoid nonuniform and/or purely adherent electroless and electroplated coating. In addition, coating porosity and eventual failure to provide corrosion protection can often be attributed to substrate precleanliness. In the laboratory as well as on the shop floor, measurements of various degrees of precision are available, are used to determine the degree of cleanliness, and may help to develop better (cleaner) formulations and systems. Testing the quality and properties of a finished product is an inefficient and uneconomical way to assess the quality of a cleaning step prior to coating. Below we discuss some methods of surface characterization that may define cleanliness in terms of a specific surface property.

Surface Tension.   Testing for surface tension is sometimes referred to as the *contact-angle test* (Fig 13.6). Surface free energy depends on surface tension, which is directly related to surface cleanliness. If an adsorbate (dirt) is present on a surface, the free-energy surface tension is reduced, as energy is spent in bonding the adsorbate to the surface. In other words, the surface has become less clean, accompanied by a reduction in surface tension. A practical method to determine surface free energy has been described in the literature (see Refs. 3 to 9). In this, solutions of different known surface tension values are dropped on a surface. Those drops that bead up have surface tension values higher than that of the surface; those that wet the surface have lower values. A systematic proper choice of test liquids provides a good measure of surface energy and with it, information regarding the cleanliness state of the surface.

Surface Carbon Analysis.   This method is based on the observation that the presence of carbon on automotive body sheet steel, for instance, can be linked to poor corrosion performance. The carbon content on the surface can be determined by subjecting the body sheet to about 500°C in an oxygen environment and determining the CO_2 thus formed.

These are only two examples of the vast number of methods available to the practitioner for the characterization of surfaces. In general, though, it should be stated that

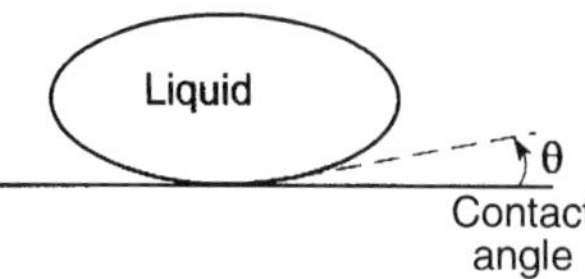

Figure 13.6. The lower the contact angle, the lower the surface tension values.

even on the shop floor, one should put more emphasis on the condition of the substrate than on the degree of dirt (soil) removal.

A final point to remember is that after the proper (i.e., the one fitting the conditions and requirements of a given surface) cleaning procedure has been completed, care must be taken to ensure that the surface will not be recontaminated during subsequent storage.

13.4.3. Stress in Deposits

Stresses are almost always present in the surface and bulk of deposits of all types. Such stresses may be a major influence on the characteristics of a deposit, such as resistance of the coating to wear, fatigue, and crack propagation. In addition, there exists the possibility that these residual stresses are causing diminished adhesion and even peeling of the thin film or coating (see Section 13.4.4). That is so because release of the stored elastic strain energy, as stresses get relaxed, may drive the separation of the thin film or coating from the substrate. At the same time, it should also be noted, the quantity of energy released per unit area of interface rises lineraly with coating thickness; hence, these effects become increasingly likely with increasing thickness.

Detailed theoretical treatments of the issue of residual stress indicate that for a thin deposit ($\sim$10 nm) the stress level is essentially uniform, whereas for a thick deposit ($\sim$100 nm), a through thickness variation is present. In both cases, however, the coating may be viewed thin enough so that usually no stresses are built up in a direction normal to the plane of the coating. Consequently, neglecting edge effects and assuming approximately planar specimen geometry, at any depth of the system an equal biaxial stress state is established (i.e., a single stress value).

Thin films are in widespread use in a large number of applications, such as for interconnections in semiconductor devices, as optical coatings, and as protective layers on high-temperature alloys. The stresses in these films often play an important role in the functioning and/or reliability of products. Methods are thus required for the measurement and possible quantification of these stresses. It must be understood that the well-known properties of bulk materials are often no guidance in a study of the behavior of the material in thin-film form. This is the case as a result of dimensional constraints as well as to the microstructure that develops during the growth of thin films. It is thus not even possible to extrapolate substance properties from bulk to thin films of thicknesses in the range 10 nm to 10 μm. Consequently, a number of specialized testing techniques have been developed to study the mechanical properties in small dimensions. Just as in the case of macroscopic testing, the point is to determine properties such as Young's modulus, strength, and fracture resistance. It is clear, then, that all of this requires new ways to measure stress and strain in small volumes. It is true that sometimes the principles of macroscopic testing may be transferred to small dimensions, as in the case of microtensile or nanohardness testing.

In general, for a thin film, the stress is related to the elastic strain as

$$\sigma = \frac{\varepsilon E}{1 - \nu}$$

where ε is the elastic strain, E is Young's modulus, and ν is Poisson's ratio (for the film). The sign of the stress depends on the nature of the deposition. Sputtering, for example, typically yields films in compression, since newly arriving atoms are forced into places where they do not belong. Films produced by chemical vapor deposition (CVD), on the other hand, are frequently initially in tension as a result of the departure of by-products (e.g., water) of the deposition reaction. Plasma-enhanced chemical vapor deposition (PECVD) can typically produce films in compression. This is due to ion implantation, commonly of hydrogen, from the plasma.

Since the films are often deposited at elevated temperatures, this may result in considerable magnitude of stress when the film is cooled down to room temperature after deposition. Also, for metal films, the total strain is often greater than the yield strain, so that plastic flow occurs and the stress is determined by the yield stress of the metal. This is, as a rule, much larger than the typical yield stress for the bulk metal. A second, "intrinsic" contribution to the stress may result from the deposition process itself. Specifically, the temperature of the film during deposition, although elevated, is still often far below the melting temperature of the material. The deposition then occurs under highly nonequilibrium conditions; that is, the atoms are not sufficiently mobile to attain minimum energy positions during deposition.

Finally, and rather importantly, an additional source of stress can be phase transformations occurring during thermal treatment following deposition, and a volume change normally occurs with the transformation. Some typical examples include transformation from amorphous to polycrystalline in some electrolessly deposited metal systems (10) and the formation of intermetallic compounds during the annealing of alloy interconnections.

Stress changes after deposition, even at room temperature, are also quite common. Stress initially present in a metal film may relax by plastic flow. Oxide films may gain or lose water. Even in the absence of phase transformations, complex stress changes typically occur during thermal cycling after deposition. Thus, for example, PECVD nitride films typically lose hydrogen; and CVD oxide films generally lose water, then at higher temperatures undergo densification when atomic mobility is sufficient to permit relaxation of the initial relatively open structure; and metal films undergo extensive plastic flow and changes in microstructure. To study these effects, it is necessary to carry out stress measurements as a function of time and temperature.

Among the most often used stress measurement methods for crystalline films is determination of the change in interplanar spacing of the crystallites in the film by x-ray diffraction. X-ray determination of near-surface strain and stress in bulk materials has a long history, dating back to the very early period of x-ray powder diffraction measurements and is a well-established technique.

Changes in interplanar spacing d can be used with the Bragg equation to detect elastic strain ε through knowledge of the incident wavelength λ and the change in the Bragg scattering angle $\Delta\theta$,

$$\lambda = 2d \sin \theta$$

giving

$$\varepsilon = \frac{\Delta d}{d} = -\cot\theta\ \Delta\theta$$

It is, of course, necessary to have an accurate measure of d, the stress-free spacing. The strain results can then be converted into stress using a suitable value of the stiffness (e.g., see Ref. 11).

In practice, the application of x-ray measurement techniques to thin films involves some special problems. Typical films are much thinner than the penetration depth of commonly used x-rays, so the diffracted intensity is much lower than that from bulk materials. Thin films are often strongly textured; this, on the other hand, results in improved intensity for suitable experimental conditions but complicates the measurement problem. Measurements at other than ambient temperature, not usually attempted with bulk materials, constitutes additional complexity. Since typical strains are on the order of 1×10^{-3}, measurements of interplanar spacing with a precision of the order of 1×10^{-5} are needed for reasonably accurate results; hence, potential sources of error must be kept to a low level. In particular, the *sample displacement error* can be a major source of difficulty with a heated sample. The sample surface must remain accurately on the axis of the instrument during heating.

13.4.4. Adhesion of Deposits

Adhesion is defined as a measure of the force required to separate a plated coating from the substrate. The bond between the two can be chemical (some refer to this as metallurgical), mechanical, or both. When plating metal on metal, with good chemical bonding, the adhesive strength is greater than the strength of the weaker metal. Plating on surfaces is an atom-to-atom process. Even smooth-appearing surfaces can be relatively rough on a submicroscopic scale, resulting in a physical keying/anchoring effect that adds to the adhesion. This microanchoring is the primary source of mechanical adhesion. Plating on plastics and other nonconductors depends on this microetch for adhesion.

Adhesion tests can be broken into two categories: qualitative and quantitative. They vary from a simple Scotch tape test to a complicated flyer tape test, which requires precision-machined specimens and a very expensive testing facility. Quantitative (such as peeling) tests have been developed for coatings on plastics (12), but not to the same extent for metal-to-metal systems. The quantitative testing systems in limited use, mainly in the electronics industry, are not commonly present in production plants but have been used to aid in process development. For quality control purposes, qualitative tests for metal-to-metal adhesion (13) are usually adequate. The adhesion of some plated metal parts is improved with baking for 1 to 4 h at relatively low (120 to 320°C) temperatures.

The deposition of thin metallic films carried out by electrochemical or other methods, such as classical evaporation or sputtering of a solid metal source, may all benefit from an extra intermediate layer to improve their adhesion to the substrate. Specifically in the case of the noble metals most often used in modern electronics (i.e., Pt, Au, Ag, and Ir), a layer of Si_3N_4, Ti, or Ta deposited sequentially in a single pump-down process prior to the deposition of the noble metal is the most frequently used adhesion promoter. Typical thicknesses of adhesion promoters and noble metal electrodes are 0.05 μm and 0.15 μm. Microfabrication of thin-film noble electrodes is reviewed in Ref. 14.

Many of the qualitative adhesion tests vary with plate thickness. As indicated above, adhesion is better for thinner deposits. That, it was stated, has to do with the stress present in deposited films. A specified plating thickness should therefore be a given parameter requirement for adhesion testing.

REFERENCES AND FURTHER READING

1. G. A. Somorjai, *Introduction to Surface Chemistry and Catalysis*, Wiley, New York, 1994.

2. M. Schlesinger, in *Proceedings of the Symposium on Electrocrystalization*, Vol. 81–6, Electrochemical Society, Pennington, NJ 1981, p. 221.

3. W. Vandervorst et al., *Microelectron. Eng.* **28**, 27 (1995).

4. M. Meuris et al., in *Proceedings of the 3rd International Symposium on Cleaning Technologies of Semiconductor Devices*, Vol. 93, Electrochemical Society, Pennington, NJ, 1993, p. 15.

5. R. Schild et al., *Proceedings of UCPSS'94*, Acco Leuven, 1994, p. 31.

6. M. Depas et al., *Proceedings of UCPSS'94*, Acco Leuven, 1994, p. 319.

7. B. Schueler et al., in *Proceedings of the 3rd International Symposium on Cleaning Technologies of Semiconductor Devices*, Vol. 93, Electrochemical Society, Pennington, NJ, 1993, p. 554.

8. S. P. Smith, *Proceedings of SIMS-IX* (Yokohama, Japan), Wiley, New York, 1995, p. 476.

9. L. E. Cohen, *Plat. Surf. Finish.* 58 (Nov. 1987).

10. J. P. Marton and M. Schlesinger, *J. Electrochem. Soc.* **115**, 16 (1968).

11. A. D. Krawitz, R. A. Winholtz, and C. M. Weisbrook, *Mater. Sci. Eng.* **A 206**, 176 (1996).

12. STM B533-85, *Standard Test Methods for Peel Strength of Metal Electroplated Plastics*, American Society for Testing and Materials, Philadelphia, 1998.

13. ASTM B-571-97, *Standard Methods for Adhesion of Metallic Coatings*, American Society for Testing and Materials, Philadelphia, 2003.

14. (a) R. L. McCarley, M. G. Sullivan, S. Ching, Y. Zhang, and R. W. Murray, in *Microelectrodes: Theory and Applications*, Kluwer, Dordrecht, The Netherlands, 1990, pp. 205–226; (b) R. R. Tummala and E. J. Rimaszewski, eds., *Microelectronics Packaging Handbook*, Van Nostrand Reinhold, New York, 1989; (c) M. Schlesinger and M. Paunovic, eds., *Modern Electroplating*, 4th ed., Wiley, New York, 2000.

PROBLEMS

Figure P13 depicts the notation, number of electrons, and binding energy, which is the amount of energy required to remove (ionize) an electron from the inner shell

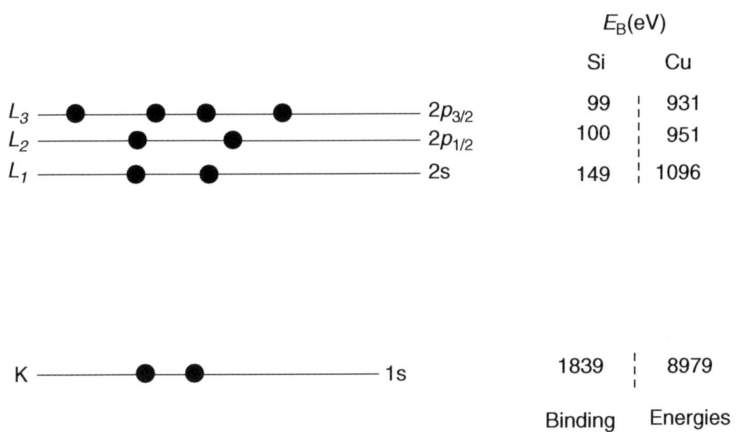

Figure P13

indicated, in silicon and copper. Three techniques in the analysis of thin-film deposits are: (1) x-ray photoelectron spectroscopy (XPS), (2) Auger electron spectroscopy (AES), and (3) electron microprobe analysis (EMA).

13.1. In XPS, an incident x-ray ejects an electron. For incident $Al\,K_\alpha$ x-rays ($E = 1.49\,keV$), what are the kinetic energies of the Si and Cu 2s electrons ejected?

13.2. For KL_1L_2 Auger electrons, an L_1 electron makes a transition to a vacancy in the K shell and an L_2 electron is ejected. What are the energies of the electrons for Si and Cu?

13.3. For $K_{\alpha 1}$ x-rays, a $2p_{3/2}$ electron makes a transition to a vacancy in the K shell. What are the energies of the Si and Cu $K_{\alpha 1}$ x-rays?

14

In Situ Characterization of Deposition

14.1. INTRODUCTION

In Chapter 13 we discussed methods available for the characterization of surfaces and thin films. These were shown to be of practical importance mainly before and/or after the plating process. It is no less important to discuss the methods that are available to characterize, study, understand, and thus control the plating operation in situ: that is, while the plating process is taking place.

The importance of an in situ study of electrochemical processes can be well illustrated by considering an issue of some considerable practical importance: optimization of the conditions of semiconductor–electrolyte interfaces. In this case, change in the quenching of photoluminescence intensity due to surface nonradiative recombination centers may indicate changes in the nature of chemical bonds in the interface region during chemical or electrochemical processing. Thus, in practice, for instance, the measurement of photoluminescence quenching is applied to the study of oxidation of HF-treated silicon surfaces in air (1). Yet another in situ method for studying HF-treated silicon or germanium surfaces is described in the literature (2). In this procedure, a strobe lamp is positioned to inject carriers into the semiconductor wafer, and the carrier density decay is then monitored by an inductively coupled radio-frequency bridge. This contactless method may be employed equally in, say, aqueous media or vacuum. The method was used for the demonstration of unusually low surface recombination velocity on high-frequency-treated silicon and germanium surfaces. In passing it should be noted here that from the point of view of electronic circuit design, covalently "satisfying" the surface bonds results in shifting the surface states into the conduction and valence bands. This is a desirable result that can be achieved using surface oxidation followed by a high-frequency etch.

As a general observation it may be stated that as far as in situ studies are concerned, the testing methods are derivatives of those discussed in Chapter 13. Experimental setups, however, require special attention. Conditions must be maintained to ensure

Fundamentals of Electrochemical Deposition, Second Edition.
By Milan Paunovic and Mordechay Schlesinger
Copyright © 2006 John Wiley & Sons, Inc.

that the processes examined remain the same as they would be if in situ studies were not attempted.

14.2. ELECTROCHEMICAL STUDIES

14.2.1. Extended X-Ray Absorption Fine Structure

Assume a monochromatic x-ray beam directed through a sample. Now, as the (wavelength) energy of the x-rays is gradually decreased or increased, it crosses a sudden absorption step or edge typical for one of the elements of interest in the sample. The part of the x-ray radiation that is transmitted will contain small variations in absorbance on the high-energy (short-wavelength) side of the absorption step or edge. The nature and characteristics of that variation provide information about the structural environment of the atoms surrounding the element whose absorption edge is being examined.

Specifically, x-ray absorption is dominated by photoelectron absorption where the x-photon is completely absorbed, creating a photoelectron–hole pair (10). Conservation-of-energy considerations require that the kinetic energy of the excited photoelectron be equal to the difference between the exciting photon and the electron's binding energy. To a rather good first approximation, the final energy state of the photoelectron is modified by a single scattering by each of the surrounding atoms. From a quantum mechanical viewpoint, the photoelectron may be viewed as a wave whose wavelength (λ) is described by the de Broglie relation, $\lambda = h/p$, where p is the momentum of the photoelectron and h is Planck's constant. For an EXAFS experiment (11), the momentum is determined by the free electron relation

$$\frac{p^2}{2m} = hf - E_0$$

where an x-ray photon of frequency f has an energy hf, E_0 is the binding energy, and m is the mass of the photoelectron.

For a given isolated atom, the photoelectron is represented as an outgoing wave. The surrounding atoms will scatter the outgoing wave. The final state is the superposition of the outgoing and scattered waves. The total amplitude of the electron wavefunction will be amplified or diminished, modulating the probability of absorption of the x-ray beam. Thus, the (wavelength) energy-dependent variation of the fine structure in EXAFS is a direct consequence of the wave nature of the photoelectron. Variations in the observed phase with (wavelength) energy of the photoelectron depends on the distance of the excited atom to the backscattering atoms. Variation of the backscattering strength as a function of the (wavelength) energy of the photoelectron depends on the atomic number of the backscattering atoms.

14.2.2. X-Ray Absorption Near-Edge Structure

It is understood that the x-ray absorption spectrum should be treated as divided into near-edge and extended fine structures. The x-ray absorption near-edge structure

(XANES) is limited to the first 30 to 40 eV past the absorption edge, whereas the extended x-ray absorption fine structure (EXAFS) covers the photon energy range from about 40 eV to about 1000 eV past the edge. This division is indeed justified, as the research community observes that the interpretation of XANES spectra is different and substantially more complicated than that of EXAFS spectra (11).

Briefly, XANES is associated with the excitation process of a core electron to bound and quasibound states, where the bound states interacting with the continuum are located below the ionization threshold (vacuum level) and the quasibound states interacting with the continuum are located above or near the threshold. Thus, XANES contains information about the electronic state of the x-ray absorbing atom and the local surrounding structure. However, as stated above, unlike EXAFS, since the excitation process essentially involves multielectron and multiple scattering interactions, interpretation of XANES data is substantially more complicated than that of EXAFS data.

14.2.3. Practical Studies

For in situ x-ray diffraction measurements, the basic construction of an electrochemical cell is a cell-type enclosure of an airtight stainless steel body. A beryllium window, which has a good x-ray transmission profile, is fixed on an opening in the cell. The cathode material can be deposited directly on the beryllium window, itself acting as a positive-electrode contact. A glass fiber separator soaked in liquid electrolyte is then positioned in contact with the cathode followed by a metal anode (3). A number of variations and improvements have been introduced to protect the beryllium window, which is subject to corrosion when the high-voltage cathode is in direct contact with it.

We next discuss x-ray absorption studies. To put matters in context, it is useful to understand that conventional studies using Auger electron spectroscopy (AES) and x-ray photoemission spectroscopy (XPS) can be carried out only ex situ in high vacuum after electrochemical treatment since the techniques involve electron detection. X-ray absorption spectroscopy can, in contrast, be used for valence and structural environment studies. As x-rays only are involved, they can be carried out in situ in an electrochemical or similar cell.

Ex situ experiments (4) have shown, for instance, that an oxide film containing Cr(IV) can be formed on AlCr alloys by polarization to high potentials in a borate buffer solution. The chromate was reduced to the $+3$ state at a low potential and then reoxidized to the $+6$ state by polarizing again to the high potential. In situ XANES (x-ray absorption near-edge structure) experiments (4) have confirmed those results. The experiments were performed using the electrochemical cell depicted in Figure 14.1. The cell allows both electrochemical control and surface sensitivity. The electrode consisted of a thin Mylar window ($\sim 6 \times 10^{-6}$ m) on which there has been sputtered about 100 Å of tantalum (to maintain electric contact) and about 20 Å of Al–12% Cr (the purpose of the alloy is preoxidation during formation of the passive layer, to minimize signal from metallic Cr). Monochromatic x-rays are incident at $45°$ to the electrode surface with a solid-state detector positioned as

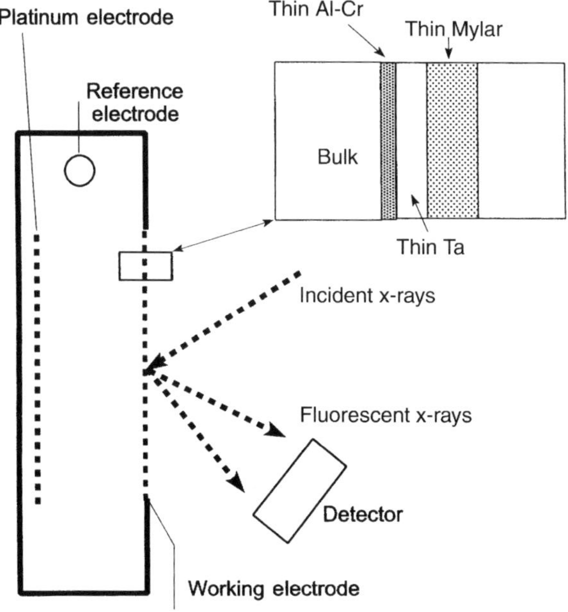

Figure 14.1. Schematic plan of experimental apparatus. (From Ref. 4, with permission from the Electrochemical Society.)

shown. It was concluded that x-ray absorption methods can be employed to make in situ valence measurements of species in passive films under electrochemical control.

An improved electrochemical cell for in situ studies is presented in Figure 14.2. In this method a platinized Pt electrode located in the anode compartment serves as the reference electrode. This cell can be installed in a test station. Such a station can have facilities for temperature and pressure control, humidification of reactant gases (e.g., hydrogen and oxygen), gas flow rate measurement, and measurement of half- and

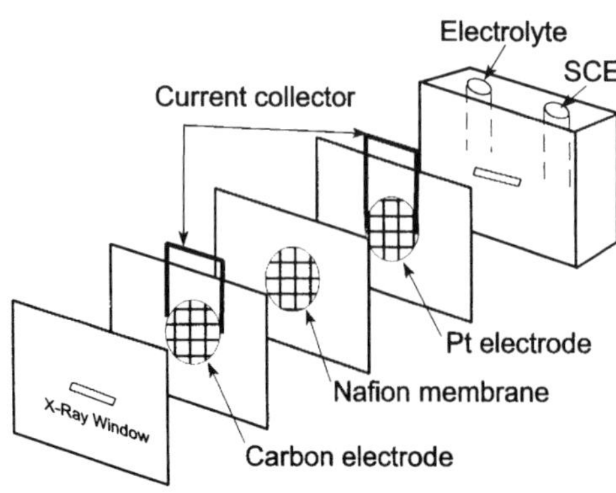

Figure 14.2. Schematic diagram of an in situ electrochemical cell used to obtain x-ray absorption spectra. (From Ref. 5, with permission from the Electrochemical Society.)

single-cell potentials as a function of current density. Finally, electrical leads from this test station could be connected to, say, power supplies.

Electrode surface images are of growing interest. Surface modifications as a result of oxidation or reduction reactions can be studied using scanning tunneling microscopy. In a new approach (6), in situ monitoring of electrode surface modification via a confocal scanning-beam laser microscopy (CSBLM) was achieved. Figure 14.3 shows the experimental setup used. Electrochemical experiments were conducted in a Plexiglas cell. A copper ring located at the top of the cell functions as the cathode, and a copper disk in the center of the bottom serves as the anode. Variation in the anode surface is monitored continuously by the CSBLM, which, in turn, can generate a two-dimensional array of pixel (picture-element) images of samples of high lateral resolution (typically, $\sim 10^{-6}$ m). The laser light, whose power is controlled and maintained at a sufficiently low level to avoid creation of photochemical side effects on the surface, is focused through the opening on top of the cell and through the electrolyte onto the anode. Images are constructed point by point by the scanning laser beam and displayed in real time on the computer terminal screen. In practice, the overall electrode may be scanned within about 1 s. Surface variations are thus captured at a sufficiently high speed to allow true image reconstruction.

Figure 14.4 gives the optical system particulars of the confocal scanning-beam laser microscope used. (The reader should keep in mind that a detailed understanding of this system requires a good knowledge of optics. Some readers may choose to skip the next section on first reading.) The laser beam passes through a spatial filter and beam expander and beamsplitter BS and is deflected by the first scanning mirror, which gives a scan in the x–y plane. Lenses L_1 and L_2 constitute a primary telescope

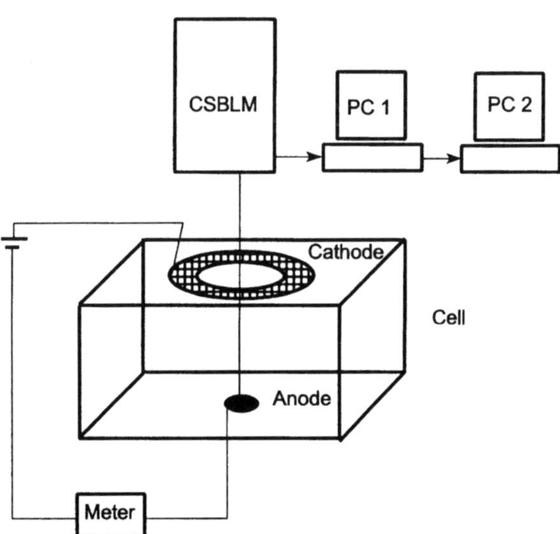

Figure 14.3. Sketch of experimental setup. The CSBLM sends a laser beam to the anode. (From Ref. 6, with permission from the Electrochemical Society.)

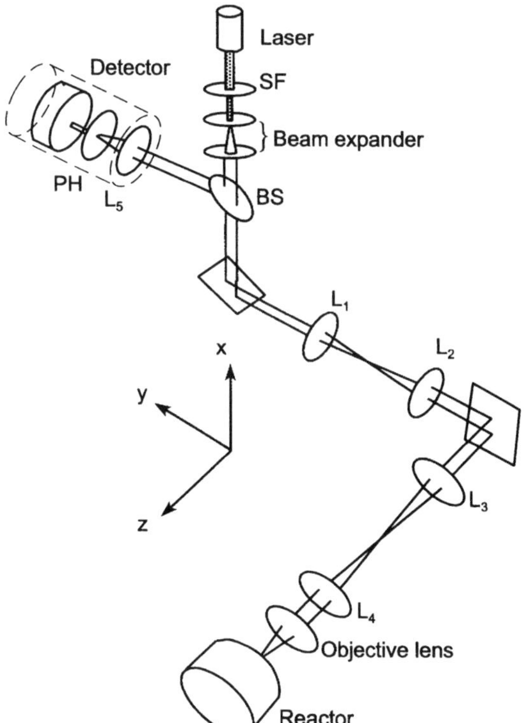

Figure 14.4. Confocal scanning-beam laser microscope. (From Ref. 6, with permission from the Electrochemical Society.)

that brings the scanning beam back to the center of the second scanning mirror, which gives a scan in the y–z plane. Lenses L_3 and L_4 constitute a second telescope that brings the scanning beam back to the axis at the entrance pupil of the microscope objective lens. The objective lens focuses the incoming beam on a spot on the electrode surface inside the reactor. Light reflected from this spot is collected by the microscope objective and passes back through the scan system, where it is descanned. The returning beam (stationary) is partially reflected by beamsplitter BS into the detection arm of the microscope, where it is focused on pinhole PH by lens L_5 and light passing through the pinhole is detected by the detector. The scan system is computer controlled to provide faster scan of the focused spot across the specimen, and the reflected light image is collected on a pixel-by-pixel basis as the scan proceeds. As pinhole PH is at the focal point of lens L_5, only light that forms a parallel beam before entering L_5 (i.e., light returning from the focal point of the objective) will pass through PH and be detected. Light coming back from specimen planes that are closer to the microscope objective than its focal plane (above the focal plane) will form an expanding beam that would focus behind PH, so that most of that light will hit the edges of PH and will not be detected; similarly for light from planes below the focal plane. Thus, its pinhole PH rejects light from above or below the plane of focus in the specimen. This action allows the confocal microscope to perform optical sectioning.

14.3. SOLID-STATE STUDIES (CHEMICAL MECHANICAL POLISHING)

Mechanical polishing of silicon wafers is a fundamentally important process in the production of flat, defect-free, highly reflective surfaces. Such samples are the starting materials in the production of integrated circuits. Techniques for polishing silicon are usually based on water solutions of colloidal silica suspension with moderate (basic) pH values. Polishing is the result of a combined effect of the reaction of the silicon with water (producing H_2 gas) and the physical removal of the reaction/polish products. Of interest are the details of the chemical process and the interaction between the silicon surface and the polishing pad. It is known that silicon is unstable and will dissolve in water. Formation of an oxide layer will prevent this from happening. Consequently, etching of silicon will require solutions that render oxidation products soluble. Indeed, if the oxide layer is removed physically, silicon is soluble in pure water (7). It is of interest to perform current–voltage measurements during (in situ) polishing to be able to learn the anodic nature of silicon while the oxide is being removed physically.

Experiments are performed (8) in the fashion shown in Figure 14.5. A polishing machine with a double-eccenter mechanism to permit random movement of the silicon wafer over the pad is set up for electrochemical measurements as shown. The working electrode is a Pyrex wafer carrier containing four contact points, on which the silicon slices are positioned with a conducting wax. The edges of these slices are sealed with lacquer to prevent contact between the wax and the polishing liquid. The rotating polishing pad carrier, which is electrically isolated from the framework of the machine, is used as a counterelectrode. A saturated calomel electrode (SCE) is the reference electrode. The polishing fluid is supplied to the polishing pad via a bowl-feed setup.

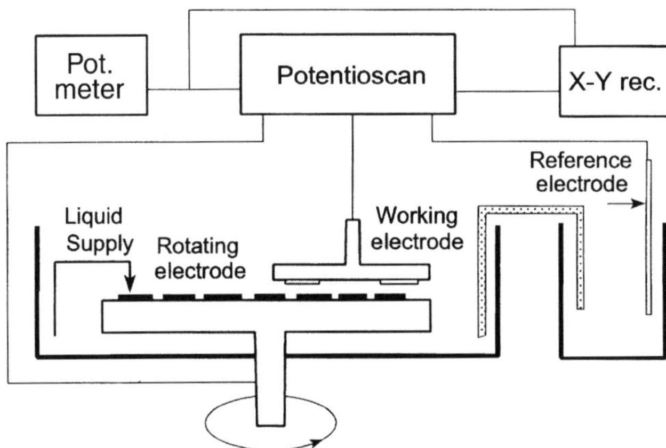

Figure 14.5. Schematic representation of a polishing system with a double-eccenter mechanism. (From Ref. 8, with permission from the Electrochemical Society.)

14.4. DISCUSSION

It should be apparent to the reader that the methods discussed in this and previous chapters are useful in both ex situ and in situ conditions. The later is more of a challenge to the experimenter, as conditions have to be maintained to avoid interference with what is being measured by the measurement itself. Although it does not deal directly with electrodeposition, the next section should be of interest, as we discuss a closely related issue: in situ experiments in surface science.

14.5. IN SITU TEST IN TRIBOLOGY

Tribology is the branch of science and engineering that deals with surfaces in relative motion. Included are issues of friction, wear, and lubrication of surfaces. Modern technology has enabled the study of these characteristics in a number of different ways. These studies have given rise to a new branch, *atomic-scale tribology,* which deals with issues and processes from atomic/molecular scale to microscale. These studies facilitate understanding of interfacial phenomena such as those observed in magnetic storage systems.

Four methods are used to size dust or other unwanted particles inside systems and/or near tribological interfaces (9). These are illustrated in Figure 14.6. These methods are of relevance to electrodeposition as well when the presence of dust particles in solution is an issue.

In Figure 14.6*a*, a sample of air is taken from a system, particles are counted, and the air sample is returned to the system. Airflow in the system is disturbed in this case, and sampling errors are possible. In Figure 14.6*b* an in situ arrangement is shown where optical counterdetector and laser are placed inside the system. Figure 14.6*c* and *d* show an alternative to the previous in situ method, in which a laser beam is passed through the system. An important problem is the nonuniformity of the light intensity across the laser beam, which would cause larger particles in the wings of the beam to give rise to the same light perturbation as smaller ones in the middle of the beam. Many commercial counterdetectors overcome this problem by concentrating the particles aerodynamically on the central area of the laser beam. Others use a second laser beam of a different color sent through the center of the first. Signals that measure scattering from the smaller beam trigger sizing by the larger beam (Fig. 14.6*c*). A single laser beam with a single detector (Fig. 14.6*d*) is the setup of choice of most workers.

14.6. HIGH-RESOLUTION ELECTRON MICROSCOPY

The past decade saw introduction of the low-energy electron microscope (LEEM) (12). It was expected potentially to be capable of moderately high resolution (6-nm) video-rate imaging of surfaces and interfaces. This would make it capable of studying dynamic processes at surfaces, thin-film growth, strain relief, etching, absorption,

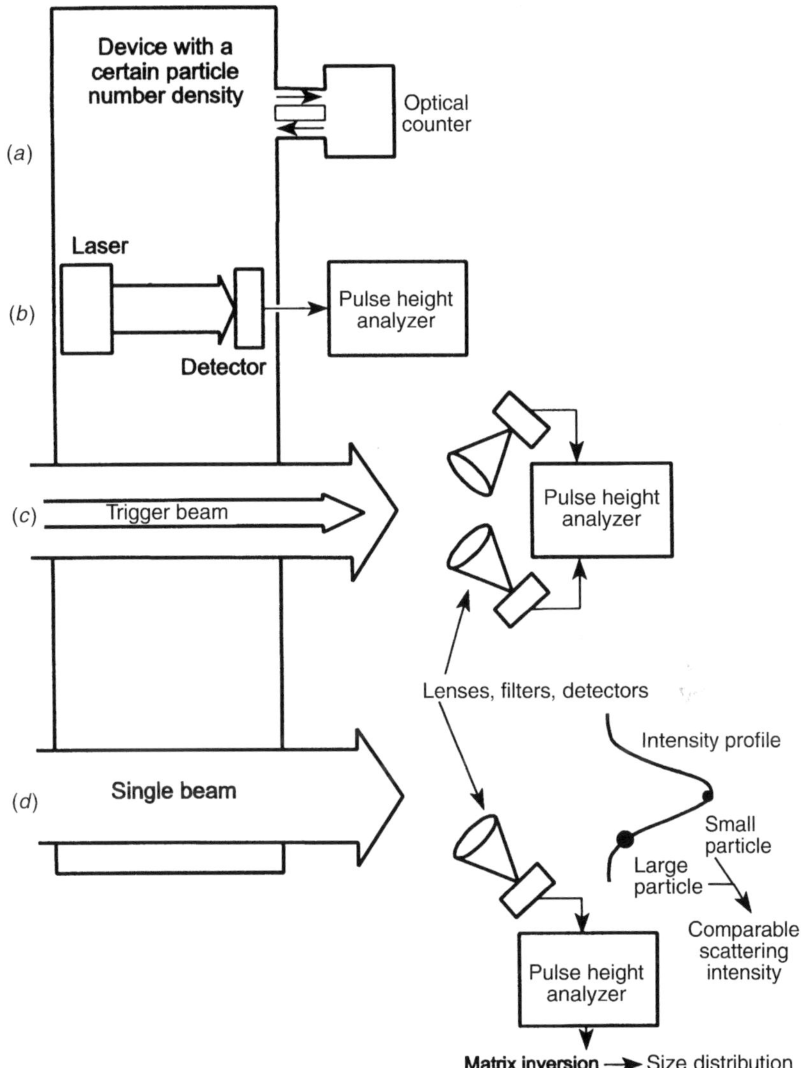

Figure 14.6. Approaches to optical in situ sizing of contaminant and wear particles with (*a*) an external counter; (*b*) an internal counter; (*c*) two concentric beams; (*d*) a single beam. (From Ref. 9, with permission from CRC Press.)

and phase transition in real time in situ as those occur. The resulting video movies, it is claimed, should contain an unprecedented amount of information that would be amenable to detailed quantitative analysis. The concept of atomic resolution with the use of an electron microscope has become reality in recent years, thanks to the fact that the high-resolution electron microscope (HREM) has evolved into a clever, sophisticated instrument capable of providing quantitative structural information on the atomic scale.

Constant improvements in design and careful attention to technical details have made possible the potential offered by the extreme short wavelength of the highly energetic electron beam used in an electron microscope compared to the wavelength of light used in an optical microscope. At the present time, instruments worldwide routinely provide images that are directly indicative or interpretable in terms of atomic structure. The HREM has an important role to play in many scientific areas. Those include such a diversity of fields as materials science, physics and chemistry of the solid state, biology, microelectronics, and many more. It is an established fact that HREM has a pivotal role in determining atomic configurations in complex materials, particularly at structural irregularities and defects, during and/or subsequent to layer and film growth. The theoretical upper limit of resolution that is provided by the ultrashort wavelength of the electron beam still eludes us. This is due to the spherical aberrations of the objective lens.

Specifically, the wavelength offered by the electron beam is determined and may be controlled by the accelerating voltage, and it may be expressed in simplified form as

$$\lambda = \left(\frac{1.5}{V}\right)^{1/2} \tag{14.1}$$

where V is the accelerating voltage of the electron beam. For example, the wavelength of an electron beam accelerated at $V = 80\,kV$ is $0.04\,\text{Å}$ or $0.004\,nm$.

It is worth remembering that, in general, the image formation in an electron microscope is the result of two processes. The first is the elastic (and inelastic) scattering of the electrons in a beam as it traverses through the specimen. This stage results in the creation of an exit surface image (ESI). In the subsequent step this wavefunction travels through the imaging lenses of the microscope. At the end of this stage the image is formed on the final viewing screen or recording medium: either a photographic negative, a luminescent screen, or a charge-coupled device (CCD) camera.

The resolution of a microscope is defined as its ability to discriminate between two close objects. Put differently, it is defined as the minimum separation of two self-luminous point sources and is given by the expression

$$D_{\text{min}} = \frac{\lambda}{2 \sin \alpha} \tag{14.2}$$

where λ is the wavelength of the light or electron beam employed in the microscope and α is the half-angle subtended at the object. This expression indicates that the resolution is dependent on the wavelength of the electron beam, and its best value may be half that of the wavelength.

In practice, however, lenses in the electron microscope have unavoidable aberrations, and a more realistic expression for the resolution is

$$\delta \approx 0.66(C_s\lambda^3)^{1/4} \tag{14.3}$$

where C_s is the spherical aberration of the objective lens. Typically, its value is between 0.5 and 4.0 mm (see Problem 14.2). This resolution represents the upper limit to which intuitive image interpretation can be made directly in terms of the projected crystal structure, with the additional assumption that the specimen must be thin (no multiple scattering of electrons). It is of interest to note that a lens system for C_s correction based on hexapole lenses has been developed and has been incorporated in commercial electron microscopes (13). Following expression (14.3), this results in improved resolution.

REFERENCES AND FURTHER READING

1. T. Konishi et al., *Jpn. J. Appl. Phys.* **31**, L1216 (1992).

2. E. Yablonovich et al., *Phys. Rev. Lett.* **57**, 249 (1986).

3. G. G. Amatucci, J. M. Tarasron, and L. C. Klein, *Electrochem. Soc.* **143**, 1114 (1996).

4. A. J. Davenport, *J. Electrochem. Soc.* **138**, 377 (1991).

5. S. Mukerjee, S. Srinivasan, M. Soriaga, and J. McBreen, *J. Electrochem. Soc.* **142**, 1409 (1995).

6. Z. H. Gu, T. Z. Fahidy, S. Damaskinos, and A. E. Dixon, *J. Electrochem. Soc.* **141**, L153 (1994).

7. I. V. Kolbanev et al., *Kinet. Catal.* **23**, 271 (1982).

8. W. L. C. M. Heyboer, G. Spierings, and J. Van den Meerakker, *J. Electrochem. Soc.* **138**, 774 (1991).

9. B. Bhusham, ed., *Handbook of Micro/Nano Tribology*, CRC Press, Boca Raton, FL, 1995.

10. R. Jenkins, R. Manne, J. Robin, and C. Senemaud, *Pure Appl. Chem.* **63**, 735 (1991).

11. See, e.g., (a) L. R. Sharpe, W. R. Heineman, and R. C. Elder, *Chem. Rev.* **90**, 705 (1990); and (b) D. C. Koningsberger and R. Prins, eds., *X-Ray Absorption: Principles, Applications, Techniques of EXAFS, SEXAFS and XANES*, Wiley, New York, 1988.

12. R. M. Tromp, *IBM J. Res. Dev.* **44**, 444 (2000).

13. See, e.g., B. Kabius, M. Haider, S. Uhleman, E. Schwan, K. Urban, and H. Rose, *J. Electron Microsc. (Suppl.)* **51**, 551–558 (2002).

PROBLEMS

14.1. X-rays are produced by directing at a target material a beam of electrons or radiation with sufficient energy. The target material most often is a metal. The table lists selected x-ray wavelengths and corresponding excitation potentials. Fill in the missing data using the relation

$$\text{energy (keV)} = \frac{12.398}{\text{wavelength (Å)}}$$

Element	Energy (keV)	Wavelength (Å)
Cu		1.54051
Mo		0.70932
Cr	5.99	
Fe	7.11	

14.2. The resolution of an electron microscope was stated in the text to be express-ible as

$$\delta \approx 0.66(C_s \lambda^3)^{1/4}$$

Calculate the values of the wavelengths and resolutions for the electron beam energies indicated. Present your results by filling in the table.

Energy (keV)	λ (Å)	C_s (mm)	Δ (Å)
100		0.7	
300		0.9	
600		1.5	
2000		4.0	

14.3. Discuss and explain the difference between the expressions relating wave-lengths to potential (voltage) values. Specifically, in Problem 14.1 you are to use a straight reciprocal relation between the two, whereas in Problem 14.2 a $(\cdot)^{-1/2}$ relation [see Eq. (14.1)] is used.

15

Mathematical Modeling in Electrochemistry

15.1. INTRODUCTION

Our purpose in this chapter is to review the nature of mathematical modeling in the context of modern electrochemistry and to describe how current and emerging trends in computer applications and system development are intended to assist practitioners. One trend is toward the merger of these two disciplines as computer-aided mathematical modeling.

It is recognized that most practitioners in the field of electrochemistry resort to the use of commercially available software packages when it comes to modeling. Most packages are rather user-friendly and do not require familiarity with their internal content and makeup for successful application. In most cases, however, such familiarity and understanding may be of invaluable benefit. Understanding the process of modeling should result in acquisition of the system that best fits the setup to be modeled. In the same way, when it comes to possible revisions and other modifications, again, using the software as "black boxes" makes it virtually impossible to obtain optimal results. It is for these reasons that readers who may want to perform modeling will find the present chapter very useful.

As indicated above, there are a large number of modeling packages on the market. Some of those are mentioned below. In the vast majority, differential equations that describe the electrochemical setup are solved using numeric methods. Two of the most common methods are the finite-difference method and the finite-elements method. These are discussed in some detail in this chapter, including example calculations in Section 15.3. We begin with a few general remarks.

The term *modeling* refers to a process of determining an appropriate description of reality that approximates its behavior to some specified degree of accuracy. Models are constructed using well-understood primitive components, or building blocks, defined by their inherent functionality and also their interaction mechanism, typically the manner by which data are communicated among them. The activity of producing models promotes a greater understanding of reality by virtue of the need to understand both the primitive components and the linkages between them. Further, to make understanding feasible, models can be adapted to provide analytic and predictive power to developers and producers alike.

A paper airplane and a plasticine car each represent models that might apply, say, in aerospace or automotive research and development. That each approximates the reality of actual aircraft and motor vehicles is intuitively obvious, yet it is immediately clear that neither suffices to explain the theory of their operation or to overcome production problems. Paper and plasticine and even fashion modeling provide types of media that appeal to visual appreciation but fail to provide deep understanding of fundamental issues. To overcome this lack, better media that allow for truer representation and insightful analyses are required.

In recent decades, computers have played an increasing role in developing models for research and applied purposes. In particular, the cost of constructing working models, or prototypes, of a research and development (R&D) product such as a plane, a car, or even a microelectronic component, has grown significantly. Thus, computer systems have been called on to produce simulations of greater accuracy, thereby reflecting reality in ways not achieved previously. The practicalities of modeling in the first decade of the twenty-first century continue to make it a rich and complex activity requiring a broad range of expertise in the use of both tools and fundamental theory.

Mathematical modeling is concerned with describing reality using mathematics (i.e., equations and relationships) and methodologies for solving these. The outcome of a mathematical solution (such as numbers or functions) does not always lend itself to straightforward comprehension or use. Computer modeling, on the other hand, uses mathematics to construct tools to aid in obtaining solutions while providing means to examine, test, and visualize the solution process, hence to simulate reality (i.e., produce a virtual reality). Current and developing generations of computer software and hardware require less fundamental expertise from the users of such systems, striving instead toward more intuitive approaches to productivity and understanding.

By way of illustration, consider determining the orbits of bodies (planets, moons, etc.) in the solar system. A stick-and-ball model might suffice to visualize planetary arrangements but fails to deal realistically with movement or the mutual (gravitational) interaction between bodies. Using Newton's theories, one can construct an elaborate system of variables and equations whose solution, in terms of various complicated (elliptic) functions and numbers, have little meaning except for the mathematically able. A suitable computer graphics program can transform the mathematical solution into pictures of planets revolving around the sun and incorporate in a straightforward way the effects of one body on the others. This last effect is of fundamental importance when considering the orbit of a human-made satellite traveling through the solar system. What is even more profound, perhaps, is that an end user of such a program can use it as a virtual laboratory to conduct research, even in the absence of mathematical expertise.

15.2. THE MATHEMATICS OF MODELING

Mathematical modeling is used extensively in electrochemistry, and as new applications arise, techniques of modeling evolve as well. A particular area of interest in electrochemistry is electrostatics. Research in electrostatics is concerned with issues relating to the properties and behaviors of static electric fields about sample element geometries, henceforth referred to as *plate (electrode) geometries.* These sample geometries contain various charged, uncharged, and neutral components, each of which may consist of a number of possible materials. Specific interest lies in the effects on the electric potential, electric fields, and currents as different geometry properties are varied. For instance, one may ask what would be the consequences of using different materials for different components, in a plating bath, or applying different electric charges to different elements in the system (bath). Also of relevance is the configuration or positioning of elements with respect to each other. These are representative of the issues that a computer modeling software system for electrostatics should allow a researcher to explore.

In many practical problems the sequence of steps to follow is: (1) geometric and physical properties specification, (2) solution method specification, (3) solution process, and (4) postprocessing and analysis. Each of these steps also involves issues of verification as well as of data storage and communication. In Figure 15.1 we have represented the previous points as a basic architecture for the design of a computer modeling system. We now describe the details of each of these steps.

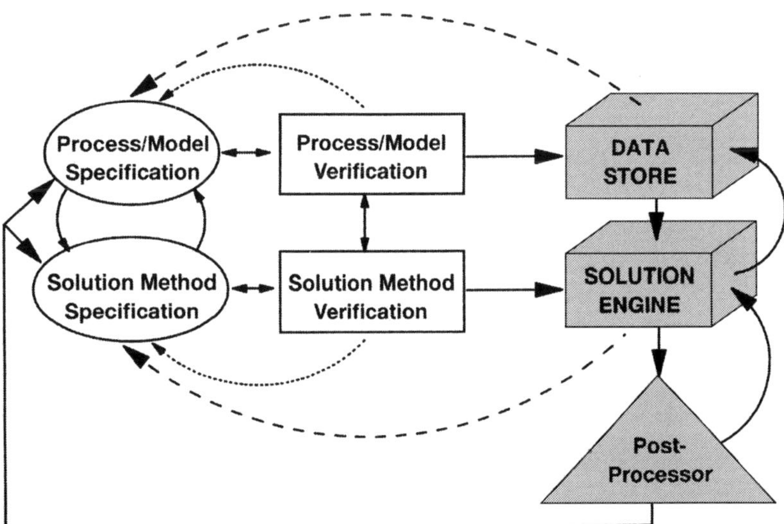

Figure 15.1. Basic system architecture in the mathematics of the modeling process required to specify, solve, and analyze a physical system. The system permits modeling to proceed linearly or iteratively.

15.2.1. Geometric and Physical Properties Specification

The first step involves *specification* of the model properties. Usually, one begins with geometric configuration: that is, the physical layout of the passive and active elements. This task is facilitated by drawing a facsimile of the system. This process is crude and may give rise to errors in specification; but, using computer drawing tools that allow for initial rapid sketching, then fine editing, an improved version of the system is readily achieved. Current computer software supports two-dimensional drawing and editing; three-dimensional applications are available for all but the extremely complicated cases.

Although the geometry can be expressed entirely mathematically in terms of formulas and relations, to do so may be awkward for those proficient in mathematics and quite incomprehensible for those who are not. It is often the case that curves and surfaces must be specified functionally, parametrically, or as piecewise sections, all of which add burden and potential error in specification.

Additionally, one must specify the physical properties and individual components of the system. Annotation of the facsimile drawing with digital or analog component properties such as voltage and resistance is accomplished in a straightforward fashion using tabular storage for quick lookup or formula storage that can be interpreted at required points or times for purposes of calculation. For instance, a set of discrete sample values may have been obtained by direct measurement in one case, while in other cases a theoretical formula could be subject to testing for self-consistency using the model.

The process of specification should always be subjected to *verification* to ensure accuracy and meaning in the data provided. Even without recourse to full-scale calculation of the solution, internal consistency of the geometry can be checked, as can closure of curves or overlap of distinct components, whereas physical properties can be matched, say, with tables of established values representing material properties, or compared against experience accrued by modellers. In Figure 15.1 each operational component is connected multiply and reversibly with other components, illustrating the practical side of modeling, where one is often required to repeat steps to correct, clarify, or modify actions taken previously.

15.2.2. Solution Method Specification

For the second step one establishes a *solution method.* The system under consideration may be static, dynamic, or both. Static cases require solving a *boundary value problem,* whereas dynamic cases involve an *initial value problem.* For the illustrative problem, we discuss the solution of a static Laplace (no sources) or Poisson (sources) equation such as

$$-\nabla \cdot \mathbf{E} = \nabla^2 U(x, y) = \left(\frac{\partial^2}{\partial x^2} + \frac{\partial^2}{\partial y^2} \right) U(x, y) = 4\pi\rho(x, y) \qquad (15.1)$$

for scalar potential $U(x, y)$, electric field vector $\mathbf{E} = \nabla U(x, y)$, and source function $\rho(x, y)$. Our ultimate goal is to obtain solutions of U, hence $\mathbf{E}$, given a specification of the source, presumably in the preceding step. As for the source term, $\rho(x, y)$ may also depend on the potential U and even derivatives of U. If these appear to first degree, that is, linear equations in U and derivatives, solutions are generally easier to obtain than with nonlinear forms of Eq. (15.1). For our purposes we will be concerned only with linear cases.

In exceptional circumstances *analytic solutions* exist that are expressible in terms of standard functions. Although numerical methods must be used to obtain solutions in general, it is useful nonetheless to briefly review analytic methods.

Consider Eq. (15.1) with no sources (Laplace) applied to a square plate with U defined everywhere on the boundary. If the problem specification is symmetric under interchange of the x and y directions, the Laplace equation may then be separable with solutions of the general form $U(x, y) = X(x)Y(y)$: namely,

$$U(x, y) = A_0 xy + B_0 x + C_0 y + D_0 + \sum_{n=1}^{\infty} (A_n \cos \lambda_n x \cosh \lambda_n y$$
$$+ B_0 \cos \lambda_n x \sinh \lambda_n y + C_n \sinh \lambda_n x \cosh \lambda_n y + D_n \sin \lambda_n x \sinh \lambda_n y)$$

$$(15.2)$$

with the various constants $(A_n, B_n, C_n, D_n, \lambda_n)$ determined by the requirement to fit the boundary conditions. At the corner $x = y = 0$, for example, it follows that $U(0, 0) = D_0 + \sum_{n=1}^{\infty} A_n$. "Fitting" the solution to the boundary and determining the (infinite) number of constants is possible only in cases where specific functional or algebraic relationships exist. For example, if the sides of the plate have potentials that vary linearly from $x = 0$ to $x = 1$ at $y = 0$ and 1 (and similarly along y), then

$$U(0 \le x \le 1, y = 0) = B_0 x + D_0 + \sum_{n=1}^{\infty} (A_n \cos \lambda_n x + C_n \sin \lambda_n x) \qquad (15.3a)$$

$$U(0 \le x \le 1, y = 1) = (A_0 + B_0)x + C_0 + D_0 + \sum_{n=1}^{\infty} [(A_n \cos \lambda_n x$$
$$+ C_n \sin \lambda_n x) \cosh \lambda_n + (B_n \cos \lambda_n + D_n \sin \lambda_n x) \sinh \lambda_n] \qquad (15.3b)$$

from which one might deduce that the complete set of A,B,C,D constants $(n > 0)$ are zero; hence, $A_0 = U(1, 1) - U(1, 0) - U(0, 1) + U(0, 0)$, $B_0 = U(1, 0) - U(0, 0)$, $C_0 = U(0, 1) - U(0, 0)$, and $D_0 = U(0, 0)$.

Often, in actual applications, a finite, discrete set of values on the boundary are available through direct measurement. Inevitably, therefore, solving for the general solution proves almost impossible. The use of discretely sampled data, however, suggests the need for specialized techniques in order to develop solutions. At the outset, then, we abandon analytic solutions as being interesting as guides, but rarely useful.

All numerical techniques require application of *sampling theory*. Briefly stated, one chooses a representative sample of points within the region of interest and at each point attempts to calculate iteratively the most accurate solution possible, guided by self-consistency of local solutions with each other and with the boundary conditions specified. We describe two seemingly contrasting techniques, finite-difference and finite-element methods (1, 2).

Finite-difference methods are based on the specific relationship between the potential at a given sample point and the potentials at nearby, or local, points; the relationship is derived using Taylor expansion, assuming that the actual potential is continuously differentiable (to at least second degree).

One category of finite-difference method uses a rectangular grid. In this approach one covers the specified layout with a grid, or mesh, as shown in Figure 15.2*a*. When curvilinear boundaries are involved, it is possible to sample the boundaries with only limited accuracy and then only by using unequal steps in the *x* and *y* directions. Using the five-point probe shown in Figure 15.2*b*, at each point one approximates Laplace's equation referring to the four neighboring points above, below, and to either side of the central point. For uniform mesh sizes *h* and *k* along the *x* and *y* axes, for instance, the second partial derivatives are approximated to second order in the derivatives, about center point potential $U_{i,j} = U(x_i, y_j)$,

$$\frac{\partial^2 U_{i,j}}{\partial x^2} = \frac{1}{h^2}(U_{i+1,j} + U_{i-1,j} - 2U_{i,j}), \quad \frac{\partial^2 U_{i,j}}{\partial y^2} = \frac{1}{k^2}(U_{i,j+1} + U_{i,j-1} - 2U_{i,j})$$

$$(15.4)$$

For nonuniform meshes or higher-order derivative approximations, the various U coefficients are more complicated algebraically, but the description that follows is essentially the same for all cases.

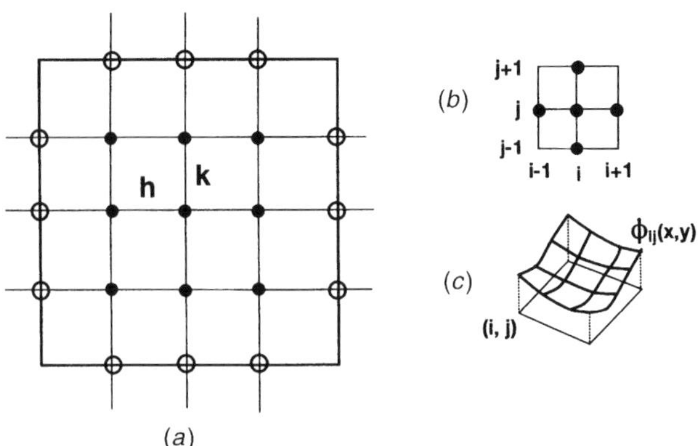

Figure 15.2. Region of interest for computing potential based on Laplace or Poisson equations, where (*a*) a complete rectangular grid is established to cover the region, which may be adapted to finite-difference techniques using (*b*) a five-point method, or (*c*) a finite-element approach based on sampling functions.

Collecting the equations for all sample points (x_i, y_j) into vectors and matrices, one recasts the problem in the form $\mathbf{AU} = \mathbf{b}$, where unknown potentials $\mathbf{U} = (U_{1,1},\ldots, U_{M,1},\ldots,U_{M,N})$ are organized by the N rows and M columns of the rectangular grid, $\mathbf{b}$ is a vector that represents the known values of U on the boundary of the region in question (as well as source information at each point in the region), and $\mathbf{A}$ is an $NM \times NM$ matrix of U-coefficients derived from the approximating equations above. The structure of $\mathbf{A}$ is determined by the U-coefficients; that is, $\mathbf{A}$ may be nonzero in most element positions or quite sparse, as shown in the following matrix for a 4×4 mesh:

$$
\mathbf{A} = \begin{bmatrix} \mathbf{D} & \mathbf{B} & \mathbf{0} & \mathbf{0} \\ \mathbf{B} & \mathbf{D} & \mathbf{B} & \mathbf{0} \\ \mathbf{0} & \mathbf{B} & \mathbf{D} & \mathbf{B} \\ \mathbf{0} & \mathbf{0} & \mathbf{B} & \mathbf{D} \end{bmatrix}, \quad
\mathbf{D} = \begin{bmatrix} 1 & \alpha & 0 & 0 \\ \alpha & 1 & \alpha & 0 \\ 0 & \alpha & 1 & \alpha \\ 0 & 0 & \alpha & 1 \end{bmatrix}, \quad
\mathbf{B} = \begin{bmatrix} \beta & 0 & 0 & 0 \\ 0 & \beta & 0 & 0 \\ 0 & 0 & \beta & 0 \\ 0 & 0 & 0 & \beta \end{bmatrix} \qquad (15.5)
$$

Matrix $\mathbf{D}$ has 1's along the diagonal, reflecting the use of a normalized discrete Laplace equation with $\alpha = \{-k^2/[2(h^2 + k^2)]\}$, and $\mathbf{B}$ is a multiple of the identity matrix with $\beta = \{-h^2/[2(h^2 + k^2)]\}$. Matrix $\mathbf{A}$ displays a sparse block structure whose off-diagonal coefficients must be less than 1 to converge to a solution.

For the approximation to be valid, the mesh must be sufficiently small: hence, the number of sample points must be large. Programming such a mathematical system is straightforward in principle but extremely difficult to compute in practice. Obtaining a solution of the system of simultaneous linear equations is time consuming and in many cases exhibits pathological behavior where the "solution" generated is patently unrealistic. In many cases of interest, the form of matrix $\mathbf{A}$ can be simplified, drastically, however. The five-point grid, for instance, results in the matrix structure shown in Eq. (15.5), which is triblock diagonal; that is, along the diagonal are $M \times M$ blocks that are tridiagonal, and additional diagonal blocks arise on the sub-and super-diagonals.

At this point, specification of the finite-difference solution method is complete in that we have chosen to utilize the finite-difference scheme and have specified the mesh properties and the sampling of points required to provide the desired approximation to the derivatives of U. Such systems can be solved efficiently even for N and M large (>1000), although time scales of typical calculations range from several minutes to hours, depending on the type of computers used, eliminating some from consideration in those time-critical situations. We defer additional discussion of achieving solutions until the next step.

Looking ahead to the issue of solutions, it is important to realize that what is being sought by the solution method is a discrete set of points, $\{x_i, y_j, U(x_i, y_j)\}$, which specify the values of the potentials at the grid locations. To obtain values of the potentials at other points lying between the sampling locations, other techniques can be employed. Straightforward linear interpolation is one such method that is simple to implement and efficient to compute, but it suffers from a lack of sufficient accuracy required in many modeling circumstances. Other forms of interpolation involve use of higher-order polynomials that increase the accuracy of approximation with increased difficulty of implementation and cost of computation. The use of polynomials to evaluate potentials at arbitrary points leads in a natural way to our next method, however.

An alternative approach to finite differencing involves *finite-element methods*. In the former approach one seeks the underlying behavior of U by solving for it numerically at all sample points in the grid. With finite elements one expresses the solution in terms of other functions (whose behavior is well known) appropriately combined to obtain U to desired accuracy throughout the region of interest. The choice of element functions reflects sampling that covers the entire region, notwithstanding curvilinear boundaries, and that are "well behaved" within each element. In many instances the sampling is more flexible and therefore more accurate than with finite difference.

Such methods start by assuming that

$$U(x, y) = \sum_{k=1}^{L} u_k \phi_k(x, y) \tag{15.6}$$

where the u terms represent a set of unknown "blending" parameters that produce a mixture (linear combination) of L known sampling, or *basis,* functions $\phi_k(x, y)$. The number of sampling functions is chosen on the basis of experience with similar problems, the existence of symmetries, adaptive analysis, or combinations of these.

One choice of basis function, based on a quadrilateral patch, is illustrated in Figure 15.2c. In the figure the element in the ith row and jth column of the mesh is assumed to have a magnitude that varies within the patch; the derivative properties may be important as well. The choice of $\phi_k(x, y)$ is not arbitrary; it is made to reflect certain mathematical qualities derived, perhaps, from prior knowledge of the general behavior of similar systems as well as properties that simplify the solution process to follow. One immediately practical constraint is that the $\phi_k(x, y)$ must satisfy the boundary conditions. Another property is that the patches meet smoothly at the intersections; this is usually obtained by continuity of $\phi_k(x, y)$ to first and second order in the derivatives. It is also convenient in many applications to choose combinations of products of functions separately dependent on x and y, reminiscent of the analytic solution, Eq. (15.2).

As with finite differences, the finite-element approach can be recast, using vectors and matrices, in the form $\mathbf{Au} = \mathbf{b}$, with $\mathbf{A}$ and $\mathbf{b}$ known and $\mathbf{u}$ to be determined. There are two basic approaches. In the first case, referred to as *collocation,* substitution of Eq. (15.6) in (15.1) leads to

$$\nabla^2 U(x_p, y_p) = \sum_{k=1}^{L} \nabla^2 \phi_k(x_p, y_p) \mathbf{u}_k = \mathbf{A}_p \cdot \mathbf{u} = \mathbf{b}(x_p, y_p) \tag{15.7}$$

where (x_p, y_p) refers to the pth sample point [for rectangular grids with constant spacing, M rows and N columns, the quadrilateral element in row r and column c, it follows that $p = N(r - 1) + c$ and $1 \leqslant p \leqslant L = NM$] and $\mathbf{A}_p$ refers to the pth row of the matrix $\mathbf{A}$ whose elements are $\mathbf{A}_{p,k} = \nabla^2\phi_k(x_p, y_p)$. Thus, L equations are generated as required to determine uniquely the L unknown u_k coefficients.

In the second approach, called the *Galerkin method*, one uses the property that the sampling functions satisfy the boundary conditions to write

$$\int \phi_m(x, y)\nabla^2 U(x, y)\, dx\, dy = \sum_{k=1}^{L} \int \phi_m(x, y)\nabla^2\phi_k(x, y)\, dx\, dy\, u_k$$
$$= \sum_{k=1}^{L} \int \nabla\phi_m(x, y)\cdot\nabla\phi_k(x, y)\, dx\, dy\, u_k = \mathbf{A}\cdot\mathbf{u} \quad (15.8)$$
$$= \int b(x, y)\phi_m(x, y)\, dx\, dy = \mathbf{b}$$

The second step in this equation involves a property called *Green's identity*. Using either method brings one to the point where the solutions of both require the same basic approaches: solving a matrix problem. As in the case of collocation, the L sample points are used to generate the rows of the $\mathbf{A}$ matrix and $\mathbf{b}$ vector whose elements are written $\mathbf{A}_{m,k} = \int\nabla\phi_m(x, y)\cdot\nabla\phi_k(x, y)\, dx\, dy$ and $\mathbf{b}_m = \int b(x, y)\phi_m(x, y)\, dx\, dy$, respectively.

Solving for $\mathbf{u}$ is greatly facilitated by choosing basis functions whose properties simplify the structure of $\mathbf{A}$, that is, functions that lead to easily computed methods of matrix inversion. In one such property, called *locality*, each function has a primary value, hence influence, within a given element region and a much smaller, or zero, value, in other regions. For example, if one chooses $\phi_k(x, y) = a_k xy + b_k x + c_k y + d_k$, which is piecewise linear in both x and y within each patch k, this leads again to a tri-block diagonal matrix structure, as in Eq. (15.5), and the method is in fact fully equivalent to the five-point finite-difference technique discussed. However, this method cannot be used for the collocation technique, since second derivatives vanish identically and lack continuity at the patch boundaries, and hence will appear bumpy instead of smooth. In many cases of interest, the method involves using *piecewise polynomials* of sufficient order (of at least cubic degree in both x and y) so as to ensure the desired patch edge continuity while allowing for matrix structure that simplifies computation.

In cases where dynamic effects must be considered, the problem is typified by $\mathbf{A}$ and $\mathbf{b}$ matrices that are parametrized, say, by time or other quantities. One such example involves modeling corrosion effects where time, acidity, and material thickness and age might be relevant dynamical parameters. In most cases one calculates a sequence of static solutions at time steps t_m that are then pieced together to fit initial (and final) conditions.

At this point what must be made clear is that although the specific interpretation of results differs among methods, the underlying set of analytic tools that one brings to bear on the problem, once it has been transformed into the language of vectors and matrices, is, of course, the same.

15.2.3. Solution Process

Many methods exist for *solving* the basic form $\mathbf{AU} = \mathbf{b}$ for the potential $\mathbf{U} = \mathbf{A}^{-1}\mathbf{b}$. The methods depend on various features exhibited by the matrices themselves, immediate by-products of how the problem was set up in the previous stage of specification. A general method, assuming that $\mathbf{A}$ is nonsingular (determinant is nonzero), is to find the inverse matrix $\mathbf{A}^{-1}$ using techniques such as Gauss elimination. However, in practice this approach is not computationally workable. Typically, one looks for features of the problem that simplify $\mathbf{A}$.

For example, uniform meshes give rise to highly structured and simplified forms of matrices, as in Eq. (15.5), which are amenable to rapid solution techniques but are very sensitive to the size of the mesh: The larger the mesh, the poorer the solution. More complicated meshes and formulations of the approximation scheme used to set up the solution scheme are used more rarely because of difficulties in programming them and their increased cost in time to achieve solution. Similarly, finite-element schemes have varying degrees of success, depending on choice of mesh, sample functions, and so on.

Relaxation methods involve iteratively seeking a convergent solution to the Laplace equation. In the present case, for instance, if we rewrite the coefficient matrix $\mathbf{A} = \mathbf{I} + \mathbf{E}$, where the latter matrix consists of elements that are all "small" compared to 1, the matrix Laplace equation takes the form $\mathbf{U}^{(r+1)} = \mathbf{E}\mathbf{U}^{(r)} + \mathbf{b}$. One begins the calculation with values $\mathbf{U}^{(1)} = \mathbf{b}$ [or, equivalently, $\mathbf{U}^{(0)} = \mathbf{0}$] and iteratively computes successive values $\mathbf{U}^{(r)}$. The calculation terminates when a specified limit of accuracy is achieved. One such measure involves calculating the proportional differences:

$$
D_{ij}^{(r)} = 2 \frac{\left| U_{i,j}^{(r)} - U_{i,j}^{(r-1)} \right|}{U_{i,j}^{(r)} + U_{i,j}^{(r-1)}}
\tag{15.9}
$$

stopping when the average is less than the tolerance. The advantages of such methods are speed of computation and circumvention of much of the need for coefficient matrix storage during the course of the computation, although two complete sets of U vectors must be maintained, the "old" one and the updated solution.

Two approaches that rely on the availability of suitable hardware involve pipelining or parallelizing the equations and solution method. In the first technique the calculation is broken into independent stages, each of which performs a complete set of operations on an input data set, producing a processed data set as output. Specialized hardware accepts values and processes them to each subsequent stage of calculation. As each stage is finished, another data set is input immediately, thus ensuring that the pipeline is always full and busy. In parallel computations the strategy is expanded even further to encompass a family of separate processors that accept independent sets of data and simultaneously apply the same or different processes.

15.2.4. Postprocessing and Analysis

The last stage of mathematical modeling consists of *interpretation of the results.* Here lies its greatest weakness, at least prior to the introduction of computer systems. To visualize the behavior of many thousands of data points, it is essential to resort to machine-assisted methods (3). In principle, however, the complete database constituting the numerical solution contains all information required to deduce other quantitative results.

Figure 15.3 outlines some of the many possible choices of postprocessing options of interest to modeling. Questions regarding, say, optimal cost or operating temperature range are expected to produce simple numerical outputs or, perhaps, simple graphical images such as pie charts or histograms. On the other hand, analysis may require intermediate stages at which the general solution properties must be studied to provide clues as to how to proceed to later stages of development.

Another significant issue in analysis of results lies in altering the parameters of the initial specification. This might be achieved in some small way, such as slight geometric deformation or subtle variation (perturbation) of one or more physical properties (such as voltage or conductivity), as shown in Figure 15.4, which illustrates a slight rotation of the triangular element to produce a *local parametric solution*, or it might be specified as part of a *global parametric solution* to the original problem. In the first case, by restriction of the primary region, within which the change is expected to manifest itself, it is possible to compute minor modifications to the initial solution in rapid fashion, indicated by the darker, solid contours in Figure 15.3. Global solutions require substantially more time, but once calculated can be used to

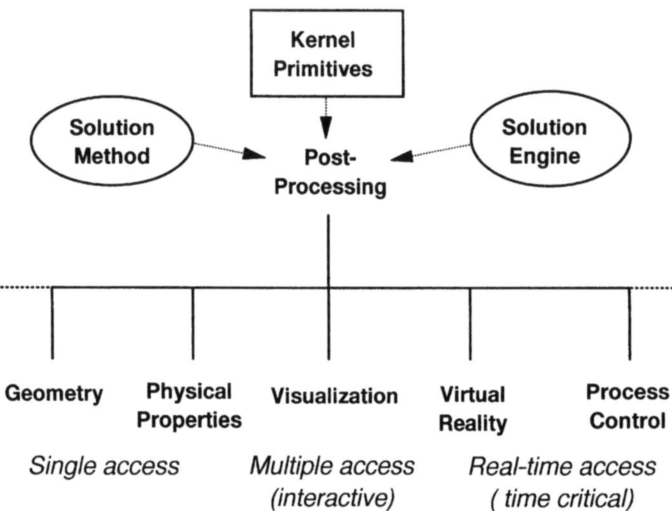

Figure 15.3. Once a solution has been obtained, the choice of postprocessing options may be quite varied, with each option requiring different means of accessing the solution data, ranging from one-time access through interactive manipulation of the solution to real-time critical applications, in latter cases requiring recomputation or updating of the solution data.

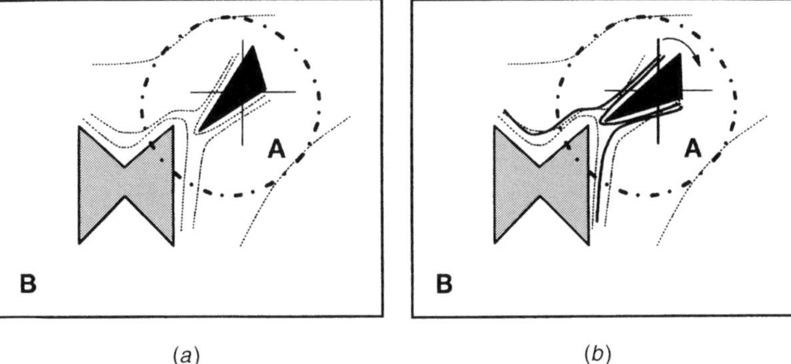

(a)   (b)

Figure 15.4. Deformation of a solution state to obtain rapid update may be obtained quickly by restricting computational effort to the region of greatest change.

perform detailed multivariable (dimensional) analysis, particularly cost/benefit analyses and the like.

15.3. EXAMPLE CALCULATIONS

In this section we present worked examples demonstrating application of the theory described earlier. In the first example we demonstrate a familiar simple electrostatic problem in which a metal sphere is held at a potential of 1 V and is enclosed in a cubical grounded metal box (Fig. 15.5). The task at hand is to find the potential distribution in volume V between the sphere and the box.

As is well known from elementary electrostatics, the equation governing the electrostatic potential, Φ, in the volume V is the Laplace equation:

$$\nabla^2\Phi = 0 \tag{15.10}$$

Another (integral) way of stating this is that for all weighting functions w, we may write

$$\int_v w\,\nabla^2\Phi\,dV = 0 \tag{15.11}$$

This form is referred to as the *weighted-residual formulation* of the problem. What is required to be determined is the potential Φ, which solves this equation while it assumes the value 1 on the surface of the sphere and the value 0 on the inner surface of the box. (One may constrain w to zero on both surfaces, since this has no effect on the equivalence of the two expressions.) Equation (15.11) may be used as the launching

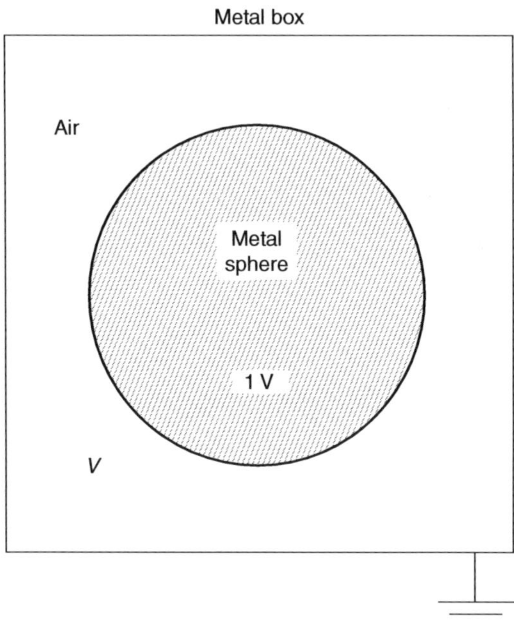

Figure 15.5. Electrostatic problem. (Adapted from Ref. 10, with permission of the Institute of Physics.)

pad of a finite-element analysis. However, as it stands, it contains second-order derivatives, and it is convenient to reduce these to first-order derivatives. This is accomplished by the use of the vector identity known as Green's identity and is akin to integration by parts. The additional term thereby produced gives a surface integral when the divergence theorem is applied to it, and the surface integral vanishes because w is zero on the surfaces. The weighted-residual formulation thus becomes

$$\int_v \nabla w \cdot \nabla \Phi \, dV = 0 \tag{15.12}$$

To solve this expression numerically, which is, after all, what the finite-element method is all about, the volume V is subdivided into a number of finite elements. For example, tetrahedral-shaped elements might be used, as shown in Fig. 15.6.

Of course, having flat faces, these elements cannot represent the curved surface on the sphere exactly. Using an adequately large number of such elements, a sufficiently accurate polyhedral approximation is possible. Inside each tetrahedron the unknown potential is approximated using a polynomial that is first order in the space coordinates (x, y, z). The polynomial can be so written that the four unknown coefficients are the values of the potential at the four nodes (vertices) of the tetrahedron, Φ_1, Φ_2, Φ_3, and Φ_4:

$$\Phi = \Phi_1 \alpha_1 + \Phi_2 \alpha_2 + \Phi_3 \alpha_3 + \Phi_4 \alpha_4 \tag{15.13}$$

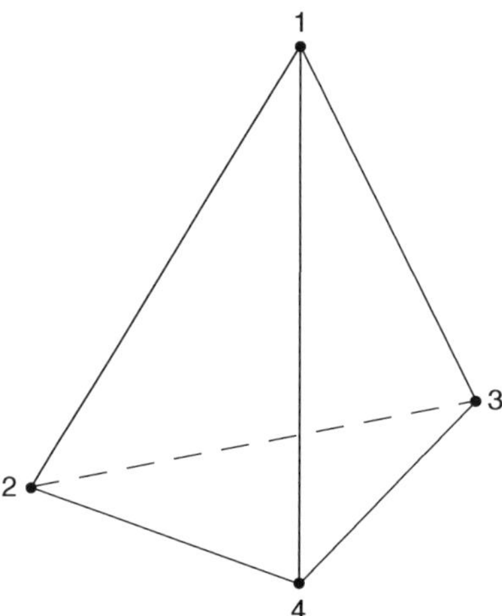

Figure 15.6. Tetrahedral element. (Adapted from Ref. 10, with permission of the Institute of Physics.)

The quantities α_i are polynomials whose dependence on x, y, z and on the size and shape of the tetrahedron is known explicitly. The polynomial α_i assumes the value 1 at node i, vanishes at the other three nodes, and varies linearly in between. The reader should note that equation (15.13) thus, provides a complete and continuous representation for the potential within the element while using physically meaningful quantities (the values at the four nodes) as unknown parameters. It also permits potential continuity to be imposed in a simple way between elements. For two elements with a common edge or face to agree in their values of Φ at their common nodes requires only that Φ be continuous everywhere over the edge or face. In actual practice this is achieved by assigning a *global numbering*, $1,\ldots, N$, to all the nodes of the finite-element mesh, such that each node has a single global number even if a given node is shared by several tetrahedra. The unknowns in the problem are then the globally numbered values of potential at the nodes Φ_i: $i = 1,\ldots, N$. Indeed, one may think of the polynomials α_i in the same global way: α_i is a function that is 1 at global node i, vanishes at the other $N - 1$ nodes, and is linear in each tetrahedron. The piecewise linear potential in V is then

$$\Phi(x, y, z) = \sum_{i=1}^{N} \Phi_i \alpha_i(x, y, z) \tag{15.14}$$

The α_i are known as *basis* or *trial functions*.

In point of fact, not all N of the Φ_i are unknown. For a node i that lies on a boundary where the potential is constrained to be 0 or 1, Φ is set to that value from the outset. For simplicity it will be assumed that the first N_f nodes are interior free nodes and the remainder are boundary nodes. The problem therefore has only N_f unknowns. These then are the *degrees of freedom* of the potential function.

Since there are a finite number of degrees of freedom with which to represent the solution, it is not generally possible to satisfy Equation (15.12) exactly for all possible weighting functions. It is desirable to have no more than N_f independent weighting functions, so that (15.12) becomes N_f equations in N_f unknowns. Although it is possible to choose any set of weighting functions, it is convenient to use the basis functions themselves (Galerkin's method). Substituting (15.14) into (15.12) with this set of weighting functions yields the following matrix equation:

$$[\mathbf{S}]\{\mathbf{\Phi}\} = \{\mathbf{b}\} \tag{15.15}$$

where $\{\mathbf{\Phi}\}$ is an N_f vector of the unknown potentials, $\{\mathbf{b}\}$ is a known N_f vector that depends on the boundary constraint, and $[\mathbf{S}]$ is a square-symmetric matrix known as the *global* or *stiffness matrix*. The term/entry i,j in $[\mathbf{S}]$ is given by the expression

$$S_{i,j} = \int_v \nabla\alpha_i \cdot \nabla\alpha_i \, dV \tag{15.16}$$

In general, there will be rather a large number of nodes, and consequently, $[\mathbf{S}]$ will be large. Even a modest problem may have $>10^4$ nodes. On the other hand, $[\mathbf{S}]$ is very sparse. The function α_i is nonzero only in those tetrahedra that share node i. If i and j are not nodes of the same tetrahedron, it is evident from (15.16) that $S_{i,j}$ will be zero. Typically, a row in $[\mathbf{S}]$ may have only 30 to 50 nonzero entries out of a total of many thousands. It is very important to exploit this sparsity both in constructing $[\mathbf{S}]$ and in solving (15.15). For example, the assembly of $[\mathbf{S}]$ is done tetrahedron by tetrahedron, not row by row. For each tetrahedron, a 4×4 *local matrix* is calculated from its four basis functions. The 16 entries of this dense local matrix are added to the global matrix in the right places, taking into account the global numbers of the four degrees of freedom, Once the solution $\{\mathbf{\Phi}\}$ has been found, the potential at any point can be determined from (15.14). If an electric field is required, it may be calculated taking the gradient of (15.14).

As a second illustration of the finite-element technique, we proceed as follows. We assume a rectangular mesh of three rows and three columns with uniform step sizes $h = k = \frac{1}{5}$ along the x and y axes, respectively. Further, assume that potentials on the boundary have values $\{U_{01}, U_{02}, U_{03}, U_{10}, U_{14}, U_{20}, U_{24}, U_{30}, U_{34}, U_{41}, U_{42}, U_{43}\}$, where by $U_{i,j}$ we refer to the top and bottom or left and right rows or columns as $i,j = 0$ and $i,j = 4$, respectively. The five-point sampling Laplace equation has the algebraic form

$$U_{i,j} = \tfrac{1}{4}(U_{i-1,j-1} + U_{i+1,j-1} + U_{i-1,j+1} + U_{i+1,j+1}) \tag{15.17}$$

which is just the average of the four neighboring potentials, and the matrix forms

$$
\begin{bmatrix}
1 & a & 0 & b & 0 & 0 & 0 & 0 & 0 \\
a & 1 & a & 0 & b & 0 & 0 & 0 & 0 \\
0 & a & 1 & 0 & 0 & b & 0 & 0 & 0 \\
b & 0 & 0 & 1 & a & 0 & b & 0 & 0 \\
0 & b & 0 & a & 1 & a & 0 & b & 0 \\
0 & 0 & b & 0 & a & 1 & 0 & 0 & b \\
0 & 0 & 0 & b & 0 & 0 & 1 & a & 0 \\
0 & 0 & 0 & 0 & b & 0 & a & 1 & a \\
0 & 0 & 0 & 0 & 0 & b & 0 & a & 1
\end{bmatrix}
\begin{bmatrix}
U_{1,1} \\ U_{2,1} \\ U_{3,1} \\ U_{1,2} \\ U_{2,2} \\ U_{3,2} \\ U_{1,3} \\ U_{2,3} \\ U_{3,3}
\end{bmatrix}
=
\begin{bmatrix}
-bU_{1,0} - aU_{0,1} \\
-bU_{2,0} \\
-bU_{3,0} - aU_{4,1} \\
-aU_{0,2} \\
0 \\
-aU_{4,2} \\
-aU_{0,3} - bU_{1,4} \\
-bU_{2,4} \\
-aU_{4,3} - bU_{3,4}
\end{bmatrix}
$$

$$
=
\begin{bmatrix}
-b & 0 & 0 & -a & 0 & 0 & 0 & 0 & 0 & 0 & 0 & 0 \\
0 & -b & 0 & 0 & 0 & 0 & 0 & 0 & 0 & 0 & 0 & 0 \\
0 & 0 & -b & 0 & -a & 0 & 0 & 0 & 0 & 0 & 0 & 0 \\
0 & 0 & 0 & 0 & 0 & -a & 0 & 0 & 0 & 0 & 0 & 0 \\
0 & 0 & 0 & 0 & 0 & 0 & 0 & 0 & 0 & 0 & 0 & 0 \\
0 & 0 & 0 & 0 & 0 & 0 & -a & 0 & 0 & 0 & 0 & 0 \\
0 & 0 & 0 & 0 & 0 & 0 & -a & 0 & -b & 0 & 0 & 0 \\
0 & 0 & 0 & 0 & 0 & 0 & 0 & 0 & 0 & 0 & -b & 0 \\
0 & 0 & 0 & 0 & 0 & 0 & 0 & 0 & -a & 0 & 0 & -b
\end{bmatrix}
\begin{bmatrix}
U_{1,0} \\ U_{2,0} \\ U_{3,0} \\ U_{0,1} \\ U_{4,1} \\ U_{0,2} \\ U_{4,2} \\ U_{0,3} \\ U_{4,3} \\ U_{1,4} \\ U_{2,4} \\ U_{3,4}
\end{bmatrix}
\quad (15.18)
$$

with $a = b = -\frac{1}{4}$. Equation (15.18) has been cast in the form $\mathbf{A}\mathbf{u} = \mathbf{b} = \mathbf{B}\mathbf{U}$, with $\mathbf{u}$ referring to unknown quantities $U_{i,j}$ at interior points, $\mathbf{b}$ the vector between equal signs that refers to the 12 boundary values of the potentials, also referred to as vector $\mathbf{U}$, and $\mathbf{B}$ the 9×12 coefficient matrix. We state this in this way to emphasize that the unknown potentials are determined using the coefficients and the known potentials and that the matrix structures reflect the differing numbers of knowns and unknowns. Applying the inverse matrix $\mathbf{A}^{-1}$ to both sides, the explicit solution to this is

$$
\begin{bmatrix} U_{1,1} \\ U_{2,1} \\ U_{3,1} \\ U_{1,2} \\ U_{2,2} \\ U_{3,2} \\ U_{1,3} \\ U_{2,3} \\ U_{3,3} \end{bmatrix} = \frac{1}{224}
\begin{bmatrix}
67 & 22 & 7 & 67 & 7 & 22 & 6 & 7 & 3 & 7 & 6 & 3 \\
22 & 74 & 22 & 22 & 22 & 14 & 14 & 6 & 6 & 6 & 10 & 6 \\
7 & 22 & 67 & 7 & 67 & 6 & 22 & 3 & 7 & 3 & 6 & 7 \\
22 & 14 & 6 & 22 & 6 & 74 & 10 & 22 & 6 & 22 & 14 & 6 \\
14 & 28 & 14 & 14 & 14 & 28 & 28 & 14 & 14 & 14 & 28 & 14 \\
6 & 14 & 22 & 6 & 22 & 10 & 74 & 6 & 22 & 6 & 14 & 22 \\
7 & 6 & 3 & 7 & 3 & 22 & 6 & 67 & 7 & 67 & 22 & 7 \\
6 & 10 & 6 & 6 & 6 & 14 & 14 & 22 & 22 & 22 & 74 & 22 \\
3 & 6 & 7 & 3 & 7 & 6 & 22 & 7 & 67 & 7 & 22 & 67
\end{bmatrix}
\begin{bmatrix} U_{1,0} \\ U_{2,0} \\ U_{3,0} \\ U_{0,1} \\ U_{4,1} \\ U_{0,2} \\ U_{4,2} \\ U_{0,3} \\ U_{4,3} \\ U_{1,4} \\ U_{2,4} \\ U_{3,4} \end{bmatrix}
$$

$$(15.19)$$

showing how each sample boundary value is probed and participates in the expression of the solution for each interior point in the grid as a weighted average.

Although the general algebraic form is desirable from a computational point of view, it is too difficult to obtain in general and is extremely costly in terms of both computation by algebraic manipulation programs such as Maple (4), used to obtain Eq. (15.18), as well as by storage of the formula expressions. As seen from comparison of the terms in the matrix, the more distant a boundary point is from an interior point, the smaller is its coefficient, consistent with intuition. As the matrix expands to include more sampling points, the coefficients decrease rapidly in value; thus, unless the relative boundary potential values increase in a manner so as to offset the decreasing coefficient values, one can usually ignore the boundary contribution from points beyond some established limit, such as that prescribed by the available floating-point arithmetic hardware.

One may require additional solution points for purposes of presentation or analysis. In many cases, *interpolating polynomials* prove to be useful. Thus, one might approach our example problem using products of quartic equations of the general form $X_j^{(4)}(x) = a_j x^4 + b_j x^3 + c_j x^2 + d_j x + e_j$ and similarly for $Y_i^{(4)}(y)$. An appropriate rationale for this lies in the fact that the quartic has five parameters (order 5, degree 4). By demanding that the curves pass through each of the five sampling points $x = j/4$ ($j = 0,\ldots,4$) for each of the five families of $y = i/4$ ($i = 0,\ldots,4$), one can always generate unique solutions to the coefficients. Additionally, the coefficients must, of course, be chosen consistent with the boundary conditions and the Laplace equation. Once obtained, the polynomial XY can be used to obtain potential values at arbitrary (x, y) as well as the positions of possible minima and maxima. This approach ultimately breaks down when the number of sample points is larger and the resulting matrix problem becomes intractable. However, this technique suggests that polynomials can be used to fashion approximate solutions that are applicable over broad areas of the interior and boundary regions.

Consistent with the notion of approximating polynomials as in the preceding paragraph, finite-element approaches attempt to simplify the solution process by carefully choosing polynomials so as to minimize the number of coefficients required to determine and simplify the matrix inversion problem. One choice is the Bezier polynomials, defined by

$$B_{t,N}(u) = \frac{N!}{t!(N-t)!}\, u^t (1-u)^{N-t}; \; B_{t,N}(0) = \delta_{t,0}; \; B_{t,N}(1) = \delta_{t,N} \qquad (15.20)$$

Curves resulting from the choice of Bezier functions blend the values of the known boundary potentials to produce interior potential values and have the appropriate smoothness properties desired in the final solution. Further, the $B_{t,N}$ have maximum values that distribute evenly through the mesh regions. For instance, for u between 0 and $\frac{1}{4}$ in Eq. (15.20), the value of $B_{0,4}$ is greatest, and all other B variables approach minimum values. Thus, $B_{0,4}$ serves to sample that particular range of u values.

Other commonly employed and related sets of approximating polynomials are *Hermite polynomials* and *B splines*. Particularly in the latter case, the functions possess the desired properties of smoothness across patch boundary intersections, strong locality leading to simplification of the **A** coefficient matrix, and efficiency of computation. In the following discussion the B functions may be viewed, up to specific values, as any of the aforementioned types.

The approximating solution function is

$$u(x, y) = \sum_{i=0}^{N} \sum_{j=0}^{M} U_{j,i} B_{i,N}(x) B_{j,M}(y) \qquad (15.21)$$

which must satisfy the Laplace equation

$$\nabla^2 u(x, y) = 0 = \sum_{i=0}^{N} \sum_{j=0}^{M} \left[\frac{d^2}{dx^2} B_{i,N}(x) B_{j,M}(y) + B_{i,N}(x) \frac{d^2}{dy^2} B_{j,M}(y) \right] U_{j,i} \qquad (15.22)$$

and the boundary conditions. For the latter we note that the known boundary potentials are sampled at $(0, j/M)$, $(1, j/M)$, $(i/N, 0)$, and $(i/N, 1)$, excluding the corners, and the Bezier functions are either 0 or 1 at the interval endpoints $x,y = 0$ and 1.

In our sample problem there exist nine interior potentials that we must obtain in terms of the 12 known boundary potentials. Thus, we apply Eq. (15.15) at each point $(x, y) = (i/4, j/4)$ with $i, j = 1, \ldots, 3$, to generate the nine equations required to obtain the nine unknown $U_{i,j}$ in terms of the 12 known boundary values. For example, at $(x, y) = (\frac{1}{4}, \frac{1}{4})$ one finds that

$$-\frac{243}{32}U_{1,1} - \frac{81}{32}U_{1,2} - \frac{27}{32}U_{1,3} - \frac{81}{32}U_{2,1} - \frac{81}{128}U_{2,2} + \frac{9}{16}U_{2,3} + \frac{27}{32}U_{3,1} + \frac{9}{16}U_{3,2} + \frac{9}{32}U_{3,3}$$

$$= -\left(\frac{3}{512}U_{4,4} + \frac{3}{64}U_{3,4} + \frac{243}{256}U_{0,2} + \frac{135}{512}U_{4,0} + \frac{2187}{512}U_{0,0} + \frac{243}{256}U_{2,0} + \frac{81}{64}U_{0,3}\right.$$

$$\left. + \frac{9}{32}U_{4,1} + \frac{135}{512}U_{0,4} + \frac{81}{64}U_{3,0} + \frac{39}{256}U_{2,4} + \frac{39}{256}U_{4,2} + \frac{3}{64}U_{4,3} + \frac{9}{32}U_{1,4}\right)$$

$$(15.23)$$

where the corner potentials are included as well. If these values are available from the specification, they are used directly; otherwise, their values can be approximated consistent with other known values. Equation (15.23), together with the remaining equations for other interior points, can be cast in matrix form, and from there a solution is deduced.

Equation (15.23) displays the feature of locality that the blending functions should possess in order to be computationally advantageous; that is, during the process of matrix inversion, one wishes the calculation to proceed quickly. As mentioned earlier, the use of linear approximation functions results in at most five terms on the left side of the equation analogous to (15.23), yielding a much cruder approximation, but one more easily calculated. The current choice of Bezier functions, on the other hand, is rapidly convergent for methods such as relaxation, possesses excellent continuity properties (the solution is guaranteed to look and behave reasonably), and does not require substantial computation.

A final note is in order. The finite-difference and finite-element techniques are entirely equivalent from a mathematical point of view. What is different about these are the conceptualization of the problem and the resulting computational techniques to be employed. One method is not better than the other, although in particular circumstances one may clearly be superior. The point is that a modeler and modeling systems should account for both methods as well as others not mentioned here.

15.4. COMPUTER ENGINEERING ASPECTS

The steps described earlier have been implemented, to varying degrees, in some commercial packages: Maple (Maplesoft, Waterloo, Ontario, Canada), Mathcad (MathSoft, Cambridge, Massachusetts), and Maxwell (Ansoft Corp., Pittsburgh), to name but a few. In recent years two software packages seem also to have become rather widespread: Matlab (the MathWorks Co., Natick, Massachusetts) and Abaqus (Abaqus UK Ltd., Warrington, UK).

A vital aspect of constructing modeling systems involves the availability of the correct software packages and tools to solve problems as they arise. In modern electrochemistry it is clear that the speed at which new problems arise may never be addressed by waiting for the next version of a favorite modeling package, not to mention issues such as cost and availability. To address the problems of new system development and construction, a new solution strategy must be developed. Figure 15.1 illustrates the basic system design and interaction of internal components for the modeling approach discussed in Section 15.2.

Since modeling is such a dynamic and largely unpredictable type of activity, it is also the case that the data flow through such a system cannot be considered using a linear model. Of necessity, the modeler makes certain progress, modifies specifications, performs trial calculations, and requires many alternative views of the same data. Data storage and solution generation are therefore fundamental to the system, with other components placed at a higher level in the system hierarchy.

Communication between logical modules must also be handled efficiently and may be analyzed in terms of data flow as well. An important point in this regard is that in designing the software, data must not become trapped so as to lead to deadlock of the entire system. This can result from a failure to deliver data or because the module cannot transfer control to another module. The system design must take into account a consistent model for the file system, data exchange, intermodule communication protocols, data security, and finally, performance and reliability. The modern discipline of software engineering (5) has developed to the extent that design and implementation of such complex software is approaching the same level of dependability required for construction, say, of a major building, although the latter is considered a much easier problem. However, modeling the modeler itself is accomplished using established and testable methods.

Developing the ability to communicate data from one application software to another is also of paramount importance for current application requirements. For example, it may be possible to calculate a certain quantity using only particular software, which is then found lacking in terms of how to process it further. The need to exchange data between applications leads to the need to develop interapplication data exchange protocols. Typically, this is currently done at the text level, which introduces the possibility of error in the output–input transformation as well as the time to perform the conversion and transfer.

An integral part of modern design approaches involves group activities, or conferencing. Although many aspects of a design may be worked on by individuals, the total product will evolve as the result of integrated efforts of a team, possibly separated across countries or the world. The possibility that team members share work and results is an increasing likelihood that must be reflected in modeling software and the supporting hardware systems; current networking capabilities have solved many of these problems, but more work is required, particularly on incorporating conferencing within the design system model.

Parts of this strategy have already been mapped out through the introduction of software products such as Java, an *object-oriented programming language* (6) suitable for immediate integration with Internet tools for distributed application across wide-area networks. In this approach, solutions to specialized problems are made available to users in the form of *applets,* or program modules, which can easily be interfaced with other modules to form higher-order programs. In short, a program can be constructed from pieces gathered from around the world; the pieces can be reused, updated, modified, and so on, to fit the needs of the specialist. The delivery system for such approaches involves the notion of clients and servers, the *client–server model,* illustrated in Figure 15.7. This model has been used for many years as the basis for operating system software and is undergoing extension and enhancement to meet the needs of modern distributed computing environments.

In a similar vein, users who otherwise might not be able to afford large, complex software packages (costing thousands of dollars, perhaps) may be able to buy time on commercial servers, thereby gaining access to powerful tools while saving precious research and development funds for other uses.

Use of the Internet as a medium for text and graphical image communication and commercial transactions is well established. However, use of the Internet as an agent in modeling is not widespread or highly developed at this time. Work (7) has demonstrated the utility of using Java applets to perform fast Fourier transformation (FFT) by downloading appropriate software to a client machine to do the task and to perform submission of data via electronic templates supplied from a server to the client, transmitted back to the server to perform calculations, and subsequent delivery to the client of graphical images, all in the form of applets (8). With such systems it is expected that client accessibility to high-performance software can be increased substantially while reducing the cost proportionately using, say, time-of-usage billing. Also, distribution from controlled servers implies that software can be maintained and features added more conveniently, ensuring constant state-of-the-art technology to customers.

To develop a distributed modeling system, it is necessary to identify the primitive components essential to the fundamental operation of the system and its various appended and programmable functions, a process referred to as *kernel identification.* Studying such systems provides much of the basis for acquiring insight into what structure and function a kernel should possess (9).

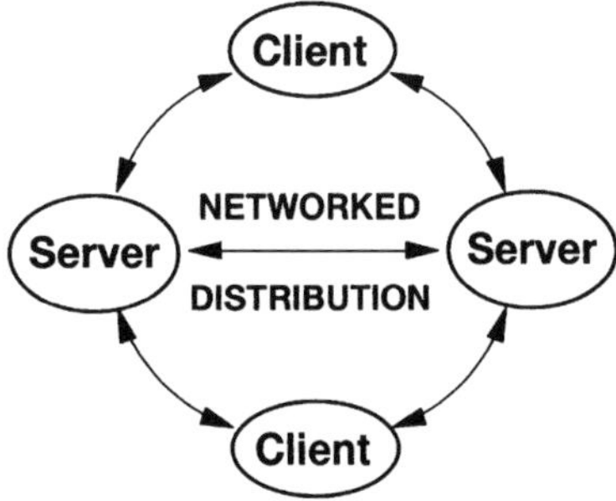

Figure 15.7. Many-client many-server model for distribution on a network. Both client–server and server–server interaction is permitted, particularly in cases where a service spawns a succession of service calls.

15.5. CONCLUSIONS

With the continuing advancement of powerful computer hardware and software systems, the nature of modeling is evolving quickly. Workers at all levels of research and development are involved increasingly with initiation, development, understanding, testing, and production of new products. In general, virtual prototypes are preferred over physical prototypes, primarily because of relative time and cost of production. It is worthwhile to train students and practitioners in the use of certain basic criteria to assist in making decisions on whether to commit resources to the modeling task. Of particular importance in this regard is an appreciation of the role of mathematics in the entire modeling process.

In this chapter we have provided an overview of mathematical modeling from inception of design through specification of solution method, production of solution, and analysis of results. Additionally, we have provided a framework for including computers, particularly current and emerging application software, as vital agents in the modeling process. With regard to both software and hardware developments, the Internet presents a great challenge and opportunity for modeling. Programming languages such as Java have emerged as suitable tools that ensure software reliability and reuse and that permit modeling to occur over a distributed set of computers.

REFERENCES AND FURTHER READING

1. M. E. Davis, *Numerical Methods and Modeling for Chemical Engineers,* Wiley, New York, 1984.

2. W. H. Press, S. A. Teukolsky, W. T. Vetterling, and B. P. Flannery, *Numerical Recipes in C: The Art of Scientific Computing*, 2nd ed., Cambridge University Press, Cambridge, 1992.

3. J. D. Foley, A. van Dam, S. K. Feiner, and J. F. Hughes, *Computer Graphics: Principles and Practice*, 2nd ed., Addison-Wesley, Reading, MA, 1994.

4. M. L. Abell and J. P. Braselton, *Differential Equations with Maple V,* Academic Press, San Diego, CA, 1994.

5. R. S. Pressman, *Software Engineering: A Practitioner's Approach.* McGraw-Hill, New York, 1987.

6. D. Flanagan, *Java in a Nutshell*, O'Reilly, Cambridge, MA, 1996.

7. I. Cidambi, M.Sc. thesis, University of Windsor, Windsor, Ontario, Canada, 1996.

8. F. Martincic, Honours Bachelor of Computer Science thesis, University of Windsor, Windsor, Ontario, Canada, 1997.

9. E. Marcuzzi, M.Sc. thesis, University of Windsor, Windsor, Ontario, Canada, 1997.

10. J. P. Webb, *Reports Prog. Phys.* **58**, 1673–1712 (1995).

PROBLEMS

15.1. Given two linear equations:

$$4x_1 + \ x_2 = 6$$
$$x_1 + 5x_2 = 6$$

(a) Express them in matrix notations as $Ax = b$.
(b) Calculate the matrix Ax.
(c) What do you require so that xA be calculated?
(d) Calculate xA. Is it identical to what you get in part (b)? Why?

15.2. Apply the finite-difference method for solving a linear boundary value problem as follows: Given the second derivative of the function y in the interval 0 to 4 as

$$y'' = y + x(x - 4), \qquad 0 \le x \le 4$$

The boundary conditions (i.e., the values of y on the two edges) are given as

$$y(0) = y(4) = 0$$

Determine the functions values at $x = 1, 2, 3$ in steps, as follows:
(a) Replace the derivative with a finite-difference approximation.
(b) Write three ordinary differential equations.
(c) Obtain a system of linear equations.
(d) Solve the equations for the (approximate) values of x_i, where $i = 1, 2, 3$.
(*Hint*: First, divide the interval of interest into $n = 4$ subintervals separated by grid points x_0, x_1, x_2, x_3, and x_4. Next, calculate the value of y on these grid points and denote them as y_i, $i = 0, 1, ..., 4$. From the boundary condition you have $y_1 = y_4 = 0$.)

15.3. To obtain higher-accuracy solutions for Problem 15.2, divide the interval into more subintervals. Increase n from 4 as in Problem 15.2 to 8, and decrease h from 1 to $\frac{1}{2}$. You will generally obtain a tridiagonal system of $n - 1$ linear equations up to 7 in the present case. For large n, solving the system of equations with paper and pencil becomes impractical, and it is necessary to find algorithms suitable for computation by computers.

16

Structure and Properties of Deposits

16.1. INTRODUCTION

Properties of materials in general and deposits in particular are strongly dependent on the structure of the material. For this reason, we discuss both in one chapter, starting with issues pertaining to structure. In the solid state, atoms are arranged in a regular manner, forming a pattern that may be described by the three-dimensional repetition of a certain pattern unit. The structure, it may be said, is periodic. When the periodicity of the pattern extends throughout a certain piece of material, we refer to it as a *single crystal.* In polycrystalline material the periodicity is interrupted at grain boundaries. The size of grains in which the structure is periodic varies markedly, as discussed below. When the size of the grains or crystallites becomes comparable to the size of the pattern unit, one then deals with an amorphous substance.

All deposits of metals are made of grains whose structural–physical nature (1) can be divided into four types: (1) columnar, (2) fine-grained, (3) fibrous, and (4) banded. In terms of their practical macroscopic physical properties, their main characteristics may be summarized as follows:

1. Those types that are of low strength and hardness but possess high degree of ductility. Examples are metals deposited under low-current-density conditions.

2. Those types are characterized by typical grain sizes of 10 to 100 nm. The deposits are relatively hard and brittle, whereas some are rather ductile. Examples are metals deposited under high-current-density conditions containing hydrated oxides as a consequence.

3. Those types that are intermediate in nature between types 1 and 2.

4. Those types that contain grains of extremely small dimensions (less than 10 nm). Typically, bright deposits (as a result of additives, for instance) such as Ni–P

Fundamentals of Electrochemical Deposition, Second Edition.
By Milan Paunovic and Mordechay Schlesinger
Copyright © 2006 John Wiley & Sons, Inc.

(electroless and electrodeposited) exhibit such structure. Those deposits can be expected to be of high strength and hardness but poor ductility.

Grains are generally understood here to be individual crystallites in a polycrystalline body of material. This definition is, however, not common to all authors. Thus, clumps of crystallites are referred to by some as *grains*, whereas others refer to such clumps as *islands*. In this discussion, ductility is related to the maximum amount of strain to which a body may be subjected without breaking.

Other authors (2) prefer a different grain structural classification. Specifically, these classifications are (1) columnar, (2) equiaxed, (3) dendritic, (4) nodular, and (5) fibrous. This classification is more in terms of overall structure than in terms of macroscopic and physical properties.

1. *Columnar.* These types of grains are most common in compact thin films. They are the result of preferred growth in certain crystal directions. Randomly oriented grains are usually small compared to film thickness.
2. *Equiaxed.* These grains grow to relatively larger sizes than in case 1.
3. *Dendritic.* In electrodeposited films, dendritic grains result from mass-transport-controlled growth, and the individual crystals may vary in shape.
4. *Nodular.* Because of their appearance, nodular structures are often referred to as "cauliflower type." Although the phenomenon has been attributed to a number of factors (such as impurities), the origin and growth mechanism of this feature are still not well understood.
5. *Fibrous.* This type of grain is the result of oriented growth of grains that cover the substrate only incompletely.

A schematic summary presentation of four of these grain types is given in Figure 16.1.

As stated above, crystalline material, whether single-crystal, polycrystalline, or even nanocrystalline–amorphous, is made such that its component (constituent base)

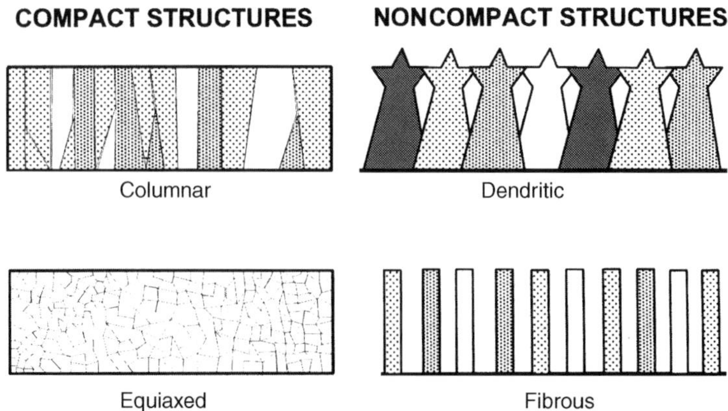

Figure 16.1. Schematic presentation of grain structures of thin films. (From Ref. 2, with permission from the Electrochemical Society.)

atoms and molecules are arranged in a three-dimensional regular repetitive pattern called a *lattice*. The commonest form among (plated) metals is the *face-centered cubic* (fcc) *lattice*, which is the lattice structure of Ag, Al, Au, Cu, Ni, and other elements. Fcc here refers to a lattice arrangement of metal atoms at each corner of a cube plus one atom each in the center of every cube face (Fig. 16.2a).

Next in frequency is the *hexagonal close-packed* (hcp) *lattice*, which is assumed by Co, Zn, and some other elements. Hcp here refers to an arrangement of metal atoms with planes of atoms placed at the corners of hexagons that are separated by planes of atoms grouped in sets of *three* between hexagons in the adjacent planes. In this way, one of the two close-packing arrangements of hard spheres is realized (Fig. 16.2b). The other method of close packing of hard spheres is realized by the fcc arrangement. That arrangement is therefore sometimes referred to as *cubic close-packed.*

Body-centered cubic (bcc) is the lattice symmetry of Fe, for instance (Fig. 16.2c). Bcc here refers to a crystal arrangement of atoms at the corners of a cube and one atom in the center of the cube equidistant from each face.

In Chapter 13 the characterization of thin films was described and the concepts of edges, steps, and similar surface sites were introduced (see Fig. 13.1). In the present chapter we discuss the crystal structure of solids, so that, for instance, the number of atoms per square centimeter on a surface N_s and the height h of a monolayer may be determined. The values of N_s and h depend on the crystallographic structure and the orientation of the surface plane.

The atomic volume can actually be calculated without recourse to, or use of, crystallography. The atomic density n of atoms per cubic centimeter is expressed as

$$n = \frac{N_{\text{Avogadro}}\, \rho}{A} \tag{16.1}$$

where N_{Avogadro} is Avogadro's number, ρ is the mass density in g/cm^3, and A is the atomic mass (number).

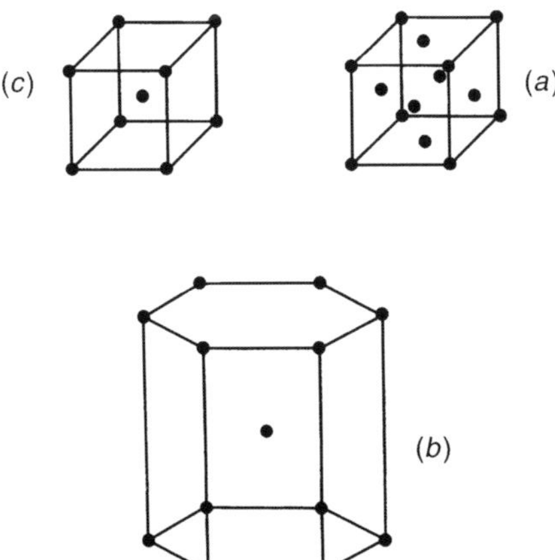

Figure 16.2. Unit cells of the three most important lattice types: (*a*) face-centered cubic; (*b*) hexagonal close-packed; (*c*) body-centered cubic. (From Ref. 1, with permission from Noyes.)

Let us look, as an example, at silicon. In this case we have

$$n = \frac{6.02 \times 10^{23} \times 2.33}{28} = 5 \times 10^{22} \text{ atoms/cm}^3 \tag{16.2}$$

Other semiconductors, such as Ge and GeAs, have atomic densities of 4.4×10^{22} atoms/cm^3, a metal such as aluminum has about 6×10^{22} atoms/cm^3, and metals such as cobalt, nickel, and copper have about 9×10^{22} atoms/cm^3.

Note also that the volume occupied by an atom in the lattice is simply

$$\Omega = \frac{1}{n} \tag{16.3}$$

The typical value of this parameter is about 2.0×10^{-23} cm^3.

In the last analysis, crystalline structure exhibited as a product of electrodeposition depends on competition between rates of new crystalline formation and existing crystal growth. Specifically, the effect of deposition conditions during plating on a deposit structure is such that deposition close to the limiting current leads to dendritic growth, as the effect of transport is more pronounced in systems exhibiting low activation overvoltage. Well below the limiting current, an increase in the activation overvoltage tends to favor formation of equiaxed smaller-grained deposits because nucleation is facilitated.

It ought to be stressed that, in fact, a large number of variables in the plating process have a bearing on structure. These include metal-ion concentration, additives (see Chapter 10), current density (see Chapter 12), temperature, agitation, and polarization. It is outside the scope of this book to discuss these in much more detail. To visualize the effects that these parameters can have on grain size, see Figure 16.3.

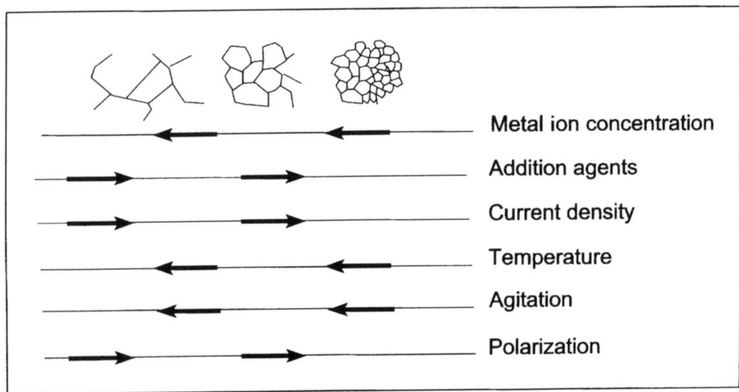

Figure 16.3. Relation of structure of electrodeposits to operating conditions of solutions. (From Ref. 3, with permission from the American Electroplaters and Surface Finishers Society.)

The arrows in the figure indicate the tendency that an increase in the given parameter will cause in determining grain sizes.

16.2. CRYSTALLOGRAPHY

The properties of periodic arrangements of atoms are of vital importance in condensed-matter engineering and science. The branch of science and engineering that deals with the enumeration and classification of the types of crystal structures as well as the methods used to determine the structure of a given specific material is called *crystallography.* It should, at this stage, be evident to the reader that the crystallographic structures are classified according to their symmetry properties. Those are operations such as rotations and reflections that leave the lattice invariant. These can be employed in simplifying theoretical calculations. They are useful in defining the number of parameters required to define the macroscopic properties of a given solid. To be able to appreciate and exploit the symmetry theories, one needs to resort to the mathematics of *group theory.* That is far outside the scope of this book. It is of interest to state, however, that there are, in three-dimensional space, 230 distinct space groups; that is, there are that many different repetitive patterns in which symmetry elements may be arranged on a space lattice.

16.3. SUBSTRATE

Yet another factor that has a substantial influence on electrodeposits, including their structure and properties, is the nature of the substrate on which the plating occurs. Two phenomena are of importance in this context: epitaxy and pseudomorphism. *Epitaxy* refers to the systematic structural "keenship" between the atomic lattices of the substrate and the deposit at or near the interface. In other words, *epitaxy* refers to the induced continuation of the morphology and structure of a substrate material into a coating applied to it. An important parameter of epitaxial growth is the substrate's temperature. For a given material being deposited on a substrate, all conditions being fixed, there exists an epitaxial temperature. That is a temperature above which epitaxial growth is possible and below which it is not. For example, for silver evaporated on NaCl, that temperature is 150°C. *Pseudomorphism* is the continuation of grain boundaries and similar geometric features of the substrate into the deposit. An alternative definition states that pseudomorphic deposits are those stressed to fit on a substrate. Also, as a working rule, pseudomorphism persists deeper (up to 10 nm into the deposit) than epitaxy.

Prior cleaning, or the state of cleanliness of the substrate, also has a great influence on the structure and adhesion of the deposit (see also Chapter 13). For instance, copper deposited on cast copper that was cleaned but not pickled prior to plating yields a fibrous deposit, while the substrate itself is coarse grained. Pickling after cleaning results in a structure such that the copper crystals are simply a continuation of those in the copper basis. By *pickling*, one means an additional cleaning step for scale and oxide removal. In general, for the sake of cleaning, both mechanical and chemical methods may be employed. It ought to be emphasized, since it is somewhat unexpected, that morphological reproduction of the basis metal itself, even in the case of electrodeposition of a different metal on the substrate with differing lattice structure, is possible in the case of pickling after cleaning.

16.4. PHASE CHANGES

Although the concept of phase is well defined thermodynamically, here *phase* refers to a mechanically separable homogeneous part of an otherwise heterogeneous system. The concept of *phase change* refers here to a change in the number present or in the nature of a phase or phases as a result of an imposed condition such as temperature or pressure. To clarify and illustrate the topic at hand, we use the specific cases of electrolessly deposited nickel and electrodeposited cobalt.

As noted as far back as the 1960s (4), electrolessly deposited nickel (Ni–P) film, when in the fresh "as deposited" stage, is a metastable system exhibiting diffuse electron-diffraction rings, typical of amorphous material, in the transmission electron microscope (TEM). It undergoes a crystalline transition when heated to about 330°C, exhibiting diffraction patterns in the TEM typical of a material of a moderate degree of crystallinity. Such patterns are illustrated in Figure 16.4. The result of this crystallization is a harder, less ductile film. The extent or the degree of crystallinity, either before or after heat treatment, is a compound function of phosphorus content, metallizing solution (bath) pH, temperature to which the sample was heated, time of exposure to the "highest" temperature, and a number of additional factors. By *degree of crystallinity* we mean here component crystalline sizes, possible preferred orientation, and the like.

Another example of phase change is the one exhibited by electrodeposited cobalt. In this case the transformation is from fcc- to hcp-type lattice structure as a result of hydrogen inclusion during deposition on the one hand and subsequent out-diffusion on the other hand.

In general, when exposed to certain types of gas, metal surfaces will undergo transformation, and their properties may undergo significant changes in the process. The rate and extent of these changes depend on the metal, the gases, and the new phase or product that will form at the interface between the two original phases. We mention here the case of Si oxidizing to SiO_2 that undergoes expansion, thus creating strong compressive stress on the interfacial surface, for instance.

As an additional example of high practical significance, we refer here to copper deposits when used in microelectronics, mirrors, and other optical applications. Those deposits have been observed to soften in time even when stored at room temperature for only 4 to 6 weeks. Also, mirrors and other precision objects made of copper will undergo surface deformation after a few months. This type of degradation can be counterbalanced by a suitable metal overcoating. Another, not always practical way is heat treatment to about 300°C. These phenomena are the direct results of microstructural instabilities, often referred to as *recrystallization* in the copper. It is worth stressing that recrystallization is not limited to copper (5).

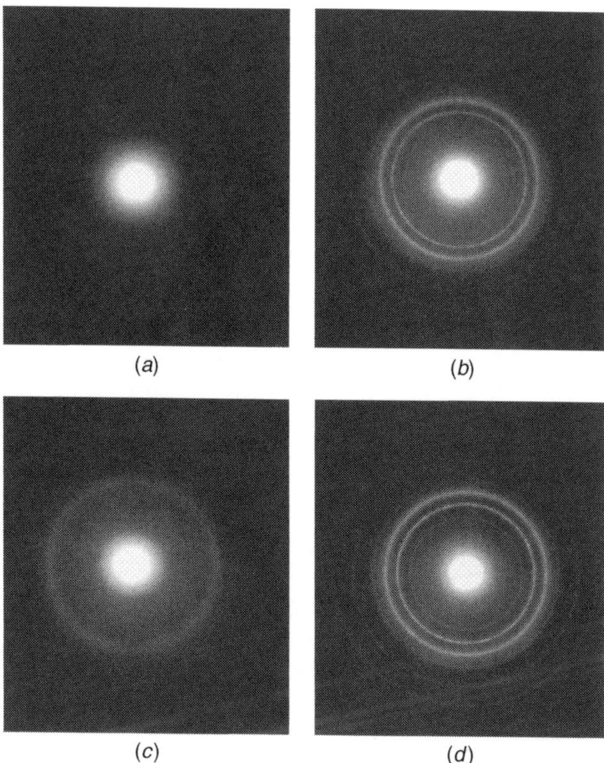

Figure 16.4. Diffraction patterns of (*a*) as deposited and (*b*) heat-treated thin electrolessly deposited Ni–P films; (*c*) as deposited and (*d*) heat-treated electrolessly deposited Cu film. Films were heated and examined in the sample holder of an electron microscope.

16.5. TEXTURE

Crystals growing on a substrate may be oriented every which way; that is, the direction axes of individual crystallites can be randomly distributed. However, where one particular axis is oriented or fixed in *nearly* one direction, we speak about a single texture. When two axes are thus fixed or oriented, we speak about double texture. *Monocrystalline orientation* refers to a scenario in which there are three such *nearly* oriented axes, including epitaxial films. Orientation here is viewed with respect to any fixed (in space) frame of reference. Crystal planes and directions are illustrated in Figure 16.5. A brief discussion of the enumeration of these elements follows.

In general, the lattice points forming a three-dimensional space lattice should be visualized as occupying various sets of parallel planes. With reference to the axes of the *unit cell* (Fig. 16.2), each set of planes has a particular orientation. To specify the orientation, it is customary to use the *Miller indices*. Those are defined in the following manner: Assume that a particular plane of a given set has intercepts p, q, and r

with the crystallographic axes. The Miller indices of the set of parallel planes are given as

$$h:k:l = 1/p:1/q:1/r \tag{16.4}$$

with the proviso that h, k, and l in (16.4) are the smallest integers satisfying the expression. In other words, the three indices have no common factor larger than 1. For a cubic lattice it is possible to show (and it may be verified by the reader) that the distance between successive planes in a set (hkl) is given by the expression

$$d_{hkl} = \frac{a}{(h^2 + k^2 + l^2)^{1/2}} \tag{16.5}$$

In Figure 16.5, different types of parentheses around Miller indices are apparent. In general, the accepted convention is that parentheses (or | · |) [e.g., (100) or |100|] signify a set of parallel planes. Braces, such as {100}, signify planes that are equivalent, such as cube faces. The directions that are normal (perpendicular) and that are equivalent are seen to be enclosed in angular brackets, <>; thus, <100> indicates directions normal to the set of (100) planes.

The texture of coatings and films deposited by various techniques do differ widely. Electrodeposited palladium serves as one of examples the best to illustrate the point. In this case and depending on current density and/or solution pH, three different deposits, in addition to no texture, can be obtained: (111), (110), or even a combination of four separate components. Here, instead of specifying the fiber axes, we indicate planes that stay *nearly* fixed in orientation. This is the practice of a number of authors.

It appears in this discussion that electrochemical parameters and not substrate properties are the main deciding factors in determining the texture of deposits. This is indeed so when a deposit's thickness is 1 μm or more. In case of thinner deposits, the substrate plays an important role as well (see the text above). Another nonelectrochemical factor may be the codeposition of particulate matter with some metal deposits. To summarize, we note that texture is influenced mostly by deposition current density, as it is itself a function of bath pH, potential, and other parameters. Not surprising, then, is the fact that in the case of physical vapor deposition (PVD), the deposition rate is the determining factor in setting the texture of the coating.

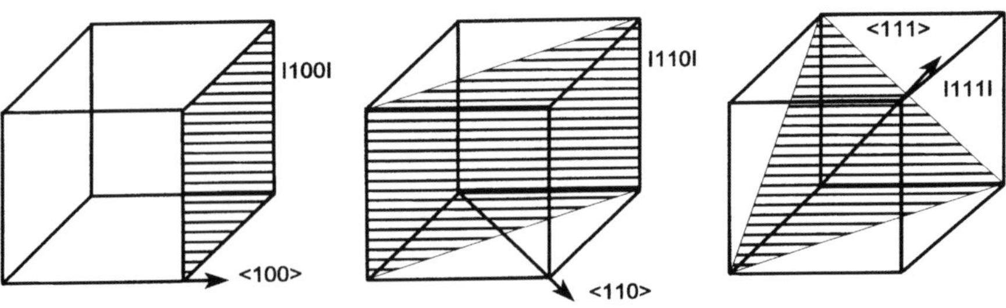

Figure 16.5. Crystal planes and directions as indicated. One plane and one direction in each cube are indicated. (From Ref. 1, with permission from Noyes.)

TABLE 16.1. Influence of Solution Hypophosphite Concentration on Preferred Orientation and Magnetic Coercivity of Electroless Cobalt–Phosphorus Films

Solution Hypophosphite Concentration (wt %)	Preferred Orientation (002)/(101)(%)	Hypophosphite Concentration (Oe)
5	0	1026
6	0	938
4	<10	930
3	<10	745
7	<10	461
2	>50	493
8	>50	200

Source: Ref. 6, with permission from the Electrochemical Society.

Texture has a rather marked influence on the properties of a given deposit. Thus, rather seemingly unrelated parameters (properties), such as corrosion resistance, hardness, magnetic properties, porosity, contact resistance, and many others, are all texture dependent. By way of illustration, we discuss (6) first the case of magnetic properties.

Electrolessly plated cobalt–phosphorus thin films are usually characterized by low- to medium-coercivity values (300 to 900 Oe). For some applications it would be desirable to have films with coercivities greater than 900 Oe. Table 16.1 shows that films with minimum preferred orientation exhibit maximum coercivity values. The ratio of the intensity of the (002) and (101) diffraction rings changes with hypophosphite concentration of the plating solution, with zero preferred orientation obtained at solution hypophosphite concentrations of 5 to 6 wt %.

As discussed in Chapter 17, the specific texture of the copper substrate on which a nanomultilayer system such as, say, Cu/Ni layer pairs, is deposited has a direct bearing on the system's texture and consequently, magnetic characteristics. Different nickel deposits show a great variety of contact resistance values. This is particularly so after the deposits have been exposed to the atmosphere for an extended period of time. The differences between these values may best be explained in terms of variations in plated texture. Nickel electrodeposits with polycrystalline nature have been observed to behave as single crystals (!) when their grains were oriented such that the (100) plane was parallel to the surface. Not surprisingly, the oxidation rate in (100)-oriented single crystals is self-limiting at ambient temperature. The optical properties of some thin-film light polarizers owe their ability to polarize to the specific texture of their structure.

16.6. PROPERTIES

In 1974 and in 1986, W. H. Safranek published two versions of texts entitled *The Properties of Electrodeposited Metals and Alloys* (7). Those two excellent volumes make it *almost* unnecessary to include here details regarding deposit properties.

16.6.1. Mechanical Properties

Material science and engineering deals with, and actually is concerned primarily with, the relationship between structure and properties of matter. Apart from the scientific interest, it is, of course, not of less practical importance. Thus, for instance, knowledge of the mechanical properties of a given substance enables prediction of its performance in a number of applications. The precise functional relationship between structure and property is rather complex, however. To be specific, the precise relationship between these two is determined by the different microscopic and macroscopic interactions that take place in the material. These include quantum mechanical interactions between constituent atoms and/or molecules as well as "classical" interactions between grains and groups of grains. The case of electrodeposition adds a measure of complication, as it is inherently a nonequilibrium process. As such, the product films exhibit a large number of imperfections, which vary in number and nature from grain to grain. In this connection it is also important to remember that the most basic imperfection of a crystal lattice is the surface itself, where the periodicity—the hallmark of bulk material—is being cut off.

For illustration purposes, we mention the case of gold, which when electrodeposited containing about 0.5% cobalt has a hardness about four to five times that of annealed gold. This degree of hardness cannot be achieved using any of the known metallurgical methods. It is widely believed that grain sizes of 20 to 30 nm in this material are responsible for this high degree of hardness. In a 1979 paper, Lo et al. (8) state that "other … mechanisms, such as solution hardening, precipitation hardening, strain hardening, and void hardening, account for only small alterations in the hardness of this coating." This seems true even today. Cobalt-hardened gold films have found a large number of practical applications in electronics and other areas.

In general, the tensile strength (defined as the tensile stress/strain ratio) of most electrodeposited metals varies over a broad range of values and is a function of the conditions of the deposition process. Here, *tensile stress* is the force acting perpendicular to the cross-sectional area of an object. The *strain* is defined as the change in length divided by the original length. The corresponding (Young's) modulus has dimensions of force per unit area. Table 16.2 illustrates the foregoing by making comparisons between strength and ductility of electrodeposited metals versus wrought metals. It would be quite correct to state that in the last analysis, fine grain size is the main reason for the higher strength of electrodeposited metal compared to their wrought forms.

In an attempt to relate the grain size in a metal to its mechanical properties quantitatively, Petch and Hall (9 and references therein) proposed an expression relating grain size d with hardness H in a metal. *Hardness* is defined in this case as the *yield stress*, the stress at which value the material experiences the onset of permanent deformation. Thus,

$$H = H_0 + K_{\mathrm{H}} \frac{1}{\sqrt{d}} \tag{16.6}$$

TABLE 16.2. Strength and Ductility Data for Electrodeposited Metals and Their Wrought Counterparts

| | | Tensile Strength | | | Wrought Metal[a] | |
| | | Minimum | Maximum | Elongation | Tensile Strength | Elongation |
Metal	Plating Bath	psi	psi	(%)	(psi)	(%)
Aluminum	Anhydrous chloride–hydride–ether	11,000	31,000	2–26	13,000	35
Cadmium	Cyanide	—	10,000	—	10,300	50
Chromium	Chromic acid	14,000	80,000	<0.1	12,000	0
Cobalt	Sulfate–chloride	76,500	172,000	<1	37,000	—
Copper	Cyanide, fluoroborate, or sulfate	25,000	93,000	3–35	50,000	45
Gold	Cyanide and cyanide citrate	18,000	30,000	22–45	19,000	45
Iron	Chloride, sulfate, or sulfamate	47,000	155,000	2–50	41,000	47
Lead	Fluoroborate	2,000	2,250	50–53	2,650–3,000	42–50
Nickel[b]	Watts and other types of baths	50,000	152,000	5–35	46,000	30
Silver	Cyanide	34,000	48,000	12–19	23,000–27,000	50–55
Zinc	Sulfate	7,000	16,000	1–51	13,000	32

Source: Adapted from Ref. 7 (2nd ed.).

[a] Annealed, worked metal.

[b] Data do not include values for nickel containing >0.005% sulfur.

where H_0 and K_H are determined experimentally for a given specific metal and where their physical meaning is as follows: H_0 is the value determined by dislocation blocking, which is, in turn, dependent on the friction stress; and K_H represents the penetrability of the moving dislocation boundary. This is related to the available slip systems.

This rather useful empirical expression is applicable to many electrodeposited materials [e.g., molybdenum, zinc, steel (10)]. The expression has been able, for instance, to provide an acceptable explanation for the phenomenon of the brittle cracking in chromium electrodeposits. It has been quite helpful in the general study and understanding of the functional connection between hardness and grain size values in many electrodeposits.

Experiments have determined (8) that mechanical blocking of dislocation motion can be achieved quite directly by introducing tiny particles into a crystal. This "process" is responsible for the hardening of steel, for instance, where particles of iron carbide are precipitated into iron.

A detailed discussion of the consequences of the Hall–Petch equation in terms of the nature of dislocations is outside the scope of this work. Readers are directed to the many excellent texts and monographs dealing with solids in general and dislocations in particular (e.g., Ref. 11).

Tench and White (12) have shown that the room-temperature tensile strength of CMA (composition-modulated alloy) Ni–Cu exhibits values around three times that of nickel itself. The hardness of the same CMA has been demonstrated by Gimunovich et al. (13) to be many times greater. This is so as long as the thickness of the CMA layers is less than 100 nm. Stress due to lattice mismatch may be the prime cause of this.

16.6.2. New Materials

The recently highly active field (see, e.g., Ref. 14) of multilayer coating from one solution employing time-dependent modulation of the cathodic potential offers great promise in the production of new materials possessing new properties hitherto unthought of. Some examples of CMAs that have been produced to date are Cu–Ni, Cu–Zn, Cu–Co, Ag–In, and Cr–Fe; some of these are expected (on the basis of ambient-temperature tensile strength values) to exhibit different degrees of superplasticity. Some show unique magnetic properties (see Chapter 17) and more. Here *superplasticity* refers to behavior similar to that of glass when heated, specifically, the ability to stretch when pulled beyond the normal breaking point of the constituent individual metals. For a material to exhibit superplasticity, it should possess a largely fine-grained equiaxial microstructure. Moreover, this structure should in addition remain stable while the material is being deformed at the superplastic temperature (which may be as high as one-half of the melting temperature). A large number of electrodeposited metals do meet those requirements and thus offer the ability to be deformed superplastically at lower than their melting temperature. This has a number of practical advantages.

16.7. IMPURITIES

Electrodeposited films almost always contain various types of inclusions or impurities. Those may be from one or more of the following origins:

1. Added "chemicals" (additives such as levelers and brighteners)
2. Added particles (for composite coating)
3. Cathodic products (complex metal ions)
4. Hydroxides (of the depositing metals)
5. Bubbles (gas, e.g., hydrogen)

By and large it may be stated that most deposits produced using low current densities possess a higher impurity content than that of those deposited under conditions of higher current densities. Then, too, small amounts [of ppm (parts per million) order] of impurities will influence strength markedly. For instance, as noted above, a small amount of cobalt in electrodeposited gold results in enhanced hardness. Also, the presence of 100 ppm of carbon in a sulfamate nickel bath increases the tensile strength of the deposit about twofold (500 MPa to ~900 MPa). Similarly, for Sn–Pb alloy, if its carbon content is increased from 100 ppm to 700 ppm, the tensile strength goes from 30 MPa to 40 MPa. Although not quite that pronounced, a change is observed in the case of cast nickel. Here an increase in carbon content from 20 ppm to as high as 800 ppm results in an increase in flow stress from about 200 MPa to only 250 MPa.

Sulfur impurities can be detrimental to nickel deposits. Specifically, an increase in sulfur content is known to reduce the fracture resistance of electroformed nickel. Since sulfur has a direct influence on the properties of electrodeposited nickel, if no other impurities are present in the deposit, hardness by itself can be used as an indicator of sulfur impurity content.

From the foregoing it should be quite clear that in practice it is virtually impossible to obtain and maintain electrodepositing baths free of impurities. This is so even in the setting of a research laboratory. Thus, in research work pertaining to electrode kinetics, for instance, careful and often complex purification procedures are a necessity to remove some key contaminants. Otherwise, no reproducible or reliable results can be obtained. The practical plating shop worker, on the other hand, deals with "highly" contaminated (from the point of view of the researcher) solutions from a variety of by-and-large unknown sources. It is for this reason that for the most part, technical practice must rely on empirical observations.

In some deposits, notably those of nickel, electrical resistance "follows" current density at low temperatures in the sense that films deposited at low current density (say, 10 mA/cm^2) show lower resistance than do those deposited at higher density. Although this is so in the low-temperature range 4 to 40 K, this difference in resistance disappears closer to room temperature.

16.8. EMBRITTLEMENT

16.8.1. Hydrogen

Hydrogen is codeposited with most metals. Because of its low atomic number, it can be expected to be readily adsorbed by the basis metal. The source of the hydrogen may vary from associated preparatory processes such as electrocleaning to the specific chemical reactions associated with the plating process itself. Regardless of its origin, the presence of hydrogen may result in embrittlement, which involves a substantial reduction in ductility. *Hydrogen embrittlement* (HE) is an expression used to describe a large variety of fracture phenomena having in common the presence of hydrogen in the alloy as a solute.

Many mechanisms have been suggested in an attempt to explain HE in various systems, but consensus has not been reached as to their validity. Steel is most susceptible to embrittlement due to absorbed hydrogen. This may be, at least in part, due to the interference of hydrogen with the "normal" flow or slip of the lattice planes under stress. If, as often is the case, a deposit contains voids on the microscopic scale, hydrogen may accumulate in molecular form, possibly developing pressures exceeding the tensile strength of the basis metal, leading to the development of blisters.

16.8.2. High Temperature

Copper as well as nickel electrodeposits change from ductile to brittle at high temperature. Nickel drops from about 90% in an area at room temperature to about 25% at 500°C. In the case of electrodeposited copper, this occurs at lower temperatures, at 200 to 300°C, depending on the conditions during electrodeposition.

Wrought nickel does not exhibit this ductile-to-brittle transformation. One possible reason for this difference is that electrodeposited nickel contains sulfur and that, in turn, forms brittle grain boundary films of nickel sulfide. Wrought (201) nickel contains enough manganese that combines with the sulfur to prevent it from becoming an embrittling factor. Indeed, with small amounts of manganese with electrodeposited nickel, the effect described above can be minimized.

Electrodeposited copper exhibits the same type of transition, due to grain boundary degradation as a result of codeposition of impurities. It is not known which impurity in particular is to blame. Either sulfur and/or oxygen may be the cause of this effect.

16.9. OXYGEN IN DEPOSITS

For the sake of illustration, the relationship between internal stress in chromium deposits and their oxygen content is shown in Figure 16.6. The scatter of points shows the results of many hundreds of experiments. This is not surprising, as residual stress in every case is related to cracking of Cr deposits. These changes, depicted in the figure, were obtained by changing the solution compositions at constant temperature and constant-current density (86°C and 75 A/dm^2 or 750 mA/cm^2, respectively).

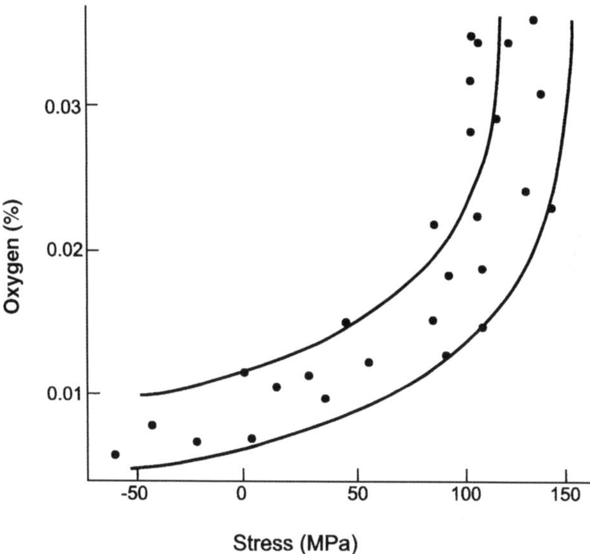

Figure 16.6. Influence of oxygen on stress in Cr electrodeposits produced at 86°C and a current density of 750 mA/cm^2.(From Ref. 15, with permission from the Institute of Metal Finishing.)

In physical vapor–deposited as well as sputter-deposited films, incorporated gases can also increase stress and raise annealing temperatures. Similar effects are present in electron beam–evaporated films.

REFERENCES AND FURTHER READING

1. J. W. Dini, *Electrodeposition: The Materials Science of Coatings and Substrates*, Noyes, Park Ridge, NJ, 1996.

2. D. Landolt, in *Electrochemically Deposited Thin Films III*, M. Paunovic and D. A. Scherson, eds., *Proceedings*, Vol. 97-26, Electrochemical Society, Pennington, NJ, 1997.

3. J. W. Dini, *Plat. Surf. Finish.* **75**, 11 (1988).

4. M. Schlesinger and J. P. Marton, J. *Phys. Chem. Solids* **29**, 188 (1968); *J. Appl. Phys.* **40**, 507 (1969); also S. L. Chow, N. E. Hedgecock, M. Schlesinger, and J. Rezek, *J. Electrochem. Soc.* **119**, 1614 (1972).

5. R. Weil and R. G. Barradas, eds., *Electrocrystallization, Proceedings*, Vol. 81–6, Electrochemical Society, Pennington, NJ, 1981.

6. M. Schlesinger, X. Meng, W. T. Evans, J. A. Saunders, and W. P. Kampert, *J. Electrochem. Soc.* **137**, 1706 (1990).

7. W. H. Safranek, *The Properties of Electrodeposited Metals and Alloys*, American Elsevier, New York, 1974; 2nd ed. published by American Electroplaters and Surface Finishers Society, Park Ridge, NJ, 1986.

8. C. C. Lo, J. A. Augis, and M. R. Pinnel, *J. Appl. Phys.* **50**, 6887 (1979).

9. See e.g., N. J. Petch, *J. Iron Steel Inst.* **174**, 25 (1953) and references therein.

10. H. McArthur, *Corrosion Prediction and Prevention in Motor Vehicles*, Ellis Horwood, Chichester, West Sessex, England, 1988.

11. See, e.g., A. J. Dekker, *Solid State Physics*, Prentice-Hall, Englewood Cliffs, NJ, 1961.

12. D. Tench and J. White, *Metall. Trans.* **A15**, 2029 (1984).

13. D. Simunovich, M. Schlesinger, and D. D. Snyder, *J. Electrochem. Soc.* **141**, L10 (1994).

14. K. D. Bird and M. Schlesinger, *J. Electrochem. Soc.* **142**, L65 (1995).

15. L. H. Esmore, *Trans. Inst. Met. Finish.* **57**, 57 (1974).

PROBLEMS

Nickel is known to have a face-centered cubic (fcc) type of crystal structure. The atomic density of the metal is 9.14×10^{22} atoms/cm^3, the atomic weight is 58.73, and the density (ρ) is 8.91 g/cm^3.

16.1. Determine the lattice parameter a and height h of a monolayer nickel metal.

16.2. Find the number of nickel atoms/cm^2 on the (110) plane in terms of the lattice parameter a.

16.3. Calculate the interplanar spacing d values for the (100), (110), and (111) planes in aluminum, the a of Al being 0.405 nm.

16.4. What are the Miller indices of a plane that intercepts the x-axis at a, the y-axis at $2a$, and the z-axis at $2a$?

17

Electrodeposited Multilayers

17.1. INTRODUCTION

Multilayers and "sandwich" systems have attracted the attention of both the scientific and engineering communities. This is hardly surprising since these types of constructs have a considerable number of practical applications as well as fundamentally explorable properties. As examples of the former, one may mention magnetically controlled read/write heads, electronic switching devices, and hard protective coatings. The latter include the study of magnetic exchange interactions, giant magnetoresistance, and many more examples. With the introduction of many reliable methods, including electrodeposition, to produce systems with nanometer-scale structural and composition variation, came the ability for greater control of material properties.

Nanostructural materials are divided into three main types: one-dimensional structures (more commonly known as multilayers) made of alternate thin layers of different composition, two-dimensional structures (wire-type elements suspended within a three-dimensional matrix), and three-dimensional constructs, which may be made of a distribution of fine particles suspended within a matrix (in either periodic or random fashion) or an aggregate of two or more phases with a nanometric grain size (these are illustrated in Fig. 17.1).

The semiconductor community has made "bandgap engineering" a reality by exploiting the ability to produce multiple-quantum-well material. (A *quantum well* is a potential well with dimensions such that quantum mechanical effects are to be taken into account. In other words, one thinks of electrons or holes as being trapped in the well, having distinct energy levels rather than a continuum.) The progress from one-dimensional to two- and three-dimensional nanostructural materials (i.e., to quantum wires and quantum dots) has provided enhanced carrier (electrons and/or holes) confinement by their presence, permitting greater control of energy-band structure.

Accurate control of microstructure on a nanometric scale makes it possible to control magnetic and mechanical properties to a hitherto unattainable degree. In particular,

Fundamentals of Electrochemical Deposition, Second Edition.
By Milan Paunovic and Mordechay Schlesinger
Copyright © 2006 John Wiley & Sons, Inc.

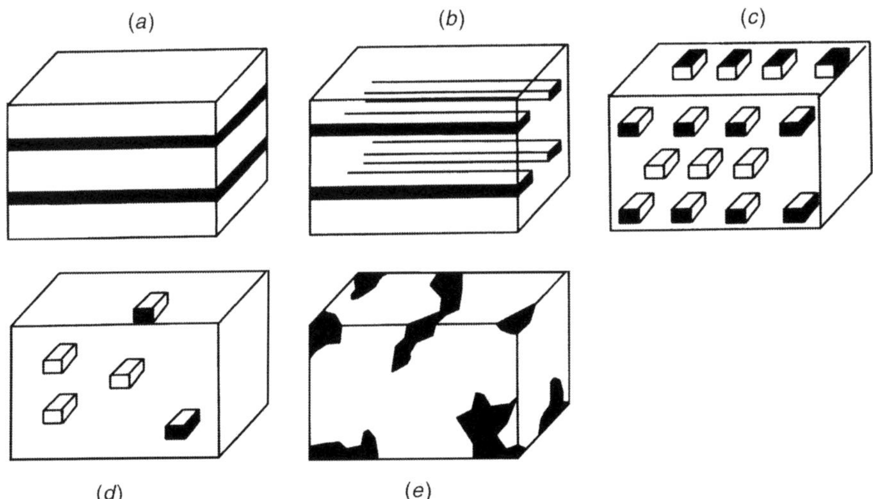

Figure 17.1. (*a*) Quantum wells, (*b*) quantum wires, (*c*) ordered arrays of quantum boxes, (*d*) random quantum dots, and (*e*) an aggregate of nanometer-size grains.

magnetic nanostructures have recently become the subject of an increasing number of experimental and theoretical studies. The materials are made of alternating layers, around 10 Å thick, of magnetic (e.g., cobalt) and nonmagnetic (e.g., copper) metals.

The magnetic layers are exchange-coupled to one another with a sign that oscillates with the thickness of the spacer (nonmagnetic) layer. This can be understood when one examines in detail the case of multilayer systems, with magnetic layers separated by nonmagnetic layers. The number of free conduction electrons is dependent on the number of nonmagnetic atoms, that is, the thickness of the nonmagnetic layer. As this number increases, subsequent orbitals in the magnetic atoms will experience alternating electron surpluses and deficits, causing a sinusoidal variation in the average magneton number per atom. Thus, the magnitude of the exchange interaction varies not only in the usual fashion, that is, inversely with the square of the distance between magnetic layers, but also in a sinusoidal manner. These results were known until recently for, say, sputtered or molecular-beam epitaxy (MBE)–grown samples.

Electrodeposition of composition-modulated films was first performed by Brenner in 1939 (1) by employing two separate baths for the two components and a "periodic" immersion of the deposit in the two baths. This is too cumbersome a method to be adopted in practice. Deposition from a single bath with the presence of salts of the two components of the multilayer is what is desired, but there was a serious problem with the deposition of two metals from one bath. Specifically, whereas a layer of the more noble member can be deposited by choosing the potential to be between the reduction potentials of the two metals, one can expect that when the potential is set to a value appropriate for reduction of the less noble member, both will be deposited, resulting in an alloy layer rather than a pure metal. Thus, to nobody's surprise, even as recently as 1983, Cohen et al. (2), were able to deposit only a layered structure of alloys rather than pure metals. In addition they cast doubt on the possibility that a modulation cycle (the thickness of the basic layers, the periodic repetition of which makes up the multilayer system) of less than 1000 Å can be achieved by means of electrodeposition.

17.2. ELECTRODEPOSITED NANOSTRUCTURES

In 1984, Tench and White (3) actually deposited composite structures with layer thickness "periods" down to hundreds of angstroms. Their nanocomposites exhibited increased tensile strength due to the harder Ni layers in a softer Cu matrix. The layers were not thin enough, however, to obtain enhancements witnessed in systems (non-electrodeposited) with layer thickness periods in the range 10 to 30 Å.

In 1986, Yahalom and Zadok (4) pointed to methods to produce composition-modulated alloys by electrodeposition, initially for the copper–nickel couple. They obtained modulation to thicknesses down to 8 Å. The principle of the method is as follows. Traces of metal A ions are introduced into a concentrated solution of metal B (assuming that metal A is nobler than B). At a sufficiently low polarization potential (point B, Fig. 17.2), the rate of reduction of metal B is high, as determined by its activation polarization. The rate of reduction of metal A is slow and controlled by diffusion. At a predetermined considerably less negative polarization potential (point A, Fig. 17.2), only metal A is reduced. The potential is simply switched between these two potential values, forming a modulated structure composed of pure A layers and layers of B with traces of A. The actual level of A alloying is given (assuming that efficiencies are about the same) by

$$\frac{i_{L_A}}{i_B + i_{L_A}}$$

where i_{L_A} is the limiting current density of metal A and i_B is the partial current density for the reduction of metal B. Proper selection of deposition parameters can yield negligible values of dilution of A in the B layers.

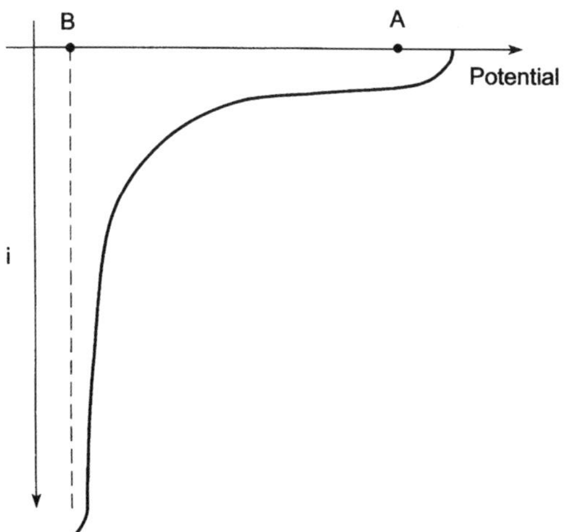

Figure 17.2. Current versus potential for a two (double)-electrolyte bath.

The actual deposition of the multilayered composite can be carried out by either current or potential control. Clearly, a pulsed polarization curve has to be constructed for the former case. The actual composition modulation cycle would be controlled coulometrically by fixing the amount of electric charge delivered while at point A in Figure 17.2 (Q_A) and the amount of charge delivered while at point B in Figure 17.2 (Q_B) via suitable input to the unit regulating the pulsing.

This method is suitable for a considerable number of metal couples. The main caveat, however, is that both metals can be deposited from similar baths. Another is that they differ sufficiently in their degree of nobility. This method (advocated by Yahalom and Zadok in Ref. 4) was first tried using a Cu–Ni system. It is still the system most studied when it comes to electrochemically produced layered structures. The reasons for this are quite compelling. There is a sufficient difference in reduction potentials between the two metals. There is a similarity in crystal structure (fcc) and proximity of lattice parameters, ensuring a good degree of coherency between the layers. Also, a considerable number of data exist in the literature concerning magnetic, mechanical, and other properties of a Cu–Ni system deposited by other than electrodeposition, enabling comparison of the modulated layers produced by the chemical method.

In 1987, Lashmore and Dariel (5) improved on Yahalom and Zadok's methods for obtaining Cu–Ni layers by:

- Substituting sulfamate electrolyte for sulfate
- Using a triple galvanostatic pulse consisting of a nickel pulse (the less noble element), an anodic pulse (to change potentials rapidly), and a copper pulse (the more noble element)
- Electropolishing the substrate; cold-rolled 150-μm-thick copper sheet or 15-mm-diameter copper single crystals

These modifications led to obtaining the following:

1. One-dimensional coherent and layered structure extending over thousands of layers.
2. Epitaxial coherency maintained for the three low-index orientations of the fcc copper lattice substrate to various degrees.
3. A strong tendency for the multilayered coating to adopt a {111} texture. Actually, an extended one-dimensional coherent {111} texture develops even on initially non-[111]-oriented Cu substrate.
4. Coherency strains, resulting in the broadening of the x-ray diffraction lines (more pronounced for the [100] and [110] than for the [111] orientation).
5. X-ray diffraction patterns of electrodeposited multilayered systems of sufficiently good quality to allow quantitative analysis of coherent and noncoherent strain features.

In what follows we delineate details of an actual (6) deposition procedure for producing, for example, Ni–Cu multilayers as a practical example. As with previous methods, artificially layered deposits may be obtained from a single chemical solution using a specially designed cell: for instance, one with adjustable anode–cathode gap (see Fig. 17.3). This two-compartment cell may be constructed from Lucite with deposition conducted in one compartment and KCl solution placed in the other. A calomel reference electrode immersed in this KCl solution should be coupled to the flat-plate cathode by a salt bridge, ending in a capillary on the deposition side. The specimen electrode is fixed, and the counterelectrode is movable using, say, a micrometer. Electrodeposition is best conducted under quiescent conditions.

Multilayer depositions may be performed using, say, an EG&G Princeton Applied Research (PAR) Model 173 potentiostat/galvanostat with a PAR Model 276 interface providing computer control capability. A PC furnished with a GPIB-PCII/IIA interface board, and an IEEE-488 general-purpose interface bus may be used for communication. The internal command set of the PAR Model 276 interface allows for waveform generation to circumvent the requirement for analog programming. The metallizing electrolyte has the following typical composition: 421 g/L nickel sulfamate, 45 g/L boric acid, 0.15 g/L antipitting agent (SNAP), and 0.25 g/L copper sulfate.

Deposition can be carried out potentiostatically, with a copper cycle deposition potential of 0.17 V versus SCE and a nickel deposition cycle of 1.19 V versus SCE. The durations of the pulses can be set to give the desired thicknesses, and for each experiment the number of coulombs passed in the copper and the nickel deposition time segments can be controlled. Deposition of samples can typically be made onto commercial polycrystalline copper sheet supplied by, say, Ventron (Alpha Products), and this copper substrate can be dissolved subsequently by immersion in an NH_4OH/H_2O_2 solution.

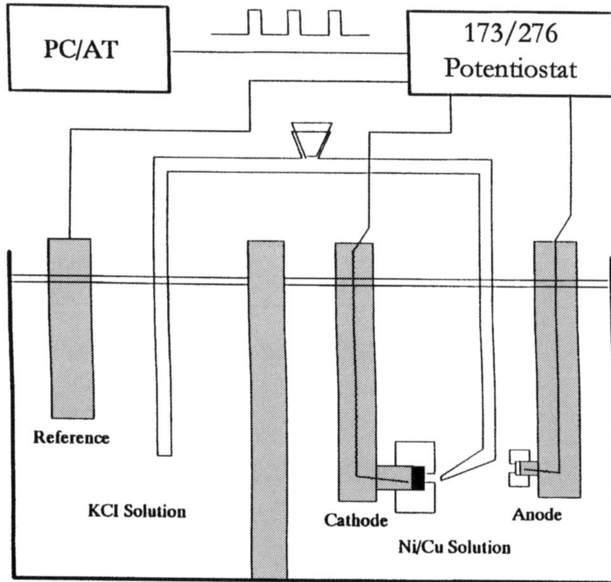

Figure 17.3. Schematic description of the electrodeposition cell for the production of superlattice multilayers. (From M. Schlesinger, Chapter 14 in *Electrochemical Technology*, T. Osaka, ed., with permission from Kodansha Ltd.)

As indicated above, a large number of metal pairs and metal alloy pairs can be deposited in a similar fashion (Cu/Co, Ag/Ni, Pd/Fe, etc.). It is also possible to electro-deposit multilayers in cylindrical pores of a suitable etched polymer membrane. Typically, wires with diameters of about 100 nm and length of 5 to 10 μm can be obtained. The deposition cycles are similar to the ones described above. Magnetoresistance [this is a term describing the relative decrease (increase) in electrical resistance of a material when subjected to a magnetic field longitudinally (transversely) to the current flow] measurements with the current perpendicular to the planes are possible. In addition, giant magnetoresistance (GMR; defined below) effects may be observed as well.

17.3. ANALYSIS OF DEPOSIT

A standard method for confirming coherence of the layers is the study of x-ray diffraction spectra. If the layers are coherent and there are enough of them to provide a relatively strong Bragg diffraction pattern, satellites due to superlattice (see Chapter 16) formation should appear on each side of the Bragg diffraction peak. Although detailed treatments can be found in the literature, we present below a simplified but rather useful formula for the determination of layer periodicity.

We may assume any of the Bragg peaks as being due to a primary beam as far as the superlattice is concerned and, further, that we are considering forward scattering by the superlattice. Then the usual formula can be used; thus, the angular shift between a Bragg peak and its satellite is given by

$$\sin \Delta\theta = \frac{n\lambda}{\Lambda} \qquad (17.1)$$

where $\Delta\theta$ is the angular shift, λ is the x-ray photon wavelength, and Λ is the periodicity of the superlattice. In practice, one does not expect that $n \gg 1$. Furthermore, since the multilayer–superlattice symmetry is one-dimensional, it is to be expected that the reflection whose k vector projection on the axis of symmetry is larger should be the more intense. In other words, the satellite that corresponds to the lower θ value is expected to be more intense than the satellite at higher θ.

In a set of experiments (6), the authors have determined the electrochemical efficiency of Ni/Cu layer deposition. It was found that the copper layers deposit with 96% efficiency and the nickel deposit with 90% efficiency. This information, together with the measured coulomb input per layer, enables one to confirm the validity of the suggested formula. Alternatively, if one accepts the arguments that lead to the formula, the electrochemical efficiency values can be viewed as confirmed. The relatively slight deviation from perfect efficiency, at least in the case of nickel, is probably connected with hydrogen evolution.

An alternative (more accurate) treatment yields a somewhat different expression for the difference between the sine values of the two satellites:

$$\sin \theta_s^+ - \sin \theta_s^- = \frac{\lambda}{\Lambda} \qquad (17.2)$$

This expression is a consequence of the analogous formula for the case of light diffraction by the optical grating.

17.4. PHYSICAL PROPERTIES

17.4.1. Hardness

We concentrate on hardness results first because of their practical implications. Since composites can be deposited from a single solution and on almost any substrate material, this property becomes important and useful. In Table 17.1 we summarize microhardness values for samples with a variety of individual layer thicknesses. The 10-Å-layer sample consists of 5000 individual layers; all the other samples contain 4000 individual layers.

The results show that a nanolayered composite structure made of two metals increase in strength, decidedly as the layer thickness is reduced to some optimal value. In epitaxial systems such as the one at hand, this strengthening may be attributed to Young's modulus and lattice parameter mismatches between adjacent layers. The modulus mismatch introduces a force between a dislocation and its image in the interface. The lattice parameter mismatch generates stresses and mismatch dislocations, and these interact with mobile dislocations. A peak in the yield stress occurs when single dislocations must overcome both barriers. The yield stress drops in thicker layers as pileups of increasing length form in the layers.

In general, one can expect systems like these to have properties different from those in bulk materials, where most ions are in a crystalline environment; that is, they do not "see" the surface. Here, most atoms "know" they are near an interface: specifically, in a noncrystalline, nonperiodic environment. As shown in Table 17.1 and Figure 17.4, the Knoop hardness of these films is increased from 2.9 to 5.6 times the value for the harder of the two components, the electrodeposited nickel.

The films are epitaxial in the sense that the lattice constant is intermediate between those of copper and nickel. As indicated above, that modulated strain is probably responsible for the increased hardness. Other authors (5) have tried to explain similar effects by stating that the layers were specifically oriented. Our example (6) demonstrates that these considerations must be reexamined since it was possible to achieve the effect in a crystalline multilayer deposited on an amorphous nickel–phosphorus underlayer. It appears that layer thickness is the important parameter here.

TABLE 17.1. Cu/Ni Knoop Hardness Data for Electrodeposited Multilayers[a]

Number of Layers	Individual Layer Thickness (Å)	Hardness
5000	10	385 ± 87
4000	20	747 ± 27
4000	40	523 ± 22
4000	100	452 ± 27

Source: Ref. 6a, with permission from the Electrochemical Society.

[a]Hardness values for electrodeposited sulfamate Ni were 133 ± 19.

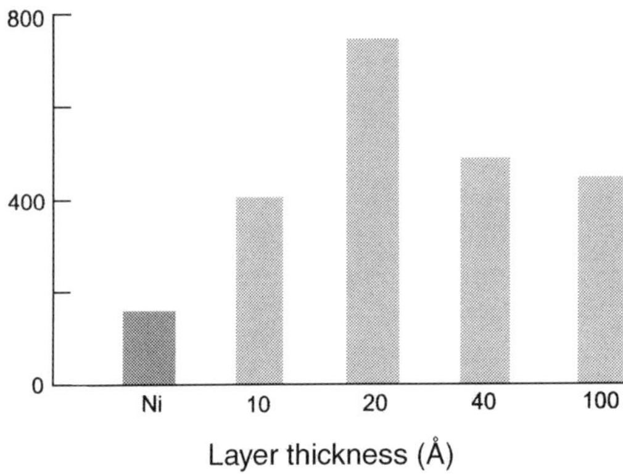

Figure 17.4. Hardness data for electrochemically grown Co/Ni superlattice multilayers as a function of individual layer thickness. (From Ref. 6a, with permission from the Electrochemical Society.)

17.4.2. Microstructure

In Figure 17.5 the main Bragg peak is seen to be intermediate between diffraction peaks for copper and nickel, not seen (in Fig. 17.5) in this angular region. The extra shoulder is a primary peak from the copper substrate, which for this (1-nm/layer) sample was not removed (by etch) following electrodeposition. Satellites at about 50° and 54° correspond to a $\Lambda/2$ value of approximately 10 Å. Note that the satellite at the lower angle is indeed stronger, as indicated above. The width of the major peaks themselves indicates clearly that the electrodeposit on an amorphous Ni–P film is crystalline, with a strained superlattice. The good definition of the satellites indicates that the layer formation is reproduced cycle to cycle, and the thicknesses of successive layers are nearly the same.

It should be pointed out also that at least in principle, higher-order ($n > 1$) satellites can be expected to be evident in superlattice multilayer x-ray spectra when the

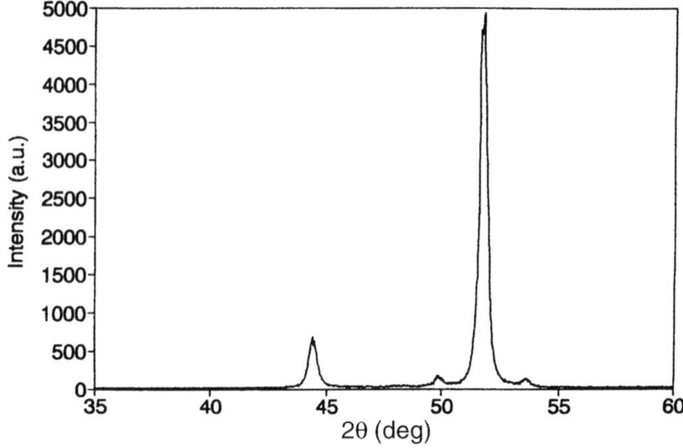

Figure 17.5. Bragg (x-ray) diffraction peak and typical satellites off a copper–nickel nanocomposite (electrochemically grown).

boundary layer between the component metals is "sharp" (i.e., the transition between layers is abrupt). This is the consequence of the satellites being the Fourier transform of the element distribution as one passes from one layer to the other.

17.5. MAGNETIC PROPERTIES (MAGNETORESISTANCE)

As stated above, *magnetoresistance* is a term describing the relative decrease (increase) in the resistance of a material when subjected to a magnetic field longitudinally (transversely) with respect to the flowing current. Typically, this effect is in the range 0 to 2%. In the past few years, a new phrase, *giant magnetoresistance* (GMR), has been coined, to describe the same effect when it increases by an order of magnitude or more. It is obvious that many applications exist for a material whose resistance is dependent on an external magnetic field. Such a substance could be used for magnetic sensing, for electronic switches, and as a memory detector in a dense storage medium.

The first material found to display GMR was a multilayer system consisting of 30-Å layers of iron interspersed with chromium layers between 9 and 18 Å. This system, reported by Baibich et al. (7), yielded almost a 50% decrease in resistivity when subjected to an external magnetic field of approximately 20 kG or 2 T. Since this groundbreaking work in 1988, several other multilayer systems, consisting of ferromagnetic layers interspersed with nonmagnetic metallic layers, have resulted in GMR values of 5 to 60%. Perhaps the most studied of these systems is again that of nickel–copper multilayers. Although the GMR effect in a nickel system is smaller, nickel has the advantage of greater stability and longevity than iron, which oxidizes rapidly and displays a large hysteresis effect. For this reason, nickel systems can be used as a benchmark as to the efficiency and accuracy of an experimental procedure, prior to the examination of another system, such as cobalt, which is the primary subject of investigation in some advanced projects. Ideally, GMR values depend on the materials used, the superlattice structure of the multilayers, and the electromagnetic conditions imposed for testing purposes in a given system, not on the method of multilayer preparation. Preparation methodology, however, bears directly on the lattice and superlattice characteristics of the sample.

All the possible methods of deposition have inherent advantages and disadvantages with regard to the quality of the multilayers they create and the ease of their production; however, electrodeposition seems best to fulfill imposed financial and temporal restrictions. Specifically, vapor deposition is rather expensive, creates quasi-amorphous interfaces, and is time consuming in terms of controlling the alternation of deposition material. Electroless deposition may suffer from the same drawback with respect to mechanical switching between solutions for alternating layer materials. Finally, sputtering requires an expensive vacuum system and cannot easily be extended to industrial use. For these reasons, electrodeposition is now the production method of choice for most practical exploitations.

In general, all metals, magnetic or otherwise, display some degree of magnetoresistivity, as noted by Kelvin almost 150 years ago. This property may be seen as a direct result of the Lorentz equation,

$$F = -e(E + v \times B) \tag{17.3}$$

where F represents the (Lorentz) force acting on the moving, at velocity v, of an electron with charge e in magnetic field B. The force is directed perpendicular to the directions of B and v. Specifically, and expressing it in the "language" of physicists, a moving charge in an external magnetic field is subject to a force whose direction is given by the cross-product of the direction vector of motion of that charge and the direction vector of the magnetic field.

Should the force be directed so as to deflect the path of the moving charge, an additional resistance, usually called the *Hall resistance*, would be observed. It should be noted that the magnitude of the effect is very small, invariably much less than 1% of the initial resistance. A secondary effect occurs in magnetic metals, however, and results in a modified magnetoresistivity curve.

The aforementioned secondary effect is the interaction of the magnetic moments of the charge carriers (e.g., each electron is said to be equivalent magnetically to a small current loop) with the magnetic domains (these are regions in a solid that act as single magnets) of the constituent metal. These interactions serve to scatter from their original path, the moving charge carrier, effectively increasing the resistance of the metal. An external magnetic field can be oriented to enhance or diminish this effect by being aligned transversely or longitudinally, respectively, with the motion of the charge carriers. Again, it should be noted that the maximum magnetoresistive change achieved in a magnetic metal, specifically Permalloy (a mixture of iron and nickel) is approximately 4%. The term *giant magnetoresistance* has been reserved for changes in resistance due to externally applied magnetic fields up to an order of magnitude (10-fold) larger than those resulting from magnetoresistance. Although GMR has been observed in systems other than compositionally modulated materials, the explanation for its mechanism given here is limited to multilayer systems. As a multilayer sample is deposited, the minimum-energy requirement for the system ensures that successive magnetic layers will contain magnetic domains that are oriented antiferromagnetically, as shown schematically in Figure 17.6.

To confirm this fact, researchers (8) have employed Lorentz-type microscopy. The main features of this method are depicted in Figure 17.7. From the description given, it is expected that brighter and darker regions will signify the magnetic domain walls, provided that the image is over or under focus. Changing from one to the other will result in interchange of the darker with the brighter domain wall images.

Shortly after finding GMR effects in artificial superlattices, a semiclassical theory using the Boltzmann transport equation was devised to explain the origin of the effect (9). The authors extended the Fuchs–Sondheimer theories to rescattering (10) and introduced electron-spin-dependent scattering, which is related to the energy-band structures of the magnetic and nonmagnetic constituents (11). They concluded that the GMR effects depend on the ratio of layer thickness to the electron mean free path (the average distance traveled) between collisions and on the asymmetry in scattering of spin-down and spin-up electrons.

The application of an external magnetic field will "bend" the orientation of the antiparallel magnetic domains toward the direction of the parallel domains. This situation creates a *spin valve*; that is, charge carriers of the "correct" magnetic moment orientation may pass through the magnetic layers unscattered by their magnetic

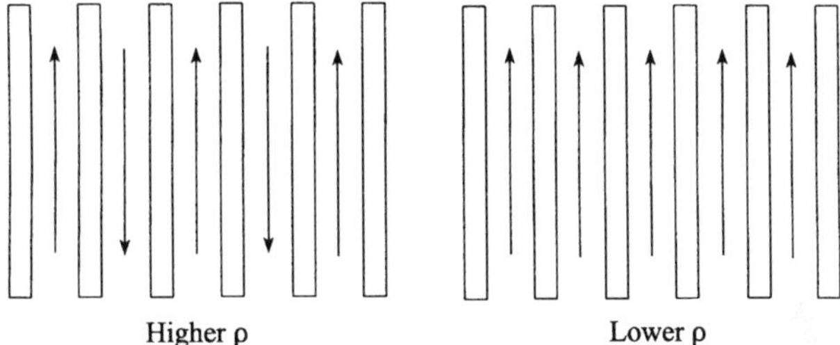

Figure 17.6. Oversimplified model of the magnetoresistance in superlattice multilayers. (From M. Schlesinger, Chapter 14 in *Electrochemical Technology*, T. Osaka, ed., with permission from Kodansha Ltd.)

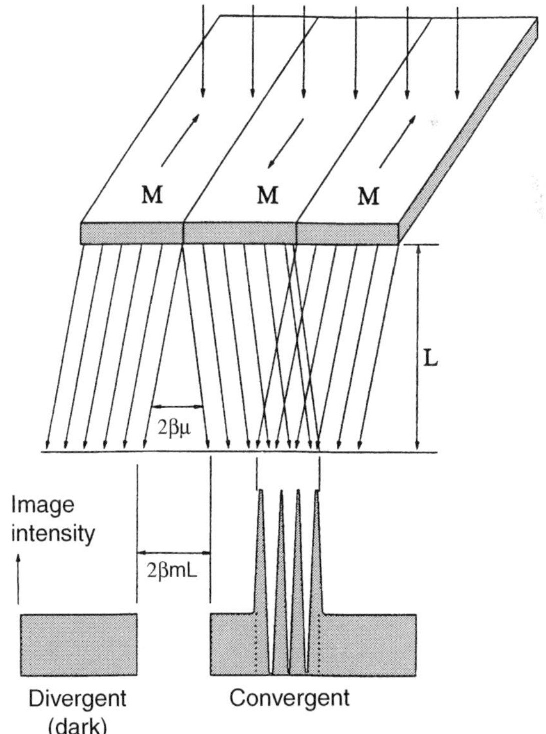

Figure 17.7. Schematic description of the main features of Lorentz microscopy. (From M. Schlesinger, Chapter 14 in *Electrochemical Technology*, T. Osaka, ed., with permission from Kodansha Ltd.)

domains. The resulting Hall/GMR/magnetoresistivity curves would be like those shown in Figure 17.8.

In actuality, the successive magnetic layers are not strictly antiparallel under zero external field, nor are they strictly parallel when under a saturating external field. The orientation of magnetic domains is dictated by minimization of the free energy, which

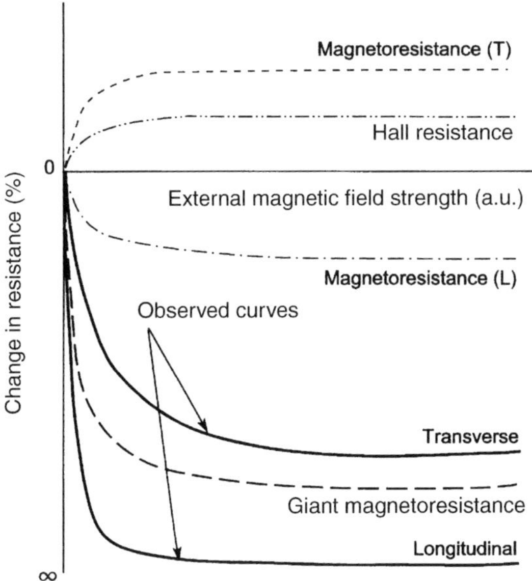

Figure 17.8. Superposition of all contributions toward observed GMR.

is composed of, among other energies, the Zeeman and interface exchange energies. The Zeeman energy is essentially linearly dependent on the number of spin layers per magnetic layer, whereas the interface exchange energy is virtually independent. Variations in GMR with respect to layer thickness exist because minimization of the former favors an antiparallel orientation, whereas minimization of the latter favors parallel orientation.

More recently, the many theoretical models proposed for an understanding of GMR effects may be classified into two types of approaches, one based on RKKY (Rudeman–Kittel–Kasuya–Yoshida)-like schemes and the other on energy-band structure calculations.

It has been demonstrated that the RKKY interaction is an accurate description of the exchange interaction of two magnetic impurities within a nonmagnetic medium. This theory has now been extended to the multilayer system, with magnetic layers separated by nonmagnetic spacer layers and yields an exchange interaction, $\mathbf{J}$, dependent on the distance between magnetic layers, z, of the form given by

$$\mathbf{J}(z) \approx \frac{A \sin 2K_F z}{z^2} \tag{17.4}$$

where A is a constant and K_F is the Fermi wavevector. Thus, it has been predicted that the saturation GMR value will vary sinusodially with the thickness of the nonmagnetic spacer layer.

In Figure 17.9 the saturation GMR in Co/Cu electrodeposited multilayers with 20-Å-layer cobalt and varying thickness of the Cu layers is given as obtained experimentally (6b) together with a fitted RKKY type of theoretical curve. For illustration

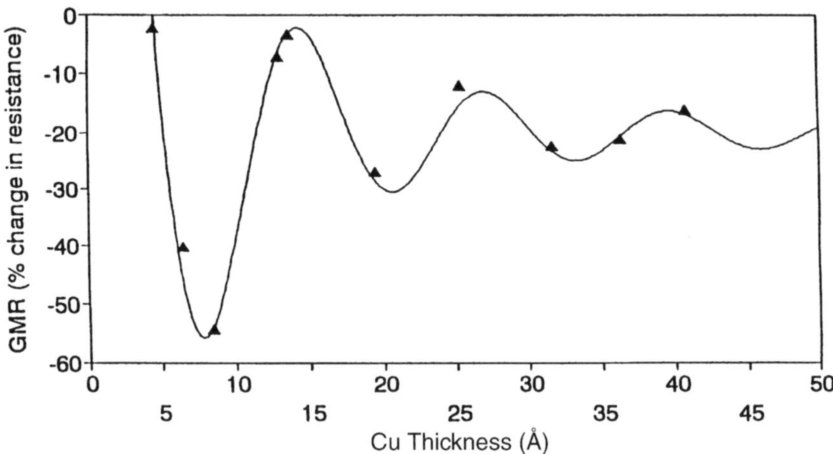

Figure 17.9. Maximum (saturation) value of the giant longitudinal magnetoresistance (GMR) in electrochemically grown Co/Cu multilayers as a function of Cu layer thickness. Cobalt layer thickness is held constant at 20 Å per layer. The continuous curve is the corresponding RKKY function. (From Ref. 6b, with permission from the Electrochemical Society.)

purposes, the GMR versus magnetic field using either dc or ac (1 kHz) for the spacer thickness giving the highest-saturation GMR value are given in Figure 17.10.

The multilayered Cu/Co systems discussed here can be grown as described next (6b). Electrolyte composition is based on a cobalt/copper ratio of 100 : 1 and consists of a solution of 0.34 M cobalt sulfate, 0.003 M copper sulfate, and 30 g/L boric acid. The pH is fixed around 3.0, and there is no forced convection while deposition is carried out. The electrodeposition may usually be carried out potentiostatically at 45°C between -1.40 V versus SCE for the cobalt and -0.65 V versus SCE for the copper with an $\approx 3\,s$ cell potential interrupt between the cobalt-to-copper transition to avoid cobalt dissolution, which can occur when there is no interrupt.

Table 17.2 summarizes typical saturation GMR results for electrochemically deposited Ni/Cu, CoNi/Cu (12), and Co/Cu multilayered constructs. Obviously, the highest GMR effects shown are by the Co/Cu system.

In general, RKKY models relate the "period" of the interlayer magnetic coupling to the topology of the Fermi surface for larger spacer thicknesses. RKKY interactions occur when a magnetic ion is introduced into a nonmagnetic crystal. The conduction electron gas in the vicinity of the magnetic ion is magnetized, and a second magnetic ion will experience an indirect exchange interaction with this magnetized electron cloud. As ferromagnetism results from a "shortage" of electrons, or holes, in a particular orbital of an ion, the conduction electron gas is supplied by the nonmagnetic metal. In the case of multilayer systems such as the one represented in Figure 17.9, the number of free conduction electrons (in the copper layer) is dependent on the number of nonmagnetic atoms, specifically the thickness of the copper layer. As this number increases, subsequent orbitals in the magnetic atoms will experience alternating electron surpluses and deficits, causing a sinusoidal variation in the average magneton number per atom. Thus, the magnitude of the exchange interaction

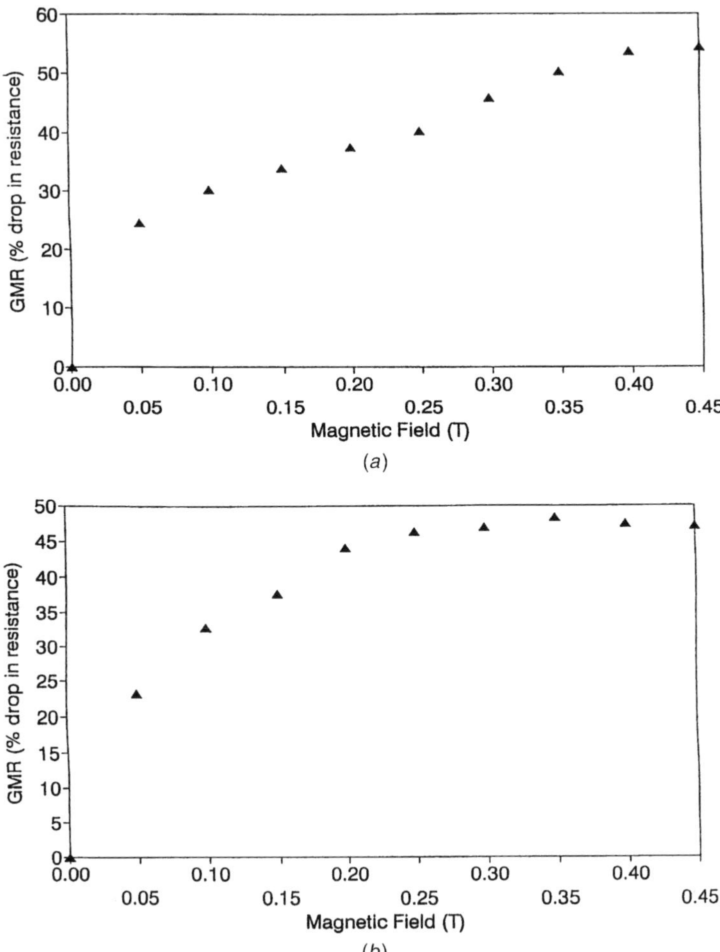

Figure 17.10. GMR value versus longitudinal magnetic field in electrochemically grown Co/Cu multilayers: (*a*) resistivity measured using direct current; (*b*) resistivity measured using alternating current (1 kHz). (From M. Schlesinger, Chapter 14 in *Electrochemical Technology*, T. Osaka, ed., with permission from Kodansha Ltd.)

TABLE 17.2. Saturation GMR Values for Three Electrochemically Deposited Compounds

	Period of Oscillation (Å)	Ratio of Peaks $P_1 : P_2 : P_3$	Position of First Peak (Å)	Thinnest Cu Layer (Å)	$\dfrac{\Delta\rho}{\rho}$ max.
Ni/Cu	12	$1 > 2 > 3$	8	6	3.5 (8 Å)
CoNi/(70%Ni)/Cu	12	$1 < 2 > 3$	10	4.6	23 (8.5 Å)
Co/Cu	12	$1 > 2 > 3$	8.5	7	60 (8.5 Å)

Source: Ref. 6b, with permission from the Electrochemical Society.

varies not only in the usual fashion (i.e., inversely with the square of the distance) but also in a sinusoidal manner (see Fig. 17.9, where this behavior is evident). These models cannot, however, explain the dependence of the "periods" on the spacer (non-magnetic) elements. They cannot be applied to systems with magnetically ordered spacers such as Fe/Mn and Fe/Cr.

The second theoretical approach uses an ab initio or "tight-binding" approach to the description of electronic structure. Determination of the interlayer magnetic coupling can be achieved by calculating the total energy difference, $\Delta E = E_{\mathrm{E}} - E_{\mathrm{AF}}$, between the two types of interlayer arrangements: ferromagnetic and antiferromagnetic coupled (13). The period and magnitude of interlayer coupling do not agree with the magnitude observed and the long period of oscillations found in practice. Clearly, none of the preceding models is capable of giving fully satisfactory explanations to all GMR phenomena as observed empirically.

17.6. MULTILAYERS IN ELECTRONICS

A discussion of current aspects of multilayer deposition should include methods and processes in the buildup and synthesis of microsystems. It is electroplating that has become the technique of choice for the production of integral components of such systems. That is due to the many advantages of electroplating methods, such as high deposition rate, high resolution, high shape fidelity, simple scalability, and good compatibility with existing processes in microelectronics. Specifically, as can be learned from the rest of this book and Ref. 14, materials with widely diverse properties, such as composition, crystallographic orientation, and grain size, are obtained with relative ease through electroplating. As an illustrative example we mention the following. High-conductivity copper or gold for interconnects and multichip applications, soldering materials based on indium or tin–lead required for flip-chip, and even soft or hard magnetic materials based on nickel, iron, and cobalt are all possible and are employed in the microelectronics industry.

A production process has recently been implemented by IBM. The aim was to reduce the electrical resistance of the interconnects in their chip to about one-third of the values attainable using aluminum and at the same time increasing the resistance against electromigration. This was made possible by employing electrodeposition of copper in a Damascene method. The manufacturing sequence is presented in Figure 17.11.

First, copper is deposited on a thin seed layer on top of an oxide layer, which, in turn, contains trenches and vias that connect to lower levels. The material is usually deposited to a thickness down to $\ll 1 \ \mu m$, both inside the features and on the rest of the surface. After plating, the wafer is polished using chemical–mechanical polishing (CMP) to remove the excess copper. Note that in this application, an electrodeposition process was introduced rather than a sputtering or evaporation step. With the latter techniques, it would be impossible to obtain properly filled features of only a few tenths of a micrometer across and aspect ratios of 1 or more. The electrodeposition process starts at the seed layer, and through the use of well-chosen additives (15) in the electrolyte (plating bath), the plating process can be adjusted to "super-fill" the cavities (i.e., the growth at the bottom of the trenches and vias proceeds more rapidly

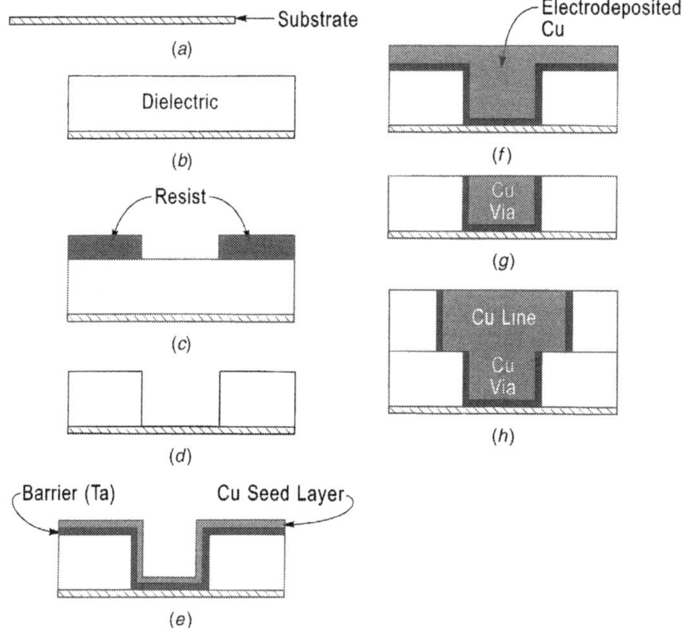

Figure 17.11. Process steps for forming Cu interconnects using the single damascene process (dielectric patterning): (*a*) planarized substrate; (*b*) dielectric deposition; (*c*) dielectric RIE through photoresist mask; (*d*) etched insulator; (*e*) deposition of diffusion barrier (Ta) and Cu seed layer; (*f*) electrodeposition of Cu into a via (vertical interconnection); (*g*) CMP of Cu excess; (*h*) patterning and deposition of Cu line (wire).

than at the top or the edges). Such control is not possible in sputtering or evaporation, where the deposits would soon start to obstruct the features and make further filling unachievable. This would result in voids within the interconnects.

17.7. CONCLUSIONS

Electrodeposition is to date the most economical and convenient path to the production of multilayered nanostructures. The contracts thus obtained can and do provide useful paradigms for fundamental studies (14) as well as for a plethora of useful applications. Some of the applications are in the arena of magnetically operated read/write elements; others are in the field of electronic switches, to name only a few.

Electrodeposition presents, in principle, several advantages for the investigation and production of layered alloys. Among these are the tendency of electrodeposited materials to grow epitaxial and thus to form materials with a texture influenced by the substrate. Electrodeposition can be used in systems that do not lend themselves to vacuum deposition. The electrodeposition process is inexpensive and can be upscaled with relative ease for use on large parts; further, it is a room-temperature technology. This last point may be important for systems in which undesirable interdiffusion between the adjacent layers may readily occur.

The studies of electrodeposited multilayer materials to date show clearly that electrodeposition is a feasible technique for the production of thin multilayered materials in systems that from an electrochemical standpoint are adaptable to the pulsed deposition technique.

Recent results were able to demonstrate that the coatings produced by electrodeposition display the same coherence and layer thickness uniformity as those of composition-modulated alloys produced by vacuum evaporation or sputter deposition.

Demonstration of the capability of electrodeposition to produce materials with predesignable, variable, and controllable composition down to practically the atomic scale constitutes an important step toward the realization of custom-tailored materials. On the theoretical side, the lack of a single satisfactory theory for the possible explanation of the different empirical results is somewhat disappointing.

REFERENCES AND FURTHER READING

1. A. Brenner, Ph.D. dissertation, University of Maryland, College Park, MD, 1939.

2. U. Cohen, F. B. Koch, and R. Sard, *J. Electrochem. Soc.* **130**, 1937 (1983).

3. D. Tench and J. White, *Metall. Trans.* **A15**, 2039 (1984).

4. J. Yahalom and O. Zadok, *J. Mater. Sci.* **22**, 499–503 (1987).

5. D. S. Lashmore and M. P. Darial, *J. Electrochem. Soc.* **135**, 1218–1221 (1988).

6. (a) D. Simunovich, M. Schlesinger, and D. Snyder, *J. Electrochem. Soc.* **141**, L10, (1994); (b) K. D. Bird and M. Schlesinger, *J. Electrochem. Soc.* **142**, L65 (1995).

7. M. Baibich, J. Broto, A. Fert, F. Nguyen Van Dav, F. Petroff, P. Eitene, G. Greuzet, A. Friedrich, and J. Chuzelus, *Phys. Rev. Lett.* **61**, 2472 (1988).

8. See, e.g., M. Schlesinger et al., *Scr. Met. Mater.* **33**, 1643 (1995).

9. R. E. Camley and B. Barnas, *Phys. Rev. Lett.* **63**, 664 (1989).

10. K. Fuchs et al., *Cambridge Philos. Soc.* **34**, 100 (1938).

11. H. Sondheimer et al., *Adv. Phys.* **1**, 1 (1952).

12. R. D. McMichael, U. Atzmony, C. Beauchamp, L. H. Bennett, L. J. Swartzendruber, D. S. Lashmore, and R. T. Romankiw, *J. Magn. Magn. Mater.* **113**, 149–154 (1992).

13. D. M. Edwards, J. Mathon, R. B. Muniz, and M. S. Phan, *Phys. Rev. Lett.* **67**, 493 (1991).

14. M. Schlesinger and M. Paunovic, *Modern Electroplating*, 4th ed., Wiley, New York, (2000).

15. For an excellent list of references, see B. A. Jones and C. B. Hanna, *Phys. Rev. Lett.* **71**, 4253–4256 (1993).

PROBLEMS

17.1. Formula (17.2) involves the difference between the sine value of the two satellites on either side of the main Bragg peak. An equivalent expression is

$$\sin\theta_i - \sin\theta_{i-1} = \frac{\lambda}{2\Lambda}$$

where the symbols are equivalent to those in Eq. (17.2). The angles are the positions of two adjacent diffraction peaks (the central and satellite peaks). With the help of the Bragg diffraction peaks and satellites shown in Figure P17.1, determine the wavelength of the x-rays used in this experiment. Note that the x-axis is labeled 2θ. This is due to the method employed in obtaining the diffraction results presented.

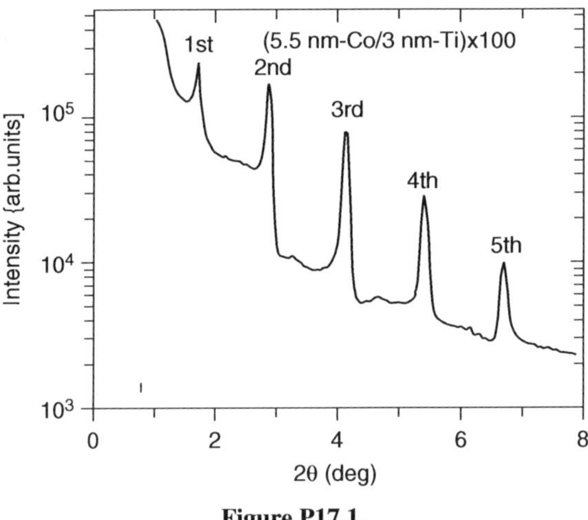

Figure P17.1.

17.2. Another set of experiments with TiN/NbN multilayers, yielded the following x-ray, reflectivity (XRR), and diffraction (XRD) curves (Figure P17.2).

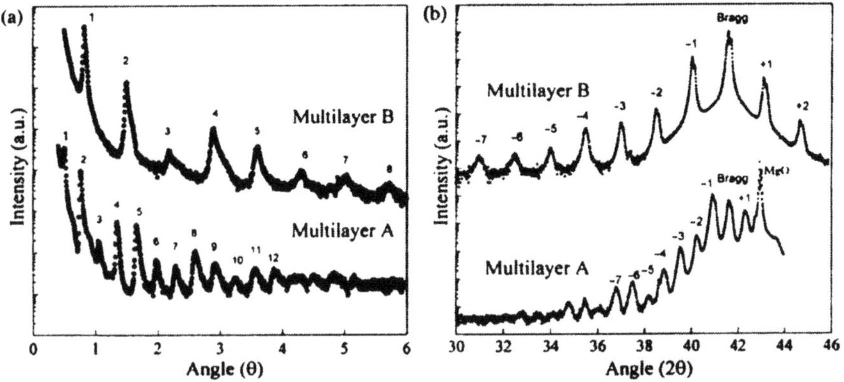

Figure P17.2. Experimental low-angle XRR (x-ray reflectivity) patterns (*a*), and experimental high-angle XRD (x-ray diffraction) patterns (*b*) from multilayers A and B.

Assume the x-ray wavelengths (λ) used are identical to the ones in Problem 17.1. Determine the values of Λ (the multilayers period) for each of the two samples.

18

Interdiffusion in Thin Films

18.1. INTRODUCTION

When two solids or films are put together, they may generally be expected to undergo interdiffusion (1). A number of parameters, such as temperature, influence the degree and rate of the process. Interdiffusion is treated here because modern integrated circuits are made of layered thin-film structures that are subject to interdiffusion during the thermal processing stage of fabrication.

Diffusion in general, not only in the case of thin films, is a thermodynamically irreversible self-driven process. It is best defined in simple terms, such as the tendency of two gases to mix when separated by a porous partition. It drives toward an equilibrium maximum-entropy state of a system. It does so by eliminating concentration gradients of, for example, impurity atoms or vacancies in a solid or between physically connected thin films. In the case of two gases separated by a porous partition, it leads eventually to perfect mixing of the two.

In equilibrium, impurities or vacancies will be distributed uniformly. Similarly, in the case of two gases, as above, once a thorough mixture has been formed on both sides of the partition, the diffusion process is complete. Also at that stage, the entropy of the system has reached its maximum value because the *information* regarding the whereabouts of the two gases has been minimized. In general, it should be remembered that entropy of a system is a measure of the information available about that system. Thus, the constant increase of entropy in the universe, it is argued, should lead eventually to an absolutely chaotic state in which absolutely no information is available.

The diffusion process in general may be viewed as the model for specific well-defined transport problems. In particle diffusion, one is concerned with the transport of particles through systems of particles in a direction perpendicular to surfaces of constant concentration; in a viscous fluid flow, with the transport of momentum by particles in a direction perpendicular to the flow; and in electrical conductivity, with the transport of charges by particles in a direction perpendicular to equal-potential surfaces.

Fundamentals of Electrochemical Deposition, Second Edition.
By Milan Paunovic and Mordechay Schlesinger
Copyright © 2006 John Wiley & Sons, Inc.

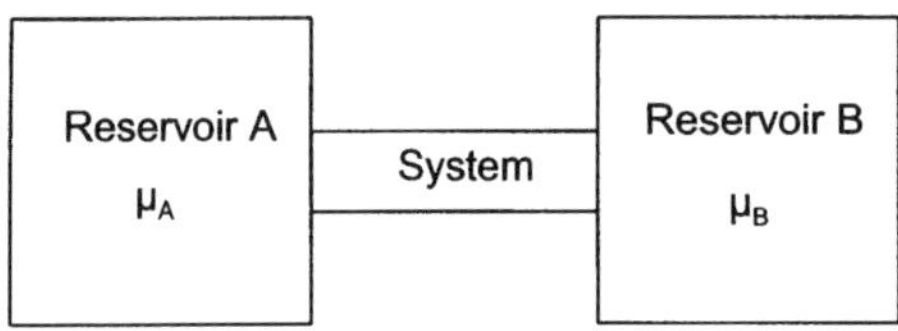

Figure 18.1. System with its ends in diffusive contact with two reservoirs of different chemical potential.

Figure 18.1 depicts the schematic–symbolic conditions for diffusion through a system (e.g., Cu lattice or similar.) Here one end of the system is in diffusive contact with a reservoir at chemical potential μ_A, with the other end in similar contact with a reservoir at chemical potential μ_B. The temperature is considered constant. In solids, for instance, the chemical potential is identified with the Fermi energy level. When two solids or thin films are brought into contact, such as in the case of a p–n junction, charged particles will undergo interdiffusion such that the chemical potentials and Fermi levels will be balanced, that is, reach the same level.

Specifically, let us assume that in Figure 18.1,

$$\mu_A > \mu_B \tag{18.1}$$

Then particles will flow through the system from reservoir A to reservoir B. This flow will result in an entropy increase of the total ensemble that is made in this case of

$$\text{reservoir A} + \text{reservoir B} + \text{system}$$

Let us assume now that particle diffusion occurs when $\mu_A \neq \mu_B$ as a result of a difference in concentration. In that case a relation known as *Fick's law* (2) is in operation, and we have

$$\text{flux} = (\text{diffusion coefficient}) \times (\text{driving force}) \tag{18.2a}$$

or using the language of mathematics,

$$J_n = -D \text{ grad } c \tag{18.2b}$$

where c is the concentration of the diffusing species (particles). The concentration gradient is acting as the "driving force."

The term *flux* here refers to the amount of the diffusing substance passing through the cross section of a unit area in unit time. The term *gradient* refers to the change in substance concentration as a function of distance. Both quantities, which have directionality in addition to a numerical value, are viewed as vector quantities.

It has been found experimentally that the diffusion constant or coefficient D in Eq. (18.2a) depends on temperature according to the expression

$$D = D_0 \exp\left(-\frac{E}{k_B T}\right) \tag{18.3}$$

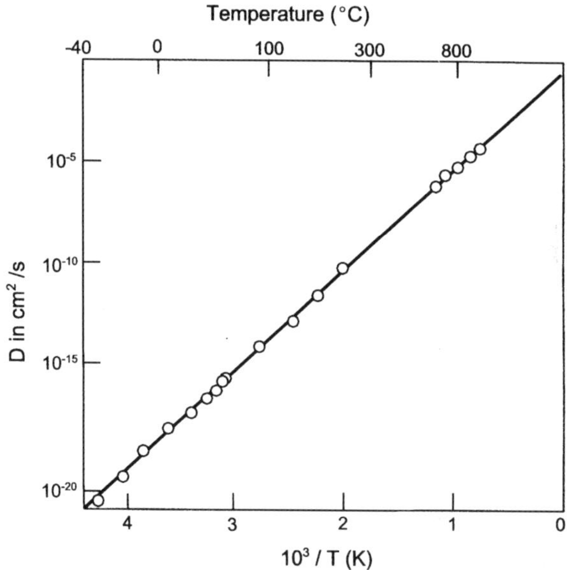

Figure 18.2. Diffusion coefficient of carbon in iron as a function of the inverse of temperature. (From C. Kittel, *Introduction to Solid State Physics*, 7th ed., Wiley, New York, 1996, with permission from Wiley.)

TABLE 18.1. Some Diffusion Constants and Activation Energies

Host Crystal	Diffusing Atom	D_0 (cm^2/s)	E (eV)
Cu	Zn	0.34	1.98
Ag	Cu	1.2	2.00
Si	Au	1×10^{-3}	1.13

where D_0 is a constant typical of the system, E is the activation energy of the diffusion process, k_B is the Boltzmann constant, and T is the absolute temperature.

In Figure 18.2 we present the diffusion coefficient of carbon in iron as a function of the inverse of the temperature ($1/T$). It is evident that the logarithm of D is directly proportional to $1/T$, as expected in view of Eq. (18.3). This type of temperature dependence is indeed typical of activation-energy-driven processes, in general. This is discussed below. For the sake of illustration, we present in Table 18.1 a few actual values of E and D.

There are three basically simple mechanisms for diffusion of atoms A in a solid AB:

1. Atoms A and B may interchange positions by rotation about a midpoint and squeezing by one another in the lattice. Actually, more than two atoms may rotate together.

2. Atoms may diffuse individually through interstitial sites.

3. Diffusion may take place with the help of vacancies, with the atoms moving only into adjacent vacant sites.

Figure 18.3 depicts these three modes of ionic motion.

If it were valid, mechanism 1 would require the activation energy for the process to be very high—so high, in fact, that it might be expected to be of the same order of magnitude as that of the cohesive energy (on the order of tens of electron volts). Values given in the literature (see, e.g., Table 18.1) are at moderate temperatures, uniformly much less than the cohesive energy, making that mechanism all but unlikely. Process 1 should thus be ruled out for most practical cases.

To actually diffuse, an atom ion must overcome a potential-energy barrier due to neighbors. To clarify the foregoing, we discuss mechanism 2 here in some detail, but it should be stressed that nearly the same arguments hold for mechanism 3.

If, as above, the potential-energy barrier height is E, statistical mechanical considerations indicate that the atom will have sufficient thermal energy to pass over the barrier a fraction $\exp(-E/k_BT)$ of the time. If f is a characteristic atomic vibrational frequency, the probability p that during unit time the atom will pass the potential-energy barrier is given by

$$p \simeq f \exp\left(-\frac{E}{k_BT} \right) \tag{18.4}$$

In other words, in unit time the atom makes f (on the order of 10^{14} s^{-1}) attempts to surmount the barrier, each time with the probability of the exponent. Thus, the quantity p is also known as the *jump frequency*.

We are now in a position to calculate an atomically meaningful value for the coefficient D_0 in Eq. (18.3). Consider two parallel planes in a lattice with impurity atoms located interstitially. The planes are a lattice constant a apart along the x-axis. Let there be n impurity atoms on one plane, with $n + [a(dn/dx)]$ on the next. Since the

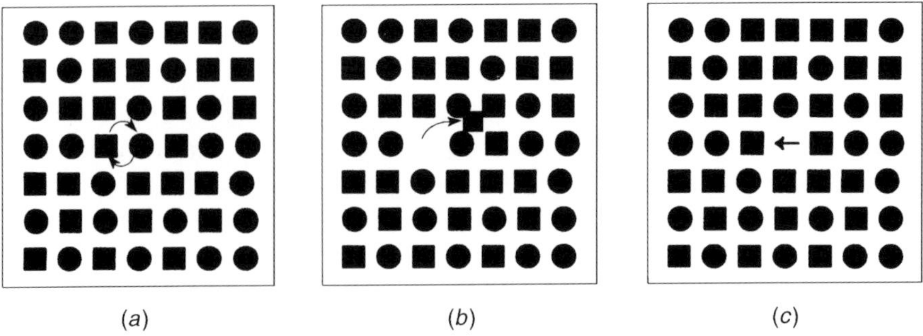

(a)   (b)   (c)

Figure 18.3. Graphical illustration of the mechanisms of diffusion: (*a*) interchange by rotation; (*b*) migration through interstitials; (*c*) atoms exchange position with vacancy. (From C. Kittel, *Introduction to Solid State Physics*, 7th ed., Wiley, New York, 1996, with permission from Wiley.)

derivative represents the change in impurity concentration per unit length along the x-axis, the number of impurity atoms passing between the planes is, in unit time, $-pa(dn/dx)$. The negative sign indicates movement from a higher to a lower concentration area. Here p is the probability as given in Eq. (18.4). If N represents the total (per volume) concentration of impurity atoms, $n = aN$ represents the concentration per unit area of a plane. The flux J_n [see Eq. (18.2a)] can then be expressed as

$$J_n = -pa^2 \frac{dN}{dx}$$

or considering Eqs. (18.3) and (18.4), we have

$$D = fa^2 \exp\left(-\frac{E}{k_B T}\right) \tag{18.5}$$

so we have the expression

$$D_0 = fa^2 \tag{18.6}$$

Thus, we are able to express this constant in terms of system characteristic quantities, making it measurable, at least, in principle. Alternatively, if we know the value of D_0 from another source, we can calculate the value of f.

18.2. DIFFUSION IN ELECTRODEPOSITS

Diffusion is liable to corrupt the properties of a deposit and defeat the purpose for which the electrodeposition was performed in the first place. This may be particularly so at the basis metal–film interface. In the case of deposits for decorative purposes, for example, diffusion of the coated underlayer (metal) to the surface will debase the intended appearance. Another example is gold plating of electronic contacts (3,4), which is often practiced to avoid corrosion of the contact areas. In these cases, the underlayer is often copper. The copper can diffuse to the surface of the gold. As a result of the copper oxidizing, the contact resistance will have been altered, markedly for the worse. As actual practical facts, we note here that a 3-μm-thick gold film deposit exposed to 300°C, will be covered with the oxide of its underlying copper within one month. If exposed to 500°C, a gold layer of 30 μm thickness will be diffused through in 4 to 5 days!

In some instances, to improve solderability, tin is deposited on nickel surfaces. In a short time, however, interdiffusion of the two metals results in the growth of an intermetallic $NiSn_3$ compound that is much less amenable to soldering. For tin over electrolessly deposited nickel surfaces, the interdiffusion results in pores in both films. Pores are to be avoided, of course, if conductivity and/or contact resistance is an issue.

Hydrogen embrittlement was mentioned in Chapter 16, where it was indicated that the presence of hydrogen in, say, steel will result in embrittlement. In that predicament we are dealing with an interstitial solid solution. The solute atoms (hydrogen in this case) move along the interstices between the solvent atoms (iron in this case), as depicted in Figure 18.3b, without having to displace them. This state of affairs is facilitated by the fact that the hydrogen atoms are much smaller than the interionic space in the lattice of the solvent (4).

Diffusion must not, however, always be viewed as being a harmful phenomenon. In some cases it is most desirable, if not essential. Such is the case in welding (5), where diffusion ensures joining of the welded parts. Steel is often coated with tin to protect it from corrosion. In this case the formation, via interdiffusion, of the intermetallic $FeSn_2$ is the key for effective protection.

Yet another positive aspect of the diffusion phenomenon is the creation of alloys by first depositing alternate layers of different coatings and then creating an alloy by heating to promote diffusion to produce an alloy. Specifically, brass deposits may be produced by first depositing copper and zinc layers alternately. Subsequent heating produces the required brass. This type of approach obviates the undesired direct method of brass deposition via a cyanide process. For jet-engine parts, which are routinely subject to temperatures close to 500°C, diffusion-alloyed nickel–cadmium is found to serve as an effective corrosion protective agent (6).

18.3. VOIDS

Sometimes interdiffusion between two metals is uneven and may lead to the creation of vacancies or voids. This type of imbalance is the result of possible unequal mobilities between a metal couple. These voids occur individually near the common interface. The voids, like bubbles, coalesce, resulting in porosity and loss of strength. Many thin-film couples exhibit this phenomenon, which is referred to as *Kirkendall void creation*. Al–Au, Cu–Pt, and Cu–Au are just a few examples. To be specific, it has been found (7), for instance, that in the case of Au–Ni, about five times more Ni atoms diffuse into Au than Au atoms diffuse into Ni.

The three main diffusion types (as opposed to mechanisms) are:

1. Bulk or lattice diffusion
2. Diffusion along defect paths (grain boundaries, dislocations)
3. Diffusion between ordered metallic phases

These are depicted schematically in Figure 18.4 in the case of metal A deposited on metal B. Bulk diffusion, as noted above, is the transfer of B into A or A into B through the crystal lattice. This is characterized by the coefficient D in the figure. Defect path diffusion is the migration along lattice defects such as grain boundaries, characterized by the coefficient D' in the figure. Ordered A_xB_y possible phases are indicated between the metals. Finally, Kirkendall void porosity is indicated and will be expected to be present if the interdiffusion rates from one metal to the other are not equal in both directions.

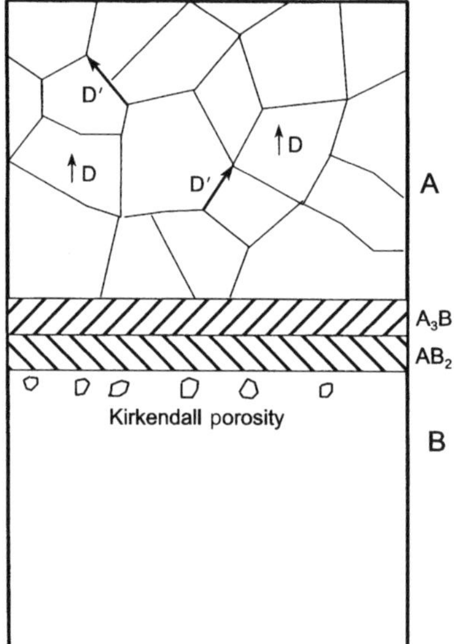

Figure 18.4. Schematic representation of the various diffusion-related reactions induced by thermal aging of a two-metal structure.

The Kirkendall effect (8) is time and temperature dependent, and with some metal couples, it takes place even at room temperature. For instance, adhesion of solder to gold is damaged by heating to about 150°C for about 5 minutes, due to the formation of Kirkendall voids. Naturally, the formation of Kirkendall voids is accelerated by increased temperature and dwelling time.

Kirkendall void formation can, however, be prevented from occurring by choosing the "right" metal species. For example, whereas platinum coating on copper is subject to the Kirkendall void creation process, the same coating on electrodeposited nickel is free of it even if heated to as high as 600°C for many hours (more than 10 hours!).

Again, this effect has useful aspects as well. It can help in creating controlled porosity for the purposes, say, of an electroformed object requiring cooling. Such, indeed, is the case in a number of electronics applications.

18.4. DIFFUSION BARRIERS

Diffusion barriers are coatings that serve in that role: specifically, protection against undesirable diffusion. One of the best examples is that of a 100-μm-thick electrodeposited copper layer that serves as an effective barrier against the diffusion of carbon. Another example is that of nickel and nickel alloys (notably, electrolessly deposited Ni–P) that block diffusion of copper into and through gold overplate. This is achieved by the deposition of a relatively thin Ni–P layer (less than 1 μm) between the copper and its overlayer. Naturally, the effectiveness of the diffusion barrier increases with its thickness. Other factors in the effectiveness of a diffusion barrier include its crystallographic properties, such as grain size and preferred crystalline orientation.

In electronic applications, where it is common to deposit copper and/or copper alloy and tin in sequence, with a nickel diffusion barrier layer, 0.5 μm thick, between the layers present, no failure occurs. Without the nickel layers between bronze/-copper/tin layers themselves, for instance, intermetallic brittle layer(s) and Kirkendall voids are formed, leading eventually to separation of the coated system and substrate.

Also, when tin-containing solder connections are made to copper, intermetallic materials are formed. Those continue to grow to render weak surfaces. Again, a nickel layer between the substrate and the solder provides a solution to this problem.

18.5. DIFFUSION WELDING

In diffusion welding, diffusion is used to create joints of high quality. In other words, here again, one deals with the practical and useful aspects of thin-film interdiffusion. In practice, clean (very clean!) cleaved or otherwise smoothed metal surfaces should be made to effect a firm mechanical contact using a strong force but one that is insufficient to cause macroscopic deformation even at an elevated temperature. Usually, this will have to be been done in vacuum or at least in an inert atmosphere. The problems of "hard to get to" (inaccessible) joints and possibly objectionable thermal conditions and the resulting undesired microstructures, such as Kirkendall voids, are minimized, if not eliminated all together. Thus, good-quality distortion-free joints requiring no additional machining or other posttreatment can be achieved.

The characteristic positive features of diffusion-welded joints are as follows:

- No change in metallurgical properties will occur.
- No change in physical properties will result.
- No cast structures will be created.
- Recrystallization and/or grain growth can be reduced or eliminated.
- Heat treatment can be included while in the "welding" process.
- Multiple joints can be welded at the same time.
- In case of precision work, no change in dimensions will occur.
- Continuous vacuum or gastight joints, even over large areas, can be achieved.
- Additional weight and possible need for machining of finished product are avoided.
- The method is recommended particularly for dissimilar metal, cermet, and composite structures.

Before diffusion welding can be performed, a special layer in the form of a thin-film coating or thin foil must often be applied to help promote joining. Such a coating can be applied by electroless or electrodeposition methods. These *intermediate layers* are used for a number of reasons, including some or all of the following: (1) plastic flow is encouraged; (2) clean surfaces can be guaranteed; (3) objectionable intermetallics are avoided; (4) diffusion is encouraged; (5) Kirkendall porosity can be minimized or even avoided altogether; (6) welding and bonding temperatures can be reduced; (7) dwell-time reduction may be achieved; (8) transient, eutectic melting in an attempt to encourage diffusion of base metals can be expected; and (9) sometimes, scavenging of undesirable elements can be eliminated.

Typically, coatings most often in use as intermediate layers are silver, nickel, copper, and gold; however, silver is used by far the most often. This is so because of the low dissociation temperature of silver oxide, making it relatively easy to obtain clean surfaces. Also, the typical thickness range of electroplates used, in practice, for diffusion welding is about 15 to 40 μm, but thicknesses as great as 130 μm must sometimes be used. A considerable variety of steel types as well as aluminum and a host of other difficult-to-join metals and even beryllium have been and continue to be diffusion bonded with the use of electroplated intermediate layers.

Essential process variables common to all diffusion-bonding techniques to be considered are (1) temperature, (2) pressure, (3) time, (4) surface condition, and (5) process atmosphere. Process temperature is usually 25 to 75% of the melting point of the lower (!)-melting-point metal in the intended weld or bond. It is to be understood that the purpose of attaining elevated temperature is to promote or accelerate interdiffusion of the atoms at the joint interface and to provide some metal softening, which, in turn, aids in surface deformation. The purpose of pressure application is to establish a firm, robust mechanical contact of the surfaces to be joined and to break up surface oxides (hence the frequent use of silver, as mentioned above). This will provide a rather clean surface for bonding. The dwell time required at elevated temperatures is determined according to metallurgical and sometimes other considerations.

18.6. ELECTROMIGRATION

In an electric cord or wire that is conducting or carrying electricity directly, electricity is conducted without transport of atoms in the conductor. The common free-electron model of electrical conductivity makes the fundamental assumption that electrons are free to move in and through a metal lattice constrained only by scattering events. Scattering is the cause of electrical resistance and of *Joule heating*. This does not, however, cause displacement of atoms or ions as long as the current density is moderate. At high current densities ($\sim 10^4$ A/cm^2), the presence of current can displace ions and thus cause transport of mass. This mass transport caused by the electric field and the charge carriers is known as *electromigration*. It is present in interconnecting lines in microelectronic devices in which the current density values are high. By way of example, a 1-μm-wide aluminum line of 0.2 μm thickness and a current of 2 mA represents a current density value of 10^5 A/cm^2, which may cause mass transport in the line even at room temperature. This constitutes a reliability failure endemic to thin-film circuits. As modern electronic circuitry tends to become smaller and smaller, the current densities become larger and the probability of circuit failure due to electromigration becomes more of a problem. It will cause both voids and extrusions. Given a line of aluminum that is subject to an electric field and current density of sufficient magnitude to cause electromigration, the line can be expected to undergo morphological change such that depletion occurs at the negative (cathode) end while extrusion will be present at the positive (anode) side. This means that the material migration is in the direction of the movement of the charge-carrying electrons. The driving force behind electromigration consists of two parts: (1) the direct action of the electrostatic field on the diffusing atoms, and (2) the momentum exchange of the moving charge carriers with the diffusing atoms. These are referred to as the *direct force* and the *electron wind force*, respectively. The latter is usually far greater than the former. There have been attempts, using quantum mechanical methods, to explain and estimate the electron wind force quantitatively. None is widely accepted, and the common practice is to adhere to semiclassical treatments (10).

The prevention of electromigration is an important challenge for the microelectronics industry. The electromigration flux is reduced in practice by adding a few atomic percent copper into aluminum. The effect of this is to slow down the grain boundary diffusion of aluminum. The second method includes the construction of a layered thin-film line such as $Al(Cu)/Al_3Hf/Al(Cu)$. The formation of a theta phase (Al_2Cu) in grain boundaries is believed to provide sources for Cu replenishment when Cu is depleted from the grain boundaries by electromigration. Thus, the desired result of a copper solution can be kept going longer. Other remedies include lowering the operating temperature or using metals other than Al, such as copper, which have a higher activation energy for diffusion. A number of review articles deal with the subject and are listed in the references and further reading section at the end of the chapter.

18.7. DIFFUSION IN ELECTRODEPOSITION

Diffusion here refers to the movement of ions and/or neutral species through the deposition bath or solution as a consequence of concentration gradients. It is primarily the result of random (Brownian) molecular motion, and it serves to produce more uniform distribution of the various component species in the bath. Depletion of ions next to the cathode will result in movement of the species from the (nearly unchanged) bulk of the bath toward the cathode.

The region near the electrode where the concentration of a species differs from that of its bulk value is referred to as the *diffusion layer*. The boundary between that region and the bulk of the bath is, naturally, not sharp. By definition, however, the region in which the concentration of a particular species differs from its bulk concentration by 1% is the diffusion layer. Figure 18.5 depicts the metal ion concentration near a cathode during electrodeposition.

The diffusion flux J, in mol/cm^2, is proportional to the concentration gradient and inversely proportional to the diffusion layer's effective thickness δ_N (also called the *Nernst thickness*). The proportionality constant D is the diffusion constant; hence,

$$J = \frac{D(c_0 - c_E)}{\delta_N}$$

where c_0 is the bulk concentration and c_E is the concentration at the electrode.

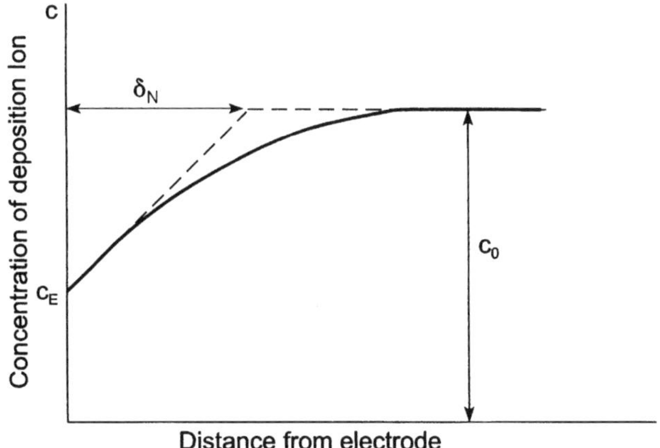

Figure 18.5. Schematic representation of metal-ion concentration at the electrode surface. (See text for definition of symbols.) (From F. Lowenheim, ed., *Modern Electroplating*, 3rd ed., Wiley, New York, 1974, with permission from Wiley.)

As can be expected, with bath agitation the effective thickness δ_N will diminish; hence the diffusion rate will increase. Typical values of δ_N are about 0.2 mm without agitation, and with a rotating-disk electrode it can be made as small as 0.02 mm. Of course, this thickness will vary from species to species.

A diffusion layer is not formed immediately on turning on the voltage (potential). It takes on the order of several seconds, depending on agitation. One consequence of pulse plating is to avoid buildup of the layer.

18.8. SURFACE WAVINESS IN THIN FILMS

The rapid developments in the microelectronics industry over the last three decades have motivated extensive studies in thin-film semiconductor materials and their implementation in electronic and optoelectronic devices. Semiconductor devices are made by depositing thin single-crystal layers of semiconductor material on the surface of single-crystal substrates. For instance, a common method of manufacturing an MOS (metal-oxide semiconductor) transistor involves the steps of forming a silicon nitride film on a central portion of a P-type silicon substrate. When the film and substrate lattice parameters differ by more than a trivial amount (1 to 2%), the mismatch can be accommodated by elastic strain in the layer as it grows. This is the basis of strained layer heteroepitaxy.

In general, surface morphological instabilities driven by stresses are an important subject to investigate in connection with microelectronic applications. In particular, the degree of surface waviness in thin films as a consequence of surface and volume diffusion is a matter of pivotal importance. This topic has attracted considerable attention in the last two or so decades (11). Although surface diffusion is an important kinetic process, other kinetic processes may affect the evolution of stressed surfaces. Indeed, a possibility at high temperatures is the diffusion of atoms through the bulk.

The chemical-potential gradient driving this volume diffusion arises due to capillarity and stress variations in the bulk brought about by the sinusoidal surface morphology. Panst et al. (11) have presented an analysis describing the evolution of wavy-sinusoidal surfaces of stressed films by considering the diffusion of atoms through the film volume and along the film surface. Volume diffusion is shown in that work to be influential at the relatively low stress levels (tens of megapascal) typically encountered in films in thermal barrier systems but not to be important at high stress levels (gigapascal range). Further, the relative importance of volume diffusion is also controlled by the ratio of the volume self-diffusivity to the product of the surface self-diffusivity times the surface defect concentration. The inclusion of volume diffusion terms in the analysis in Ref. 11 also implies different behavior of the film surfaces under tensile or compressive stress. This effect is shown to affect the stability only at low stress levels. The study shows that at the dominant instability wavelength, and under low-stress and high-temperature conditions, the only important destabilizing mechanism is volume diffusion and the only important stabilizing mechanism is surface diffusion. Furthermore, the contribution of volume diffusion depends on the sign of the film stress, with compressive stress promoting surface roughening and tensile stress promoting surface smoothing.

REFERENCES AND FURTHER READING

1. E. L. Owen, in *Properties of Electrodeposits: Their Measurements and Significance*, R. Sard, H. Leidheiser, and F. Ogburn, eds., Electrochemical Society, Pennington, NJ, 1981.

2. A. J. Dekker, *Solid State Physics*, Prentice-Hall, Englewood Cliffs, NJ, 1961; *Introduction to Solid State Physics*, 7th ed., Wiley, New York, 1996.

3. M. R. Pinnel, *Gold Bull.* **12**(2), 62 (1979).

4. A. S. Norwick, in *Encyclopedia of Materials Science and Engineering*, M. B. Bever, ed., Pergamon Press, Elmsford, NY, 1986, p. 1180.

5. J. W. Dini, in *Electrodeposition Technology Theory and Practice*, Vol. 87-17, L. T. Romankiw and D. R. Turner, eds., Electrochemical Society, Pennington, NJ, 1987, p. 639.

6. R. W. Moeller and W. A. Snell, *Plating* **42**, 1537 (1979).

7. M. Schlesinger, in *Electrocrystallization*, Vol. 81-6, R. Weil and R. G. Barradas, eds., Electrochemical Society, 1981, p. 221.

8. E. O. Kirkendall, *Trans. Metall. Soc. AIME* **147**, 104 (1942).

9. H. B. Huntington and A. R. Grone, *J. Phys. Chem. Solids* **20**, 76 (1961).

10. K. N. Tu, J. W. Mayer, and L. C. Feldman, *Electronic Thin Film Science*, Macmillan, New York, 1992.

11. R. Panat, K. J. Hsia, and David G. Cahill, *J. Appl. Phys.* **97**, 013521 (2005).

PROBLEMS

18.1. Using simple arguments, show that the coefficient of diffusion D (the net number of particles that flow in unit time through a unit plane perpendicular to a unit concentration gradient) is given by $\frac{1}{3} vs$, where v is the average velocity of the particles and s is the mean free path.

18.2. Starting with Eq. (18.6), arrive at a useful expression for the distance that a particle migrates by random walk during time t as $x = \sqrt{Dt}$, where x is the migration distance and D is the diffusion constant, as given in Eq. (18.3).

18.3. For the interstitial diffusion of carbon in iron $D = 10^{-21}\,\mathrm{m^2/s}$ (see Fig. 18.2), calculate how far a carbon atom diffuses in $2 \times 10^8\,\mathrm{s}$ (6.6 years!).

18.4. Calculate the coefficients of diffusion D at 300 K for the following cases:

Diffusing Metal	Matrix	D_0 (m^2/s)	Q (kcal/mol)
1. Copper	Aluminum	2×10^4	33.9
2. Silver	Silver (volume diffusion)	0.72×10^4	45.0
3. Silver	Silver	0.14×10^4	21.5

19

Applications in Semiconductor Technology

19.1. INTRODUCTION

Some of the most active areas in modern applications of electrochemical deposition are semiconductor technology and magnetic recording. One major recent advance in the silicon-based semiconductor industry is the development of copper interconnects on chips. This new technology replaces aluminum or aluminum alloy (e.g., Al–Cu) conductors produced by vacuum-based deposition techniques with copper conductors manufactured by electrodeposition. Vacuum-based deposition techniques include physical vapor deposition (PVD) and chemical vapor deposition (CVD). The most frequently used PVD technique is *sputtering*. In the sputtering process of Al, a target surface of Al is bombarded by energetic ions of an inert gas (e.g., Ar, Kr) (1,2). The result of this bombardment, collisions between the incident energetic particles and the Al target surface, is the ejection of one or more Al atoms from the target. These ejected Al atoms are then deposited on the silicon-based semiconductor. Sputtering tools operate under glow discharge conditions (1,2). CVD deposition process of Al involves surface chemical reaction between the silicon wafer and the reactant-gas species containing an Al compound. This surface reaction results in the formation of Al thin film (1).

Since 1999, copper has been replacing aluminum, due to its low-bulk electrical resistivity and superior electromigration resistance. The electrical resistivities of pure Al and Cu are 2.9 and $1.7\,\mu\Omega \cdot cm$, respectively, and that for Al alloys is 3 to $4\,\mu\Omega \cdot cm$. Activation energies for electromigration, using identical structures and experimental conditions, are 0.81 (±0.03) and 1.1 (±0.1) eV for Al (0.5 wt % Cu) and Cu, respectively (3–6). Lower electrical resistivity results in higher-speed devices. Higher electromigration resistance (higher activation energy for electromigration) results in higher reliability and thus a lower interconnect failure rate (7).

The number of transistors per chip is increasing continuously (for microprocessors, the total number of transistors per chip was 11, 21, and 40 million in the years

1997, 1999, and 2001, respectively) and the physical feature size of transistors is decreasing, so chip interconnect dimensions are being scaled down (8). For example, line widths are 0.25, 0.18, 0.15, and 0.10 μm for the years 1997, 1999, 2001, and 2006, respectively. This reduction in size (miniaturization) of the interconnect line width and thickness is accompanied by a change in the ratio of surface to bulk atoms. Thus, an understanding of the physical properties of the evolving submicrometer thin films requires new understanding of the physics and chemistry of both bulk and surface atoms (9).

The required degree of understanding of the physical properties of metal thin films used for interconnects on chips is illustrated by the following example. It was found that the performance of conductors on chips, Al or Cu, depends on the structure of the conductor metal. For example, Vaidya and Sinha (10) reported that the measured median time to failure (MTF) of Al–0.5% Cu thin films is a function of three microstructural variables (attributes): median grain size, statistical variance (σ^2) of the grain size distribution, and degree of [111] fiber texture in the film.

19.2. DEPOSITION OF Cu INTERCONNECTIONS ON CHIPS

The change from Al to Cu interconnects required a change in the fabrication process, from metal reactive ion etching (RIE) to dielectric RIE. The metal-RIE process is used in the fabrication of Al interconnects on chips. This process is shown in four steps in Figure 19.1. The first step in the metal-RIE process is sputter deposition of blanket thin film of Al (or Al alloys, e.g., Al–Cu, Al–Si) over planarized dielectric (e.g., silicon dioxide). In the next step the unwanted metal is etched away by reactive ion etching through a photoresist mask. The features produced in this way are separated, electrically isolated metal Al conductor lines. In the RIE process, chemically active ions such as F or Cl bombard the Al surface and form volatile aluminum fluorides or chlorides, which are then pumped away in the vacuum system (11,12). After etching, a dielectric is deposited so that it fills the gaps between the lines as well above them. In the last step, dielectric is planarized by a chemical–mechanical polishing (CMP) technique (1).

At present, multilevel Cu interconnections on chip are fabricated using a dielectric-RIE process, since it is difficult to pattern Cu by RIE. That is so because the vapor pressure of halides is very low at room temperature (13,14). In this process, blanket Cu deposition is followed by chemical–mechanical polishing of the Cu (1). This approach is known as a *damascene process* (process used in Damascus for centuries to form inlaid metal features on jewelry). Figure 19.2 illustrates the fabrication process sequence for the single damascene process. The process starts with the deposition of a dielectric layer on an Si wafer, patterning it using photolithography and a dielectric-RIE process. After patterning the dielectric, barrier metal and a Cu seed layer are deposited using a PVD or CVD technique. Finally, Cu is electrodeposited into the recesses, trenches (lines), and holes (vias). The excess of Cu deposited on the upper surface is removed by CMP (1). In the dual damascene process, vias and trenches are patterned and filled with Cu at the same time (Fig. 19.3).

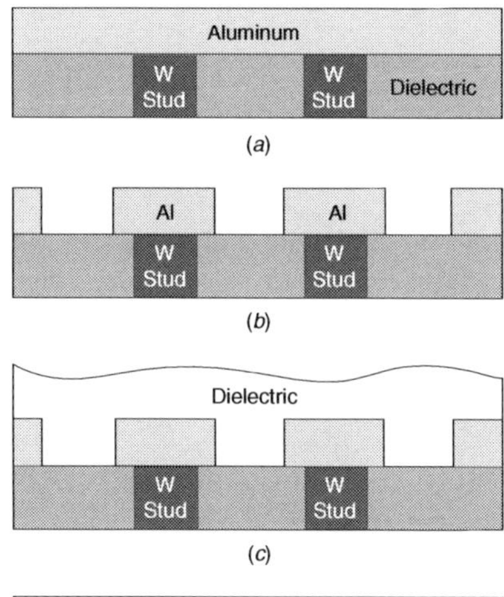

Figure 19.1. Process steps for forming Al interconnects using the metal-RIE process: (*a*) Al deposition; (*b*) Al RIE through a photoresist mask; (*c*) dielectric (e.g., SiO$_2$) deposition; (*d*) dielectric planarization by CMP.

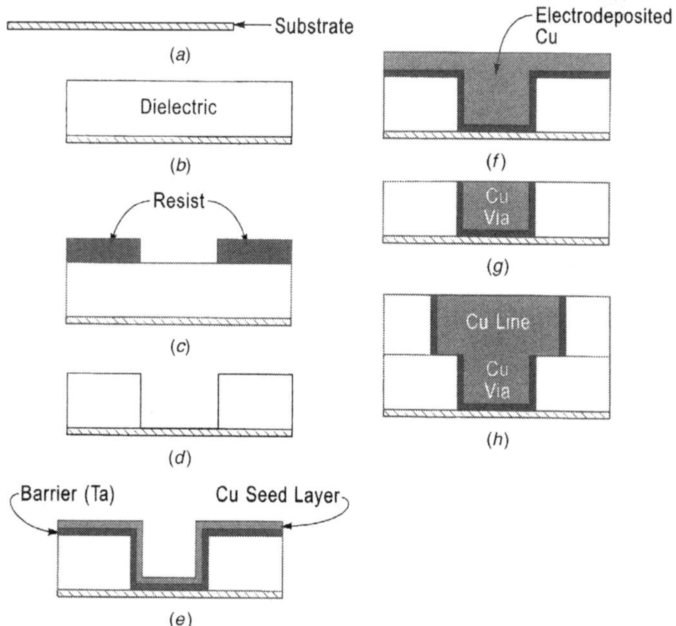

Figure 19.2. Process steps for forming Cu interconnects using the single damascene process (dielectric patterning): (*a*) planarized substrate; (*b*) dielectric deposition; (*c*) dielectric RIE through photoresist mask; (*d*) etched insulator; (*e*) deposition of diffusion barrier (Ta) and Cu seed layer; (*f*) electrodeposition of Cu into a via (vertical interconnection); (*g*) CMP of Cu excess; (*h*) patterning and deposition of Cu line (wire).

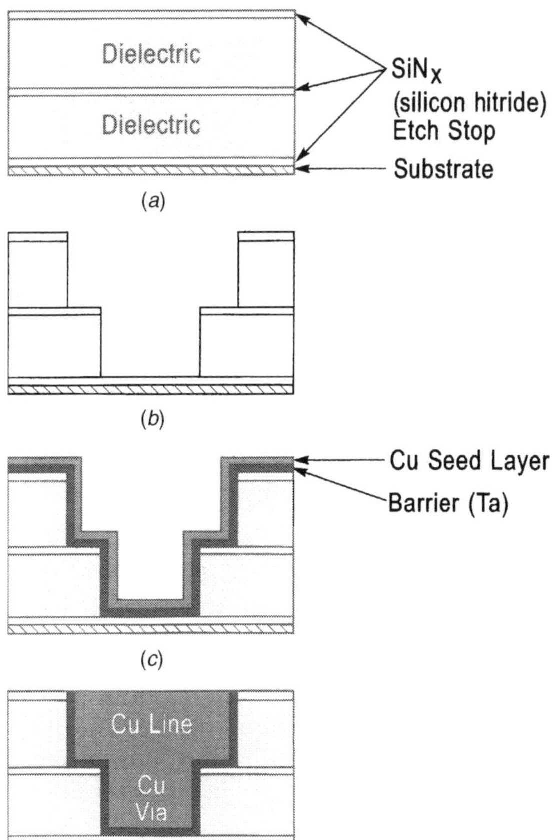

Figure 19.3. Process steps for the dual damascene process: (*a*) deposition of dielectric; (*b*) dielectric RIE to define via and line; (*c*) deposition of diffusion barrier and Cu seed layer; (*d*) electrodeposition of Cu into via and trenches followed by Cu CMP.

The selective Cu deposition process was suggested by Ting and Paunovic (13) as an alternative means of fabricating multilevel Cu interconnections (Fig. 19.4). The first step in this *through-mask deposition process* (14) is the deposition of a Cu seed layer on a Si wafer, and then a resist mask is deposited and patterned to expose the underlying seed layers in vias and trenches. In the next step, Cu is deposited to fill the pattern. After the Cu deposition mask is removed, the surrounding seed layer is etched and dielectric is deposited. Electroless Cu deposition has been suggested for the blanket and selective deposition processes (15).

A variation of the selective Cu deposition process, limited to electroless Cu deposition, is the *lift-off process*, a *planarized metallization process* (16). Figure 19.5 shows a stepwise process sequence for this technology.

There is a basic difference between the damascene and through-mask plating processes in the way the trenches and vias are filled with electrochemically deposited Cu, through either an electrodeposition or an electroless technique. In multilevel metal structures, vias provide a path for connecting two conductive regions separated

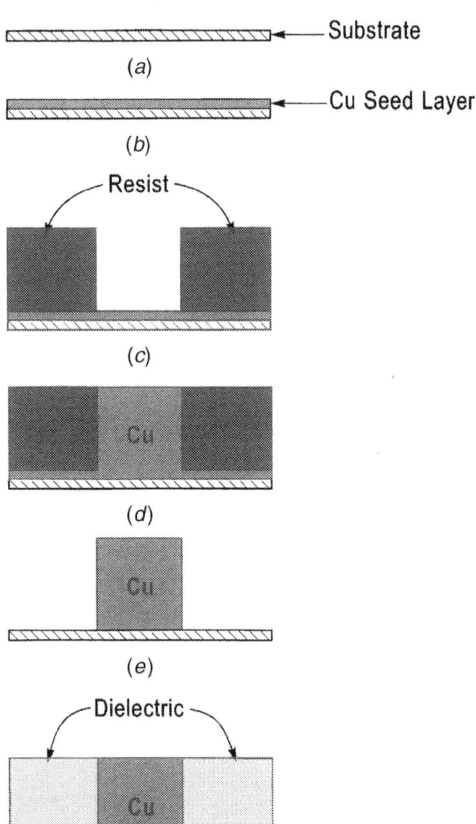

Figure 19.4. Through-mask deposition process: (*a*) Si substrate; (*b*) Cu seed layer deposition; (*c*) photoresist deposition and patterning; (*d*) through-mask electroless deposition of Cu; (*e*) stripping of photoresist and etching of Cu seed layer outside line; (*f*) dielectric deposition.

by interlevel dielectric (ILD). In a damascene process, the Cu deposit grows from the active bottom and sidewalls (Fig. 19.6*a*). Preferred growth from the bottom may be achieved by the addition of suitable additives (17,18). In the through-mask plating process, only the bottom is active; the sidewalls are inactive and the Cu deposit grows from the bottom (Fig. 19.6*b*) (19–21).

One of the challenges in interconnection fabrication on a chip is the electrodeposition of Cu in vias of small diameter ($<0.2\,\mu$m). Modeling of these processes shows that new Cu electrodeposition solutions and new deposition techniques are necessary to solve the problems introduced by the development of new integrated circuits (22).

19.3. DIFFUSION BARRIERS AND SEED LAYER

Copper introduces new problems in the fabrication of interconnects on chips, the most important of which is the diffusion of Cu into Si, SiO_2, and other dielectrics (4),

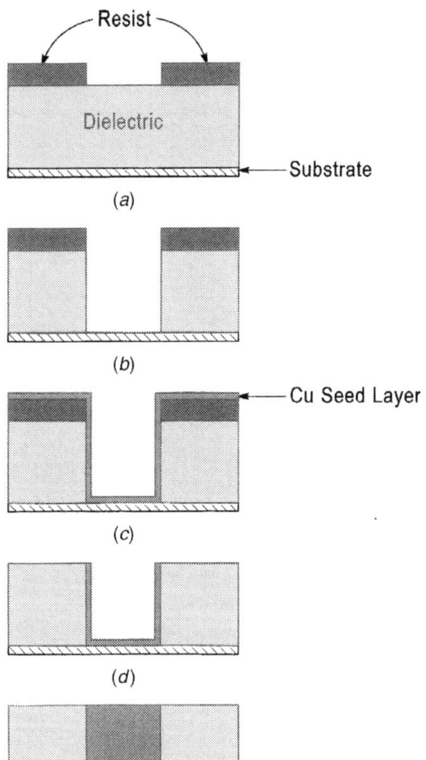

Figure 19.5. Process sequence for the lift-off process (the planarized metallization process): (*a*) a resist film is patterned on a dielectric film; (*b*) dielectric patterning; (*c*) a thin catalytic film layer (PVD or CVD Ti, Al) is deposited; (*d*) a lift-off technique removes the excess material, leaving the catalytic layer in the trench only; (*e*) electroless Cu deposition.

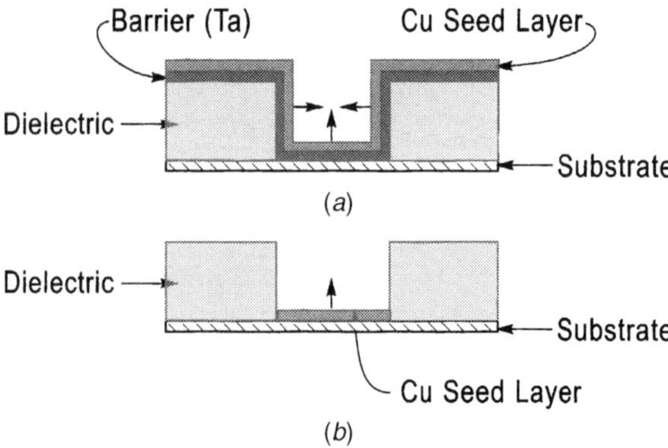

Figure 19.6. Growth of deposit in vias and trenches during Cu electrodeposition in (*a*) damascene and (*b*) deposition through- mask process.

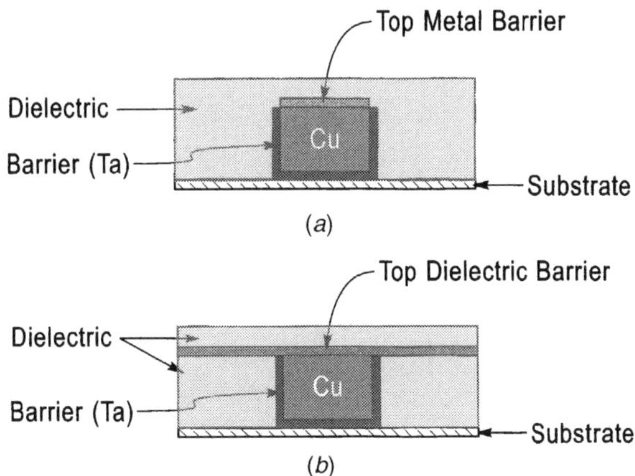

Figure 19.7. Fully encapsulated Cu line: (*a*) metal (alloy) top barrier; (*b*) dielectric top barrier.

and the reaction of Cu with Si, forming silicides (23). Diffusion of Cu through Si results in the poisoning of devices (transistors), and diffusion through SiO_2 leads to degradation of dielectrics. Thus, diffusion barrier layers are an integral part of the fabrication of copper interconnects (Figs. 19.2 and 19.3). Barrier films isolate (encapsulate) Cu interconnects from adjacent dielectric material. The diffusion barriers most studied are Ta (22,23), Ti, and TiN (24,25).

Since barrier metals have relatively high electrical resistivity (Ta, $12.4\,\mu\Omega \cdot$ cm; Ti, $80\,\mu\Omega \cdot$ cm), it is necessary to cover the barrier layer with a conductive metal layer. This conductive metal layer is a *Cu seed layer* that is deposited using PVD or CVD techniques (Figs. 19.2 and 19.3). When the electrodeposition of the Cu on a barrier/ Cu seed layer bilayer is finished, vias and trenches are filled with Cu, and excess Cu is removed using CMP (Figs. 19.2 *g* and *h* and 19.4*d*). The exposed Cu in lines needs *capping* with a barrier material to prevent diffusion. Two types of capping material may be used: metals (alloys) and dielectrics. The selectively deposited metal barriers (Fig. 19.7*a*) suggested are electroless Co(P), Co-Ni(P), Co-W(P), and selective CVD W (26–30). The preferred dielectric barrier material (Fig. 19.7*b*) is blanket SiN_x. The result of capping is a fully encapsulated Cu line (Fig. 19.7).

It is seen from the discussion above that Cu is electrodeposited in vias and trenches on a bilayer: a barrier metal/Cu seed layer. When the barrier layer is composed of two layers (e.g., TiN/Ti), Cu is electrodeposited as a trilayer: a barrier bilayer/Cu seed layer. This type of underlayer for electrodeposition of Cu raises a series of interesting theoretical and practical questions of considerable significance regarding the reliability of interconnects on chips. In Section 19.1 we have noted that interconnect reliability depends on the microstructural attributes of electrodeposited Cu (for Cu-based interconnects). These microstructural attributes, such as grain size, grain size distribution, and texture, determine the mechanical and physical properties of the thin films. Thus, one basic question in the foregoing series of questions is the problem of the influence of the underlayer barrier metal on the microstructure of the Cu seed layer. The second question is the influence of the microstructure of the Cu seed layer on the structure

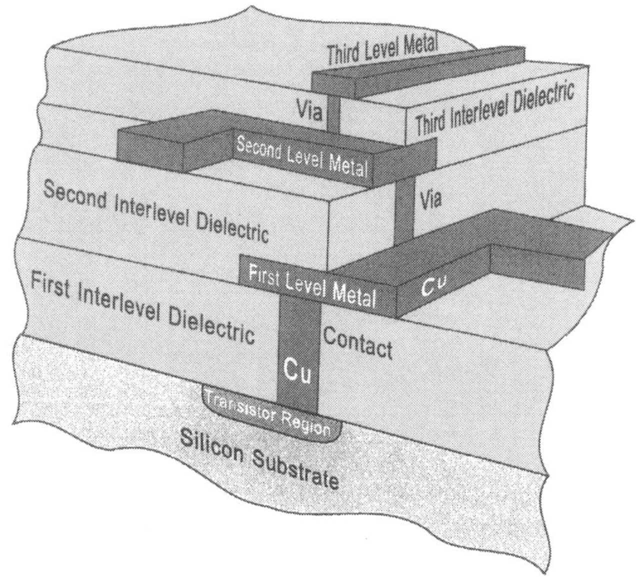

Figure 19.8. Three-level Cu interconnect IC structure.

and properties of electrodeposited Cu. Zielinski (31) showed that Cu microstructure is sensitive to the texture and microstructure of the barrier metal. For example, textural inheritance was observed on Ta underlayer. In this case there is quasi-grain-to-grain epitaxy. In addition, the presence of a barrier layer can further influence the microstructural evolution upon annealing. Tracy et al. (32) determined that a strongly textured underlayer such as Ti or Ti/TiN induces a similarly strong texture in the Cu. A strong Ti texture is passed to the TiN, which passes the texture to the Cu (or to Al–Cu). Tracy and Knorr (33) determined that a typical copper film consists of three texture components: [111], [200], and random. Indeed, [200] and [511] texture components are possible under some deposition conditions.

Integration of Cu with a dielectric introduces new problems and challenges [1(b), 34–36] (Fig. 19.8). For example, if polyimides are used as intermetal dielectrics, reliability concerns are corrosion of underlying metal and adhesion of metal films to polyimide underlayers [1(b)].

The discussion above on the influence of the underlayer on microstructure and the reliability of interconnects in ICs illustrates why there is a need for in-depth understanding of the process of deposition and the physical and mechanical properties of the electrodeposited Cu used in IC fabrication. These are great and interesting challenges and great opportunities for high-level studies of electrochemical deposition.

19.4. SUPERCONFORMAL ELECTRODEPOSITION OF COPPER INTO NANOMETER VIAS AND TRENCHES

Defect-free deposits in vias and trenches, superconformal electrodeposition, may be achieved in the presence of additives. In superconformal deposition, electrodeposition inside vias and trenches occurs preferentially at the bottom. Subconformal and conformal electrodeposition of copper in vias and trenches, on the other hand, occurs in an additive-free acid sulfate solution. These deposits have two types of defects: voids and seams (18, Fig. 3).

Mechanism of Superconformal Electrodeposition.   Two mechanisms for superconformal deposition in the presence of additives were proposed:

1. *The differential inhibition mechanism.* This mechanism, proposed by Andricacos et al. (18), considers a one-additive system. It is noted in Sections 10.4 and 10.5 that additive (inhibitor) adsorbed at the cathode affects the kinetics and growth mechanism of electrodeposition. The surface coverage of the additive (inhibitor), θ, is a function of the diffusion-controlled rates of adsorption–desorption processes. In the differential inhibition mechanism it is assumed that there is a very wide range of additive fluxes over the microprofile (vias and trenches): extremely low flux in deep interior corners, low flux at the bottom center, moderate flux at the sidewalls, and high flux at the shoulders. This type of concentration variation of the inhibitor (i.e., θ) results in (a) variation in the inhibition of electrodeposition kinetics in vias and trenches over a very wide range (i.e., several orders of magnitude), and (b) preferential electrodeposition on the bottom of vias and trenches, leading to void-free deposits. The inhibition factor used in the interpretation of leveling cannot describe shape-change behavior in superconformal electrodeposition of copper (18; and Section 10.6 herein).

2. *The inhibition–acceleration mechanism.* Moffat et al. (37) proposed the inhibition–acceleration mechanism to explain the experimentally observed corner rounding (inversion of curvature, Fig. 19 in Ref. 37) and general shape evolution in superconformal electrodeposition of copper in vias and trenches of nanometer dimensions (37,38). These authors also studied a three-additive system composed of two inhibitors and one accelerator. They concluded that superconformal deposition and corner rounding may be attributed to competitive adsorption of inhibitor and accelerator. This model is based on the assumption of curvature (in vias and trenches) – enhanced accelerator coverage.

Modeling.   Two types of modeling were used to interpret and optimize superconformal electrodeposition of copper:

1. *Deterministic modeling.* The first model in this category is based on the differential inhibition mechanism. Andricacos et al. (18) modified Dukovic's computer program, which was designed to follow the shape evolution in through-mask electrodeposition of copper (39). The essential characteristics of the modified program are as follows: (a) the flux of the inhibitor along the profile of the features (vias, trenches) varies over a very wide range (i.e., several orders of magnitude); (b) this variation is determined by a dynamic balance between the arrival of the fresh additive and its consumption (reduction at the cathode). This mathematical model is able to predict the best combination of current density and inhibitor concentration that will produce superconformal deposition. The second model in this category is based

on the inhibition–acceleration mechanism. The basic assumption of this model is that the accelerator is strongly adsorbed on the surface, thereby displacing the inhibitor (38,40).

2. *Stochastic modeling*. Pricer et al. (41) used Monte Carlo stochastic methods for modeling the filling of vias and trenches by electrodeposition in the presence of a hypothetical blocking additive. Random numbers were used to determine when and if a particle will make a certain move (42). Each move for a species has an associated probability value, depending on the time step. This Monte Carlo computer program had 14 parameters related to additive, Cu^{2+}, and Cu^+ ions. An analysis of Monte Carlo "snapshots" during trench filling resulted in a series of conclusions. Here we present only two. (a) Void size decreases as the concentration of additive increases. This decrease occurs because the additive is adsorbed on the upper part of the trench, blocking Cu^{2+} deposition, thus allowing the Cu^{2+} to diffuse farther into the trench without being consumed. (b) At low adsorption rates (e.g., 10^3 nM/s), large voids can be seen; at higher adsorption rates (e.g., 10^7 nM/s), the trench fills faster from the bottom and no voids can be seen. Faster copper deposition at the bottom is caused by more of the additive being able to get adsorbed on the upper part of the trench, resulting in an increased diffusion of Cu^{2+} ions onto the bottom of the trench.

REFERENCES

1. (a) S. Wolf and R. N. Tauber, *Silicon Processing for the VLSI Era*, Vol. 1, *Process Technology*, 2nd ed., Lattice Press, Sunset Beach, CA, 2000; (b) S. Wolf, *Silicon Processing for the VLSI Era*, Vol. 2, *Process Integration*, Lattice Press, Sunset Beach, CA, 1990.

2. R. A. Powell and S. M. Rossnagel, *PVD for Microelectronics: Sputter Deposition Applied to Semiconductor Manufacturing*, Vol. 26, *Thin Films*, Academic Press, San Diego, CA, 1999.

3. J. M. E. Harper, E. G. Colgan, C.-K. Hu, J. P. Hummel, L. P. Buchwalter, and C. E. Uzoh, *MRS Bull.* **19**(8), 23 (1994).

4. D. S. Gardner, J. Omuki, K. Kudoo, Y. Misawa, and Q. T. Vu, *Thin Solid Films* **262**, 104 (1995).

5. A. J. Learn, *J. Electrochem. Soc.* **123**, 894 (1976).

6. C.-K. Hu, K. P. Rodbell, T. D. Sullivan, K. Y. Lee, and D. P. Bouldin, *IBM J. Res. Dev.* **39**, 465 (1995).

7. M. Ohring, *Reliability and Failure of Electronic Materials and Devices*, Academic Press, San Diego, CA, 1998, Chap. 5.

8. Y. Taur and T. H. Ning, *Fundamentals of Modern VLSI Devices*, Cambridge University Press, New York, 1998, Chaps. 4 and 5.

9. C. Vickerman, ed., *Surface Analysis*, Wiley, New York, 1998.

10. S. Vaidya and A. K. Sinha, *Thin Solid Films* **75**, 253 (1981).

11. W. M. Moreau, *Semiconductor Lithography: Principles, Practices, and Materials*, Plenum Press, New York, 1988.

12. S. P. Murarka, R. J. Gutmann, A. E. Kaloyeros, and W. A. Lanford, *Thin Solid Films* **236**, 257 (1993).

13. C. H. Ting and M. Paunovic, *J. Electrochem. Soc.* **136**, 456 (1989).

14. L. T. Romankiw, *Electrochim. Acta* **42**, 2985 (1997).

15. V. M. Dubin, Y. Shacham-Diamand, B. Zhao, P. K. Kasudev, and C. H. Ting, *J. Electrochem. Soc.* **144**, 898 (1997).

16. C. H. Ting, M. Paunovic, P. L. Pai, and G. Chiu, *J. Electrochem. Soc.* **136**, 462 (1989).

17. Y. Shacham-Diamand and V. M. Dubin, *Microelectron. Eng.* **33**, 47 (1997).

18. P. C. Andricacos, C. Uzoh, D. O. Dukovic, J. Horkans, and H. Deligianni, *IBM J. Res. Dev.* **42**, 567 (1998).

19. E. K. Yung, L. T. Romankiw, and R. C. Alkire, *J. Electrochem. Soc.* **136**, 206 (1989).

20. S. Mehdizadeh, J. O. Dukovic, P. C. Andricacos, L. T. Romankiw, and H. Y. Cheh, *J. Electrochem. Soc.* **139**, 78 (1992).

21. J. O. Dukovic, *IBM J. Res. Dev.* **34**, 693 (1990).

22. S. Mehdizadeh, J. Dukovic, P. C. Andricacos, L. T. Romankiw, and H. Y. Cheh, *J. Electrochem. Soc.* **137**, 110 (1990).

23. R. Stolt, A. Charai, F. M. D'eurle, P. M. Fryer, and J. M. E. Harper, *J. Vac. Sci. Technol.* **A9**, 1501 (1991).

24. L. A. Clevenger, N. A. Bojarczuk, K. Holloway, J. M. E. Harper, C. Cabral, Jr., R. G. Schad, F. Cardone, and L. Stolt, *J. Appl. Phys.* **73**, 300 (1993).

25. M. T. Wang, Y. C. Lin, and M. C. Chen, *J. Electrochem. Soc.* **145**, 2538 (1998).

26. T. Kouno, H. Niwa, and M. Yamada, *J. Electrochem. Soc.* **145**, 2164 (1998).

27. S.-Q. Wang, *MRS Bull.* **19**, 30 (1994).

28. M. Paunovic, P. J. Bailey, R. G. Schad, and D. A. Smith, *J. Electrochem. Soc.* **141**, 1843 (1994).

29. V. Dubin, Y. Shacham-Diamand, B. Zhao, P. K. Vasudev, and C. H. Ting, *Mater. Res. Soc. Symp. Proc.* **427**, 179 (1996).

30. E. G. Colgan, *Thin Solid Films* **262**, 120 (1995).

31. E. M. Zielinski, *J. Electron. Mater.* **24**, 1485 (1995).

32. D. P. Tracy, D. B. Knorr, and K. P. Rodbell, *J. Appl. Phys.* **76**, 2671 (1994).

33. D. P. Tracy and D. B. Knorr, *J. Electron. Mater.* **22**, 611 (1993).

34. P. A. Flinn, *J. Mater. Res.* **16**, 1498 (1991).

35. Y. Morand, *Microelectron. Eng.* **50**, 391 (2000).

36. P. V. Zant, *Microchip Fabrication*, 4th ed., McGraw-Hill, New York, 2000.

37. T. P. Moffat et al., *J. Electrochem. Soc.* **147**, 4524 (2000).

38. D. Josell, D. Wheeler, W. H. Huber, J. E. Bonevich, and T. P. Moffat, *J. Electrochem. Soc.* **148**, C767 (2001).

39. J. O. Dukovic, *IBM J. Res. Dev.* **37**, 125 (1993).

40. D. Josel, D. Wheeler, W. H. Huber, and T. P. Moffat, *Phys. Rev. Lett.* **87**, 016102 (2001).

41. T. J. Pricer, M. J. Kushner, and R. C. Alkire, *J. Electrochem. Soc.* **149**, C406 (2002).

42. T. J. Pricer, M. J. Kushner, and R. C. Alkire, *J. Electrochem. Soc.* **149**, C396 (2002).

20

Applications in the Fields of Magnetism and Microelectronics

20.1. INTRODUCTION

Many materials, including iron, steel, cobalt, and nickel, display spontaneous magnetization and thus are known as *ferromagnetic materials*. The spontaneous magnetization is the result, among other things, of magnetic dipoles, due to unpaired atomic electrons in their incomplete (unclosed) orbitals. The magnetic dipoles are coupled in parallel by the exchange interaction between electrons. It is interesting to note that what is referred to today as *exchange interaction* was the *molecular field* assumed by Weiss a long time before the discovery of spin. Weiss found that a molecular field of close to 1000 T was required to align magnetic dipoles in parallel against the kT thermal energy (k is Boltzmann's constant and T is the Kelvin temperature). In the absence of exchange interaction, thermal energy aligns magnetic dipoles randomly, resulting in paramagnetic materials that do not have spontaneous magnetization. A ferromagnet will become a paramagnet above a certain transition temperature called the *Curie temperature, T_c*, when the thermal energy overcomes the exchange interaction. Below the Curie temperature, a ferromagnet has temperature-dependent spontaneous magnetization. The Curie temperatures of iron (Fe), cobalt (Co), and nickel (Ni) are much higher than room temperature, so their room-temperature magnetization is close to that at absolute zero (0 K). Notwithstanding the above, the presence of strong exchange interaction in a ferromagnetic material such as iron does not itself lead to a large spontaneous magnetization. Indeed, a ferromagnet with a large spontaneous magnetization can be demagnetized, that is, made to have no spontaneous magnetization. The reason for these was proposed by Weiss in terms of the ferromagnet being divided into magnetic domains. Each domain is itself uniformly magnetized and possesses the same magnetization, but the domains orient themselves along different directions. Thus, the spontaneous magnetization of a ferromagnet may be much smaller than the saturation magnetization. The saturation magnetization of a ferromagnet, which is equal to the magnetization of all the

Fundamentals of Electrochemical Deposition, Second Edition.
By Milan Paunovic and Mordechay Schlesinger
Copyright © 2006 John Wiley & Sons, Inc.

domains combined, is reached when the multidomain state is transformed into a single domain state.

The boundary between neighboring domains is called a *domain wall* or *magnetic transition*. Such walls have finite width within which spins rotate gradually from one direction to the other. As the spins have to overcome the exchange interaction (which would have them all oriented in one direction, or to put it differently, in the direction of the other domain) and magnetic anisotropy, a domain wall possesses certain potential energy. The basic origin of magnetic domains is nature's tendency to assume minimum energy configurations. A single domain state is generally not favorable, as a result of its large demagnetizing energy. The balance between the demagnetizing energy, domain wall energy, magnetic anisotropy energy, and so on, will determine which domain configuration is stable. This is illustrated in Figure 20.1.

Specifically, when a multidomain ferromagnet is subjected to an external magnetic field, the domains with lower energy will expand while those with higher energy will shrink. This will result in domain wall displacement. In addition, magnetization in domains can rotate in the direction of the applied field without displacing domain walls. This is what Figure 20.1 depicts. (Note that in the figure, M_s is the saturation/maximum magnetization and H_d is the demagnetization field.) It is important to note that in actual fact, the magnetization processes are more complicated than the description presented here makes clear. Thus, a ferromagnetic material is usually not perfect; it contains grain boundaries, defects, impurity inclusions, and so on, referred to as *domain wall pinning sites*. Therefore, the domain wall energy is a function of location in a rather complicated fashion.

In 1919, Barkhausen performed an experiment by connecting a solenoid with a magnetic core to an amplifier and loudspeaker. When the magnet was subjected to full hysteresis (1, pp. 19–27), he heard a cracking noise in the speaker. When such a speaker is replaced by an oscilloscope, irregular voltage spikes may be observed. Those

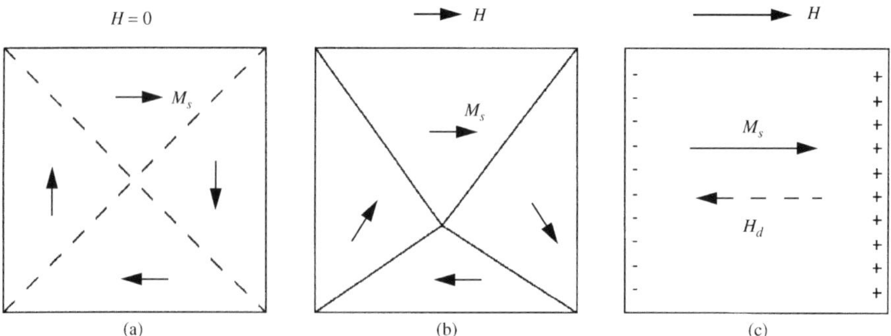

Figure 20.1. The magnetization process: (*a*) demagnetized state; (*b*) unsaturated state; (*c*) saturated state. When an external magnetic field is applied, domain rotation and domain wall motion occur simultaneously or sequentially. Domain configuration can be found by minimizing the total energy related to magnetization. (From Ref. 1, with permission from Elsevier.)

noises and spikes, later termed *Barkhausen noise*, are recognized to be due to discontinuous irreversible domain motion, often called *Barkhausen jumps*.

Coherent rotation of magnetic dipoles is yet another common mode of magnetization process. This simply means that all the spins rotate together.

20.2. MAGNETIC INFORMATION STORAGE

Ours is the Information Age. Consequently, the demand for high-performance low-cost nonvolatile information storage systems increases continuously. There are a great variety of information storage systems with varying degrees of development and commercialization: magnetic tape drives, hard disk drives, magnetic floppy disk drives, magnetooptic disk drives, phase-change optic disk drives, semiconductor flush memory, magnetic random access memory, and holographic optical storage, among others. Magnetic information storage technology (hard disk, floppy disk, and tape drives) are most widely used. Electrochemical deposition techniques are essential in the production of these systems. Figure 20.2 depicts the hierarchy of the information storage in a typical PC in terms of cost versus performance speed and efficiency. In other words, it is most efficient for a microprocessor to access and store in the semiconductor main memory (DRAM) alone because DRAM access time is less than 100 ns. However, DRAM is volatile; any information in the DRAM is lost once the computer is turned off. Thus, a lower level of information storage devices is essential. The latter may include magnetic hard disk, magnetic floppy drives, and others. As a rule, the higher-level storage media are superior in performance but inferior in cost (i.e., cost more). The lower level of the memory hierarchy is not volatile and is cheaper. In all of these, as noted above, electrodeposition techniques play an important role.

To illustrate the point, we present Figure 20.3, a schematic cross section of a thin-film hard disk. The Ni–P layer, whose purpose is to render a nearly perfectly smooth, rigid,

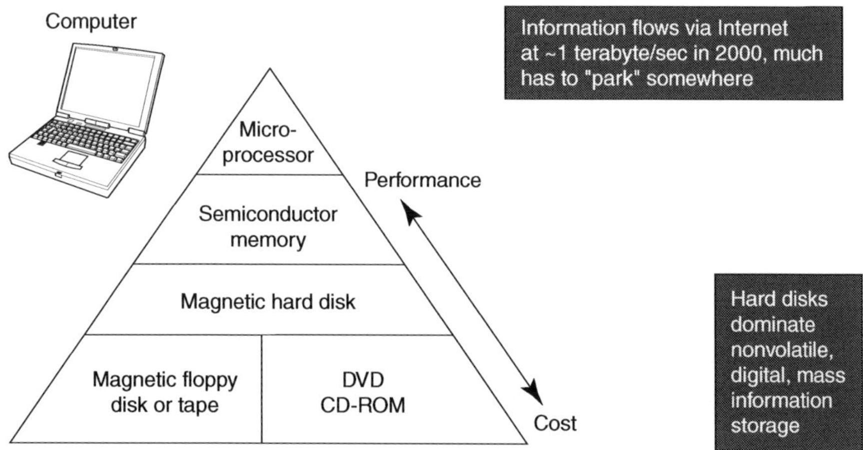

Figure 20.2. Information storage hierarchy. (From Ref. 1, by permission from Elsevier.)

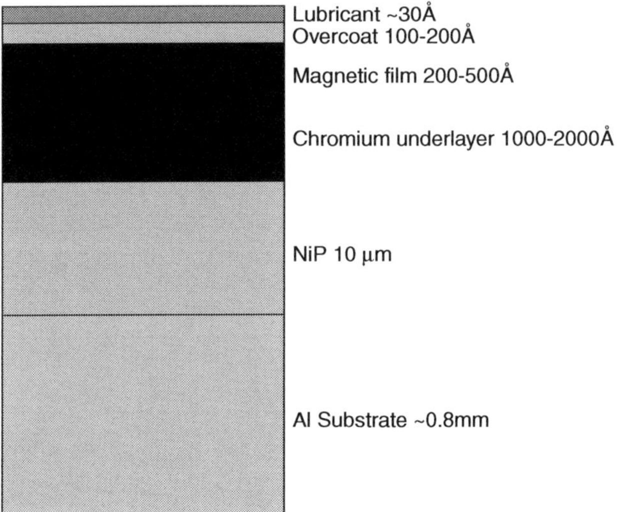

Figure 20.3. Schematic cross section of a thin-film hard disk. Mobile hard disk drives use glass substrates and smaller diameters (2.5 in. and 1 in.). (From Ref. 1, by permission from Elsevier.)

and properly textured undersurface, is almost always made of an electrolessly deposited thin-film. The chromium underlayer, which controls magnetic properties and microstructures of the magnetic recording layer, was produced in the past via electrodeposition. The magnetic layer itself, typically a cobalt-based alloy, was also produced via electroless deposition methods. Those methods required the addition of a number of mostly proprietary additives (see Chapter 10) to the deposition bath in order to have the disk possess the correct magnetic properties. Also, note that the smoothness provided by the Ni–P layer is essential if one is to maintain the read/write head at a fly height below 40 nm. More recently, though, the chromium underlayer, the magnetic recording layer, and the overcoat are most often produced via vacuum-deposition methods.

20.3. READ/WRITE HEADS

Actually, read/write heads in combination represent somewhat older technology; it should thus be remembered that in newer technologies, there is a clear trend toward using *separate* heads for reading and writing (2).

Metal in gap (MIG) or ferrite heads are produced with a combination of machining, bonding, and thin-film processes. Thin-film inductive heads are manufactured using thin-film processes similar those of semiconductor IC technology (discussed in Chapter 19). The thin-film head production process is rather unusual, as it involves both very thin and very thick films. We choose to present here a detailed summary of the fabrication process of thin-film inductive heads with a single-layer spiral coil. This may serve, once again to, illustrate the centrally important role of electrochemical deposition in connection with modern information technology.

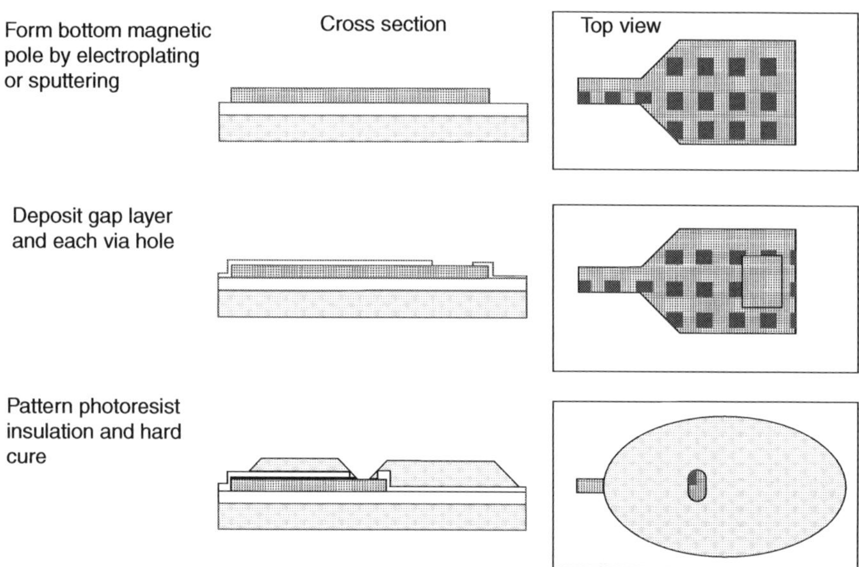

Figure 20.4. Thin-film head fabrication stage 1. (From Ref. 1, by permission from Elsevier.)

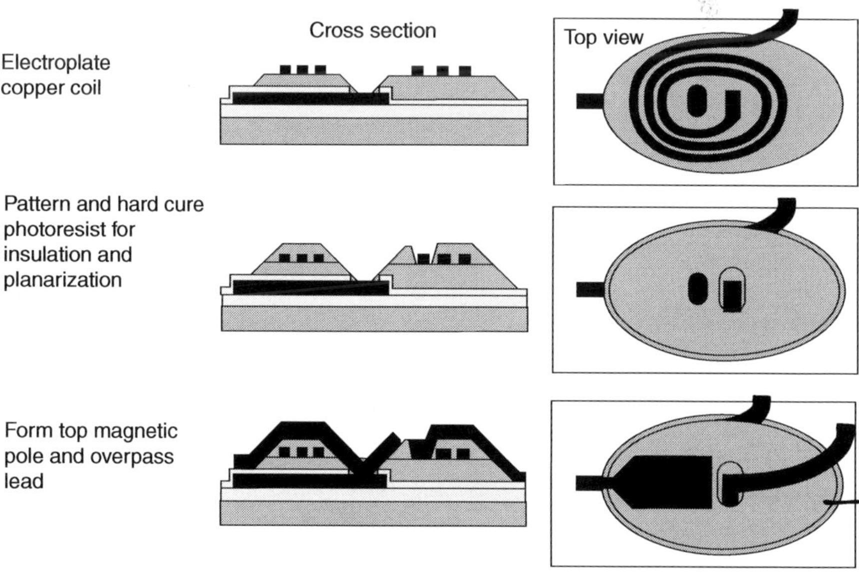

Figure 20.5. Thin-film head fabrication stage 2. (From Ref. 1, by permission from Elsevier.)

Substrates most commonly used for thin-film heads are ceramic alumina TiC wafers. The production steps are as follows (see Figs 20.4 to 20.6 and Ref. 3):

1. The substrate, made of aluminum titanium carbide (ALTIC), will become the aerodynamic slider body of the flying head. The first step in the fabrication is the

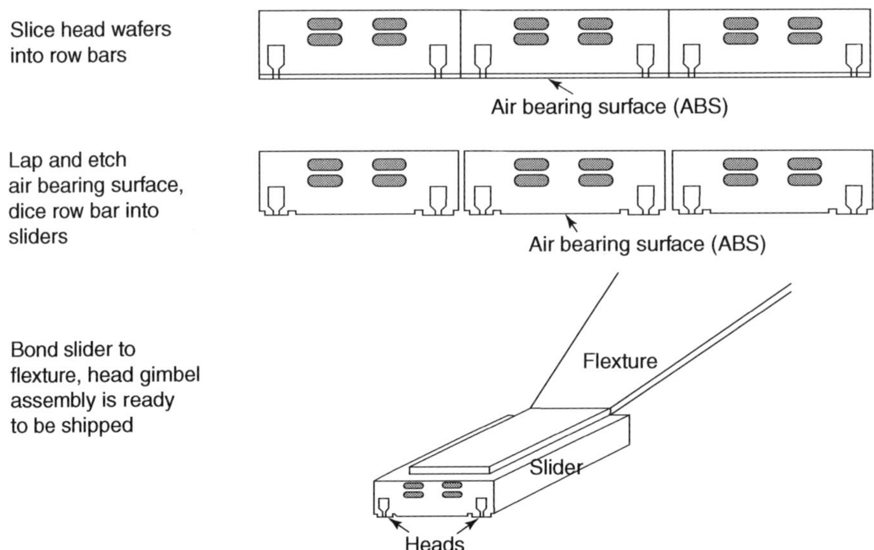

Figure 20.6. Thin-film head fabrication stage 3. (From Ref. 1, by permission from Elsevier.)

sputtering of thick alumina ($\sim 15\,\mu$m) on the polished substrate. This layer acts as an insulating layer between the substrate and the bottom magnetic layer. A thin under-coat NiFe layer (50 to 100 nm) for electroplating is deposited, often by sputtering.

2. Deposition and patterning of the bottom magnetic pole follow. The pole is usually electroplated with a through-photoresist window frame mask to a thickness level of 2 to 4 μm. Note that whereas the magnetic pole is made into a pancake shape to increase the head efficiency, it is the narrow pole tip's dimension that determines the narrow track width. As stated, the widely used Co-based alloy magnetic poles are electrodeposited (wet process). Nanocrystalline FeN-based alloys are sputter-deposited in a vacuum chamber (dry process).

3. Next comes the sputter deposition of the alumina gap layer, made about 0.1 to 0.5 μm thick. The gap layer thickness greatly affects the linear resolution and side read/write effects of the recording heads.

4. As insulation between the coil and the magnetic core, a hard-cured (to 200°C) photoresist insulator is patterned. It is a novolak polymer or polyimide about 5 μm thick, which is popular for its high insulator and photolithographic properties. This provides electrical insulation as well as a planar surface for subsequent deposition of copper coils.

5. Spiral copper coils 3 μm thick, 3 to 4 μm wide, and 2 μm apart are electro-deposited in the region above the magnetic pole. A seed layer must be put down first either through sputtering or by any other suitable means. Next, the seed layer must be selectively etched. The coils have to be made wider outside the magnetic poles to reduce overall coil resistance. It is rather noteworthy that around 1997 (shortly before the first edition of this book was published), the semiconductor IC industry began to

electrodeposit copper interconnects on-chip instead of sputtering aluminum interconnects although electrodeposited Cu had long been used in magnetic recording heads.

6. As in step 4, hard-cured photoresist insulator about 5 μm thick is patterned to provide electrical insulation for the copper coils and planarized surface for top magnetic pole deposition. Leveling of the coated surface is essential to preserve the soft magnetic properties of the top magnetic layer. Annealing above the softening temperature of the photoresist is effective for leveling. For a multilayered coil, steps 5 and 6 are repeated.

7. The top magnetic pole, 2 to 4 μm thick, and overpass lead, which provides electrical connection to the central tap of the coils are now deposited and patterned (as in step 2). The back regions of the magnetic poles are, as a rule, made thicker than the pole tips, to achieve high head efficiency and avoid magnetic saturation.

8. Copper studs, 20 to 40 μm thick are electrodeposited, a 10–15-μm-thick alumina overcoat is deposited, and the open copper studs are lapped. The thick alumina is required to protect the magnetic and coil structures during the machining and lapping processes and to prevent corrosion.

9. Soft gold bonding pads with a seed layer are electrodeposited as above. The gold pads are required to protect the underlying devices and to facilitate wire bonding. One must note that wafer-level testing of thin-film heads may be done at this point. This is useful for diagnosis of possible problems during the production process.

10. The wafer is scribed and sliced into row bars as shown in Figure 20.6. The row bars are bonded to tooling bars that hold them during mechanical processing.

11. Heads are randomly sampled for pole tip geometry measurements after lapping and good individual sliders are sorted out.

12. Good sliders are bonded to flexures as shown in Figure 20.6. Most thin-film inductive heads employ multilayer spiral coils to achieve large read-back signals. This means that the copper plating and the hard-cured photoresist steps must be repeated. Electrical connection between the neighboring coil layers must also be implemented. Naturally, these steps add greatly to the complexity of thin-film head production.

20.4. HIGH-FREQUENCY MAGNETICS

The past two to three decades have witnessed significant increases in switching frequencies. Specifically put, switching frequencies have risen from the 75-kHz range to the 1000-kHz (1-MHz) range and beyond, which has facilitated decreases in the physical size of electronics objects (e.g., ICs, board areas). Such trends have been more visible in the power module arena, where very low profile and minimal board area are required.

The physical density of modules is increasing for a number of reasons, including improved semiconductor devices, which allow increased frequency of operation and thus less energy storage. Thus, a dominating magnetic and electric energy storage element may be reduced in size, allowing further size reduction of the modules (such as storage disks) themselves. Naturally, with the elevated frequencies of operation come new challenges and developments required of the magnetic components (e.g., inductors, transformers). These challenges and developments are concerned primarily with the increase in losses as well as with the desire to minimize volume, while managing the mechanical structure to permit efficient heat removal and cost-effective manufacturing techniques.

Generally speaking, magnetic components may be called upon to perform many functions in a power conversion system (e.g., a radio receiving unit, where EM waves are converted to acoustic waves). These functions are broadly divided between power- and signal-handling magnetics. We focus here on power-handling magnetics since these are of unique concern to power electronic systems, primarily power inductors and power transformers.

To put matters in context with electrochemical deposition, our subject in this book, we observe that traditional magnetic devices have usually been made of a set of wirewound windings on an insulating bobbin/cylinder with insulating tape systems, varnishes, and so on, with a magnetic core mounted over the winding set. These construction methods have always led to limitations on size, manufacturability, performance, and repeatability. Recently, a strong movement took place toward planar, low-profile magnetics with windings fabricated into stand-alone or embedded printed wiring boards (PWBs). These offer many advantages, such as very low profiles not possible with wirewound magnetics, reduced leakage inductance, higher degrees of interleaving at no added cost, higher operating frequencies (into the megahertz regime), and further cost reductions by embedding directly into the power train circuit board. However, some issues arise in connection with this movement, including maintaining insulation systems, providing sufficient copper metal weights, minimizing line spaces for maximum packing, and tight control on tolerances on heavy copper metal weights.

Copper weights pose issues altogether new to the PWB industry. Magnetic designs typically hold tight tolerances ($\sim$10%) on dc resistance. Since heavy weights are achieved by electroplating extra copper over an existing laminate, the tolerances and uniformity that can be obtained are strongly dependent on the specific conditions of the electroplating process and master artwork. Bath chemistry, current density, and deposition rates affect the final weight, with a nonuniform thickness often formed across an entire PWB panel, being invariably heavier toward the panel's outer edges and lighter in the center (platers refer to this as the dogbone effect). Current thieves, chemistry, agitation, and cathode–anode geometry design can be used to improve this discrepancy in deposit thickness. Ultimately, depending on the PWB manufacturer, a certain tolerance toward copper thickness, and hence dc resistance, will have to exist from part to part on the same panel and must be taken into account in the design and specification stage of the magnetic device. This effect may lead to the need to specify according to PWB copper thickness (in mils or millimeters) rather than as typically done, by weight (e.g., in ounces or grams). This will be necessary to achieve tighter dc resistance tolerance, possibly at the risk of higher cost. The etching process to form the conductor patterns also produces minimum conductor-to-conductor spacing that is highly dependent on the copper thickness at hand.

20.5. SPINTRONICS

Electronic devices that operate using the spin of the electron and not just its electric charge are on the way to becoming a multibillion-dollar industry—and may lead to quantum microchips (4). As progress in the miniaturization of semiconductor electronic devices leads toward chip features smaller than 100 nm in size, device engineers and physicists are inevitably faced with the fast-approaching presence of quantum mechanics—that counterintuitive, and to some mysterious, realm of physics wherein wavelike properties control the behavior of electrons.

Practitioners in the semiconductor device world are busy coming up with ingenious ways to avoid the quantum world by redesigning the semiconductor chip within the confines of "classical" electronics. Yet it is a fact that we are being offered an unprecedented opportunity to define a radically new type of device that would exploit the quantum world to provide unique advantages over existing information technologies. One such quantum property of the electron is known as *spin*, which is its magnetism. Devices that rely on an electron's spin to perform their functions form the foundation of *spintronics* (short for *spin-based electronics*), also called *magneto-electronics*. Information-processing technology has thus far relied on purely charge-based devices, ranging from the now-outdated vacuum tube to today's million-transistor microchips. Conventional electronic devices move electric charges around, ignoring the spin of each electron. Magnetism (and hence, electron spin) has nonetheless always been important for information storage. For instance, even the earliest computer hard disk drives used magnetoresistance—which is, as discussed in Chapter 17, a change in electrical resistance caused by a magnetic field—to read data stored in magnetic domains. It is a fact that the information storage industry has provided the initial successes in spintronics technology. More sophisticated storage technologies based on spintronics are already at an advanced stage.

Magnetic random-access memory (MRAM) is a new type of computer memory. MRAMs retain their state of magnetization even with the power off, but unlike present forms of nonvolatile memory, they have switching and rewritability rates that challenge (are faster than) those of conventional RAM. In today's read heads as well as those of MRAMs, key features are made of ferromagnetic metallic alloys. Such metal-based devices make up the first—and most mature—of the various categories of spintronics.

REFERENCES AND FURTHER READING

1. S. X. Wang and A. M. Taratorin, *Magnetic Information Storage Technology*, Academic Press, San Diego, CA, 1999.
2. C. D. Mee and E. D. Daniel, *Magnetic Recording Technology*, 2nd ed., McGraw-Hill, New York, 1996, p. 6.2.
3. J. Mallinson, *Magnetoresistive Heads: Fundamentals and Applications*, Academic Press, San Diego, CA, 1997.
4. S. A. Wolf, D. D. Awschalom, R. A. Buhrman, J. M. Daughton, S. von Molnár, M. L. Roukes, A. Y. Chtchelkanova, and D. M. Treger, *Science* **294** (5546), 1488–1495 (Nov. 16, 2001).

5. B. Beschoten et al., *Phys. Rev.* **B63**, R121202 (2001).

6. S. A. Crooker et al., *Phys. Rev. Lett.* **75**, 505 (1995).

7. R. Fiederling et al., *Nature* **402**, 787, (1999).

8. B. T. Jonker et al., *Phys. Rev.* **B62**, 8180 (2000).

9. Y. Ohno et al., *Nature* **402**, 790 (1999).

20.6. APPENDIX: ELECTROMAGNETIC QUANTITIES AND UNITS

Readers may be aware that the equations and numerical values of a magnetic field change with the system used. It thus becomes necessary to review the basics of unit systems here. The units in magnetics have always been somewhat confusing. Both SI and Gaussian unit systems are widely used in the literature. This state of affairs may be understood in terms of the following:

1. If one keeps using the same set of units and a certain set of equations, one is free to choose the system of one's preference.

2. One should use the units that most closely fit the phenomenon being treated. Thus, in astronomy one uses light years ($= 9.46 \times 10^{15}$ m), whereas in atomic physics the unit of choice would be the angstrom ($= 10^{-10}$ m).

In the present book, for magnetism we use the SI unit that is based on the MKSA (meter, kilogram, second, ampere) system. In accordance with that, the tesla ($1\,\mathrm{T} = 10^4$ gauss) was presented as the magnetic unit in Chapter 17 (see Fig. 17.10a and b). It is useful to know both the SI and Gaussian systems and be able to convert between them. Thus, when one attempts to solve a magnetics problem, to avoid errors one is well advised to stick to a single convenient unit system. A useful conversion table of

TABLE A20.1. Comparison of the SI and Gaussian Unit Systems

SI (MKSA)	Gaussian (CGS)
$\mathbf{H} = \dfrac{\mathbf{B}}{\mu_0} - \mathbf{M}$	$\mathbf{H} = \mathbf{B} - 4\pi\mathbf{M}$
$m = IA$	$m = \dfrac{IA}{c}$
$\nabla \cdot \mathbf{H} = -\nabla \cdot \mathbf{M} \equiv \rho_m$	$\nabla \cdot \mathbf{H} = -4\pi\nabla \cdot \mathbf{M} \equiv 4\pi\rho_m$
$\nabla \times \mathbf{H} = \mathbf{J}$	$\nabla \times \mathbf{H} = \dfrac{4\pi}{c}\mathbf{J}$
$\mathbf{B}$: tesla (T)	gauss (G), $1\,\mathrm{G} = 10^{-4}\,\mathrm{T}$
$\mathbf{H}$: ampere/meter (A/m)	oersted (Oe), $1\,\mathrm{Oe} = 1\,\mathrm{G} = 80\,\mathrm{A/m}$
$\mathbf{M}$: ampere/meter (A/m)	emu/cm^3, $1\,\mathrm{emu/cm}^3 = 1\,\mathrm{kA/m}$
$\mathbf{I}$: ampere (A)	statamp, $1\,\mathrm{A} = 3 \times 10^9$ statamp

Source: Adapted from Ref. 1, by permission from Elsevier.

selected formulas and physical quantities is presented in Table A20.1. The constant c is the speed of light, $c = 3 \times 10^8$ m/s.

> **B**: magnetic induction or magnetic flux intensity
>
> **Φ**: magnetic flux, $\boldsymbol{\Phi} = BA$
>
> **H**: magnetic field
>
> **m**: magnetic moment
>
> ϕ: magnetic potential, $\nabla^2 \phi = -\rho_m$
>
> **M**: magnetization
>
> ρ_m: magnetic charge density

21

Frontiers in Applications: The Field of Medicine

21.1. INTRODUCTION

Most readers may not appreciate the impact of electrochemistry and/or electrochemical deposition techniques in medicine. In this chapter we discuss these topics as they relate to medical devices. Emphasis is placed on the often overlooked materials science and surface chemistry aspects of medical devices rather than on the topics, described extensively in the literature, of electrochemical sensors in medical applications. This chapter is intended to provide the reader with a view of the role in medical devices of electrochemistry in general and electrochemical deposition in particular. It is also intended that the reader gain an appreciation of the future potential role of electrochemistry in devices, particularly in the creation of biomimetic (i.e., biology mimicking) medical devices.

Medical devices, in general, are human-made structures or machines that function inside or outside the body and have a role in human functioning either in sensing or manipulating a physiological variable. If they are implanted in a host, they may be a temporary or a permanent device.

Among well-known medical devices are the mechanical heart valve and the pacemaker. Although medical devices are generally perceived as being either macroscopic or permanent, many, and in some cases all, of the effects (wanted and unwanted) of the devices are derived from interactions at the surface. Indeed, when the devices are made of metals, many, if not most, of the surface effects are electrochemical in nature. This fact is the main subject of this chapter whose role is to outline briefly the manner in which electrochemistry and thus electrochemical deposition are crucial to both the mechanical and materials stability of medical devices. In some cases, entirely new industries have been created around advances in our understanding of the electrochemical processes occurring at surfaces. An additional aim of this chapter is to touch on the ways in which electrochemistry can be used to modify surfaces actively to create a more amicable

Fundamentals of Electrochemical Deposition, Second Edition.
By Milan Paunovic and Mordechay Schlesinger
Copyright © 2006 John Wiley & Sons, Inc.

interaction between the medical device and its host. In at least one example below, electrochemical deposition itself (called *electroforming*) is used to actually create a medical device.

21.2. ELECTROCHEMICAL DEPOSITION OF MEDICAL DEVICES

Not much has been done in discussing ways to improve medical device surfaces using electrochemical coating methods: in particular, with the aim of accomplishing the medically relevant goal, aside from discussion in the relevant literature, of the processing of nitinol (an alloy of nickel and titanium; see Section 21.4).

One of the most commonly used medical devices is the stent, (Fig. 21.1), small metallic structures that are expanded in blood vessels, functioning to maintain the patency (freedom from obstruction) of the vessel in which it is placed. Although the first use of stents was in vasculature (blood vessel systems), more recent applications include, for example, implantation between two vertebrae to increase the rigidity of the spine. A typical vascular stent is placed in its anatomical location and then either plastically deformed/expanded (stainless steel) or allowed to expand to a predetermined size, as a consequence of shape memory (nitinol).

Recent research and development work in the field of medical devices has focused on creating *biomimetic surfaces*, surfaces created with an understanding of the pathophysiology surrounding the surfaces, typically created with the intention to somehow manipulate the local environment to make it more accommodating to the implant. As an illustration of a biomimetic surface, one may consider the process of restenosis and its prevention. *Restenosis* is the process wherein scar-type tissue in-growth compromises the arterial passage space, or lumen, after angioplasty and stenting. This process typically occurs in 30 to 50% of patients who receive stents in the coronary (heart) blood vessels. These patients often require subsequent surgery and/or further stenting. Several explanations for restenosis have been put forth, including a foreign body type of reaction to the stent materials, an inflammatory reaction due to the implantation trauma, or a failure of the normal blood vessel lining (endothelial cells) of the blood vessels to renormalize after being perturbed by the stent implantation.

Gold and, more recently, iridium have been considered ideal biomaterials for stents. They are highly radiopaque, that is, show clearly in x-ray-type analysis and are considered inert in that corrosion should not be expected of the surface, due to their "noble" characteristics. Indeed, a gold electrodeposited stainless steel stent made it through clinical trials and was actually approved for clinical use (1). It was, nonetheless, a failure because its rate of restenosis was higher than that of stents already on the market. When the production process for the stent was reviewed, it turned out that the plated surface was not properly heat treated. Early animal data (2) show that changing this one variable reduces the restenosis rate to that of historical data for restenosis. This is yet another example where failure to appreciate surface electrochemistry caused clinical failure of a device. More recently, work has been done to couple organic surfaces to electroplated surfaces. It has been well known for many years that organic molecules

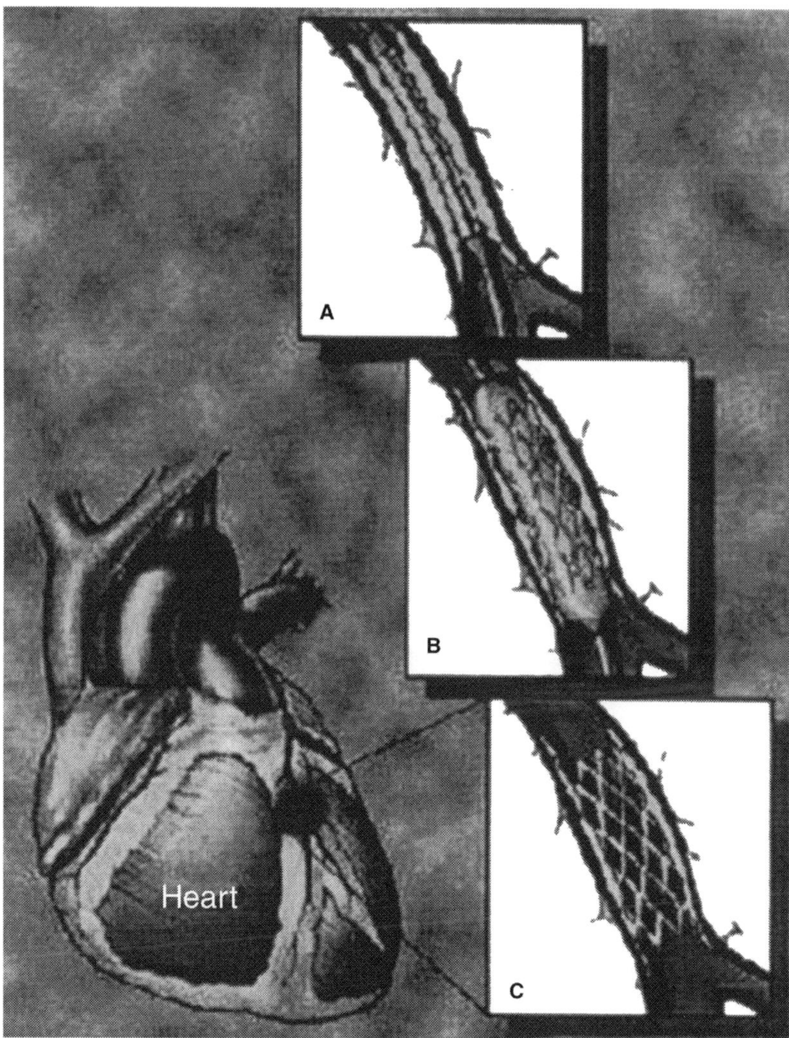

Figure 21.1. Stents are used to open arteries of the heart blocked by atherosclerotic plaques: (A) a balloon and stent are placed across the plaque; (B) the balloon is expanded, leaving the stent to prop open the artery; (C) restenosis is the process wherein scar tissue builds up around the stent, again causing a flow restriction. A balloon is required for stainless steel, whereas a nitinol stent will expand on its own, due to the shape memory property of nitinol. (From Ref. 11, with permission.)

containing a thiol group at one end will strongly adhere to gold surfaces and form molecular monolayers.

Recent work has expanded this knowledge by first electrodepositing a gold layer on the device surface and then applying an organic layer to the electrodeposited surface (3). In this case, the device was a pacemaker lead. The organic layer was utilized to prevent scarring around the lead and to decrease the resistance that traditionally develops after electrode implantation. As is well known in the field of electrochemistry, deposition can be accomplished by electroless and/or electrodeposition. With regard to medical devices, which often are not planar and often have submillimeter features, electroless deposition offers many advantages. *Electroless deposition* (see Chapter 8) is a wet electrochemical method in which deposition occurs evenly along the surface of the device, including submillimeter features and nonplanar surfaces. Electroless deposition can be used to coat nanoparticles evenly and to fill submicrometer holes. Electroplating, on the other hand, is very sensitive to the electric field lines between the anode and the cathode. As a result, the coating derived from an electroplating process can be highly irregular (4). As such, electroless deposition may meet the biocompatability goals. Recent work (5) utilized the inherent porosity of electrolessly deposited nickel to incorporate drugs and provide for their storage and release, thereby opening new doors in device design and surface chemistry.

Electroforming is a process in which electrodeposition is performed on a mandrel in a given pattern. When the desired thickness is achieved, the mandrel is etched away from the electroformed stent, leaving a free-standing structure, in this case a fully functional stent. In such a process, an entire stent, including the structure and surface, can now be formed by a low-energy, low-capital electrochemical process. Indeed, Hines (6) has developed a process to electroform stents from a gold electroplating solution. Such a process raises the intriguing possibility of having only a monolithic process to both form a stent and modify its surface for optimal biocompatability.

21.3. SURFACE ELECTROCHEMISTRY IN THE PROCESSING OF BIOMATERIALS

More recent work has utilized advanced electrochemical methods to create coatings that can provide drug to the local environment as well as provide for more biocompatible and corrosion-resistant surfaces (5). In this study, the final processing of the device surface included an electrochemical method to create voids on the surface. Pharmaceuticals were loaded into the voids and released over a long period of time. The coating also enhanced corrosion resistance. There is no doubt that an understanding of surface electrochemistry, specifically at the metal–blood interface, will allow for further biomimetic enhancement.

The success of nitinol as a medical biomaterial can be attributed to an understanding of the electrochemistry at its surface. Although previously unexploited, electrochemical (coating) methods can accomplish many of the desirable biomimetic effects described above. Electrochemistry has the potential to alter surface morphology, release drugs, enhance radiopacity, and prevent corrosion, all with the same electrochemical coating. Electrochemical methods are also highly economical, with low capital costs. In this context, electrochemistry should be considered a nanotechnology in terms of its efficiency, size scale, and self-assembly properties. Electrochemical coatings can be applied at scales of tens to hundreds of nanometers, allowing for the manufacture of nanocomposite coatings. As an example of a composite, the first 100-mm span increases radiopacity while the next 100-mm span releases the drug, and the third hides the first two from the physiological environment. Because electrochemical (electroless) deposition is an evenly applied process that can be turned on and off quickly, such composite coatings can be created on the order of 1 μm in thickness, a property not amenable to other methods for providing the same effects. Furthermore, and importantly, such a composite will not greatly affect the bulk material properties.

Extensive research regarding electrochemical processes at the surface of metallic implants has been performed in the past. This has focused almost exclusively on the corrosive processes occurring at the surface of the device and its prevention. "Bare" metals have been scrutinized intensely with regard to corrosion. Surprisingly, specific devices such as stents have not been studied systematically and certainly have not been studied under conditions of mechanical stress and strain. The first stents were implanted in the early 1990s and were made from 316L grade stainless steel. As would be expected, these stents show excellent resistance to corrosion under physiological conditions but not under conditions of static stress. However, corrosion processes are highly sensitive to shape and surface defects. Interestingly, to date, there have not been any studies on the effects of the polymer coating on corrosion at the stent surfaces. Polymer coatings can potentially be both a detriment to corrosion protection or may enhance corrosion protection. High-magnification scanning electron micrographs (SEMs) of commercially available nitinol stents are telling. Notable are micrometer-size crevices on the surface. Note that this is evident prior to expansion and prior to stresses placed on it.

Some time ago, the Johnson & Johnson Corporation began marketing a stent with a polymer-based coating containing a drug that inhibits the restenosis process. The drug is released from the polymer at a slow pace over time, whereas the polymer remains on the stent permanently. Early results have been rather dramatic, with a reduction in the restenosis rate from 40% to 10%. (7). This is one of the most dramatic examples of a biomimetic coating in clinical use. Whether polymer-coated metal is the best coating for the purpose, however, remains to be seen. There has been extensive research into the biomimetic effects of surface morphology on the attachment, spread, and even second messenger systems of cells. For example, in the field of orthopedic surgery, implants used to replace joints or fix fractures can have a sintered surface or a surface roughened by fixing 1-mm metallic beads to it. With the roughened porous surface, the activity of osteoblasts, or bone-synthesizing cells, is dramatically up-regulated, so much so, in fact, that bone cement is not required to maintain fixation of some implants. Recent work has begun to evaluate the potential for nitinol as an orthopedic material (8). The advantages cited for a nitinol-based implant include its damping properties as well as its shape memory. The surface processing used in this study was simple mechanical abrasion, the one that the supply company provided. Adequate bone in-growth was obtained. However, if prior surface electrochemistry research were to be considered, the bony in-growth could be greatly enhanced.

Although the titanium oxide layer at the surface of the nitinol is highly biocompatible and protects the underlying substrate from electrochemical corrosion, the titanium oxide layer itself is mechanically very brittle. Under mechanical stress, such as the shear of blood flow in the aorta or under the bending moments of aortic pulsations, the titanium oxide surface layer can fracture, exposing the underlying metal to corrosion. Not only is corrosion undesirable in terms of biocompatibility (i.e., leaching of nickel and its oxides), but corrosion can lead to deterioration of the mechanical properties at the specific location where the corrosion occurs. To summarize, surface defects are inherent in the processing of nitinol; the surface defects cause stress discontinuities in the surface of the titanium oxide film; the stress discontinuities lead to film breakdown and localized pitting corrosion; and the pitting corrosion leads in turn to bulk mechanical breakdown at the site as well.

In an optical micrograph of a commercially available nitinol stent's surface examined prior to implantation, surface craters can readily be discerned. These large surface defects are on the order of 1 to 10 μm and are probably formed secondary to surface heating during laser cutting. As mentioned above, these defects link the macro and micro scales because crevices promote electrochemical corrosion as well as mechanical instability, each of which is linked to the other. Once implanted, as the nitinol is stressed and bent, the region around the pits experiences tremendous, disproportionate strain. It is here that the titanium oxide layer can fracture and expose the underlying surface to corrosion (9).

In an excellent study and review, Sun et al. (10) revealed the heterogeneity of nitinol under various temperature conditions, even in a simple lactated Ringer's solution. Lactated Ringer's solution is a mixture of salts and water meant to simulate the tonicity of blood. In this set of experiments, the surfaces were obtained from commercial sources, and each sample had undergone similar surface processing prior to experimentation. When a nitinol sample was simply placed in the Ringer's solution at constant potential and a given temperature, current transients were seen, which represent breakdown and repassivation of the oxide film.

Such a study has yet to be done under dynamic (or conditions of bending) of the nitinol. Such dynamic stress would undoubtedly increase the number of passivation and repassivation events. Each event has the potential to release nickel into the surroundings and further the pit formation. Nitinol, it should be noted, achieves its shape memory properties after a series of heat treatment steps, which leaves a thick oxide residue on the surface. The oxide layer is removed by etching, mechanical abrasion, or electropolishing. After the heat-induced oxide layer is removed, a fine layer of titanium oxide (a ceramic) remains on the surface. The titanium oxide layer is extremely important for the biocompatibility of nitinol, as the passivation layer prevents corrosion of the nitinol alloy. Electrochemical processes are involved in several ways at the surface of the nitinol. Electropolishing, a preparation step, is an electrochemical process wherein the substrate is made the anode and the surface is evened out on a microscopic scale. Once the device is implanted, electrochemical corrosion occurs along the device surface. When covered by a titanium oxide layer, pits such as the ones mentioned above, do not undergo corrosion. However, the titanium oxide layer covering the pits is highly susceptible to mechanical stress and cracking when bending moments are applied.

21.4. MATERIALS SCIENCE OF BIOMATERIALS

As stated, medical devices comprise a rapidly growing industry where the basic technology changes very quickly; many of these changes come from the ever-expanding basic knowledge of materials science. The majority of metals and metal alloys used in the device industry today include titanium, nickel–titanium (nitinol), cobalt–chrome, and stainless steel. Significant advances have been made over the years in polymer science (e.g., conducting polymers) and in metallurgy, which enable more than incremental improvements in the devices and sometimes allow for entirely new classes of devices. This is, in fact, the case for nitinol. However, as will become apparent below, the understanding and consequent exploitation of the chemistry, stability, and durability of nitinol leave a great deal of room for future improvements before they can be manipulated easily.

As discussed above, nitinol is the name given to the alloy consisting of approximately equiatomic concentration of nickel and titanium. Thus, this alloy is approximately 50% nickel and 50% titanium. It was created in the 1960s and it was quickly discovered to possess a property called *shape memory*. This means that it will return to a preferred (preset) shape upon heating above a critical temperature (T_c). The relative percentages of nickel and titanium determine the actual T_c value. This property has been exploited in medical devices where T_c is set at the human body temperature, 37°C. When a device made from nitinol is inserted into the body, it reexpands to its preformed shape. The advantages of nitinol's shape memory are severalfold. For example, a preshaped device can be compressed into an "introducer" sheath, which is many times smaller than the space the device will ultimately fill (e.g., a stent in a blood vessel). After implantation, the device expands to its original shape within minutes. An additional advantage is that the device tends to exert a continuous positive force on the blood vessel, so that if the vessel is compressed externally, the stent will return to its original position. This is extremely important when a stent is placed in a peripheral vessel (e.g., a carotid artery).

The lag between the time that nitinol, was first produced and the time it was used commercially in medical devices was due in part to the fear that nickel would leach from the metal and not be tolerable as a human implant. As it turns out, with a correct understanding of the surface electrochemistry and subsequent processing, a passivating surface layer can be induced by an anodizing process to form on the nitinol surface. It is comprised of titanium oxide approximately 20 nm thick. This layer actually acts as a barrier to prevent the electrochemical corrosion of the nitinol itself. Without an appreciation for the electrochemistry at its surface, nitinol would not be an FDA-approved biocompatible metal and an entire generation of medical devices would not have evolved. This is really a tribute to the understanding of surface electrochemistry within the context of implanted medical devices.

21.5. VARIOUS APPLICATIONS IN THE FIELD OF MEDICINE

This chapter is devoted to the subject matter of electrochemistry in the service of, and how it relates to, medicine. It is indeed timely for this kind of topic to be discussed in a book such as this one. Medicine is the second-oldest profession, and as such, alone deserves our closer attention. Further, in the ongoing process of globalization, we are witnessing not only the tendency of commercial unification of the globe but the rapidly emerging interdependency of different scientific and technical disciplines as well.

Indeed the much bandied-about "n" word—nanotechnology—may well evolve as the underlying thread in this technological melding process.

With the advent of relativity and quantum mechanics in the early part of the twentieth century and the development of molecular biology in the second half of the century, it is now accepted that mathematics, physics, chemistry, and biology constitute but different parts of the same broader scientific discipline. Such interdependency, and subsequent confluence, of various disciplines may yet become one of the hallmarks of the twenty-first century. Despite these facts, it is often the case that the impact a given branch of science has upon another requires a rather special vehicle to allow it to become common knowledge. Given that, it is hardly surprising that the impact of electrochemistry in medicine is not yet properly recognized by all. This chapter is designed to focus on electrochemistry as it relates to medical devices. What is, by necessity, not treated thoroughly here are the materials science aspects of medical devices and related power sources (such as fuel cells), which may make them "tick." Another example of what is not included is the use of scanning electrochemical microscopy for imaging pathways of molecular transport across skin tissues, in addition to a good number of other topics. What *has* often been discussed in the literature of late is the topic of electrochemical sensors in medical devices.

It is hoped that this chapter will give readers a broader view and appreciation of the tremendous role that electrochemistry plays in medicine and medical devices, as well as a glimpse into the future possibilities as both of these now-related disciplines develop in time. None of these developments could have become reality if not for the collaboration of device engineers, medical professionals, electrochemists, and battery scientists. Such collaborative efforts require the ability to communicate ideas to professionals outside one's own discipline as well as to understand and appreciate input from disciplines other than one's own. It is reassuring to see the emergence in a number of universities of such interdisciplinary and crosscutting programs. This type of trend must clearly be encouraged for the process of scientific globalization to flourish.

21.6. CONCLUSION

In conclusion, electrochemistry and electrochemical coating play a crucial role in biomaterial–host interactions (11) and thus constitute integral components in medical device design. Electrochemistry should thus be viewed as a tool that can work for or against the medical device engineer and biochemist. Electrochemical methods, including electroplating and electroless plating methods, can be used to create micrometer-scale surface morphologies to induce specific types of cellular differentiation or to attract and/or repel a specific type of cell that can be used to increase the radio-opacity of the device, to produce nanocomposite coatings on a device, to enhance or control drug delivery from a device, or to accommodate a nonmetallic coating. They can even be used to create a complete medical device. These merits have to be balanced by the fact that electrochemical corrosive processes tend to destabilize surfaces and can undermine the mechanical stability of a device. There is no doubt that the field of electrochemistry and its continual progress will have a substantial impact on the future of medical devices. Devices continue to be scaled down in size, which will necessitate a greater understanding of corrosion processes. As analytical tools for the study of surface chemistry improve and become more widespread, and as nanoarchitectured control permeates the medical world, electrochemistry will be viewed as an economical, simple, yet powerful technique to modify and create biomimetic surfaces and medical devices.

REFERENCES AND FURTHER READING

1. J. Dahl, P. K. Haagar, E. Grube, M. Gross, C. Beythien, E. P. Kromer, N. Cattelaens, C. W. Hamm, R. Hoffmann, T. Reineke, and H. G. Klues, *Am. J. Cardiol.* **89**, 801 (2002).

2. E. R. Edelman, P. Seifert, A. Groothuis, A. Morss, D. Bornstein, and C. Rogers, *Circulation* **103**, 429 (2001).

3. M. H. Schoenfisch, M. Ovadia, and J. E. Pemberton, *J. Biomed. Mater. Res.* **51**, 209 (1999).

4. C. Heintz, G. Riepe, L. Birken, E. Kaiser, N. Chakfé, M. Morlock, G. Delling, and H. Imig, *J. Endovasc. Ther.* **8**, 248 (2001).

5. M. Gertner and M. Schlesinger, *Electrochem. Solid-State Lett.* **6**, J4 (2003).

6. R. Hines, in *Process for Making Electroformed Stents,* Electroformed Stents, Inc., Stillwell, KS, 2000.

7. M. C. Morice, P. W. Serruys, J. E. Sousa, J. Fajadet, E. B. Hayashi, M. Perin, A. Colombo, G. Schuler, P. Barragan, G. Guagliumi, F. Molnàr, and R. Falotico, *N. Engl. J. Med.* **346**, 1773 (2002).

8. A. Kapanen, J. Ryhanen, et al., *Biomaterials*, **22**, 2475 (2001).

9. M. Schlesinger and M. Paunovic, *Modern Electroplating*, 4th ed., Wiley, New York, 2000.

10. E. Sun, S. Fine, et al., *J. Mater, Sci. Mater. Med.* **13**, 959 (2002).

11. M. E. Gertner and M. Schlesinger, *Interface* **12**(3), 20 (2003).

INDEX

Fundamentals of Electrochemical Deposition, Second Edition.
By Milan Paunovic and Mordechay Schlesinger
Copyright © 2006 John Wiley & Sons, Inc.

THE ELECTROCHEMICAL SOCIETY SERIES

Corrosion Handbook
Edited by Herbert H. Uhlig

Modern Electroplating, Third Edition
Edited by Frederick A. Lowenheim

Modern Electroplating, Fourth Edition
Edited by Mordechay Schlesinger and Milan Paunovic

The Electron Microprobe
Edited by T. D. McKinley, K. F. J. Heinrich, and D. B. Wittry

Chemical Physics of Ionic Solutions
Edited by B. E. Conway and R. G. Barradas

High-Temperature Materials and Technology
Edited by Ivor E. Campbell and Edwin M. Sherwood

Alkaline Storage Batteries
S. Uno Falk and Alvin J. Salkind

The Primary Battery (in Two Volumes)
Volume I *Edited by* George W. Heise and N. Corey Cahoon
Volume II *Edited by* N. Corey Cahoon and George W. Heise

Zinc-Silver Oxide Batteries
Edited by Arthur Fleischer and J. J. Lander

Lead-Acid Batteries
Hans Bode
Translated by R. J. Brodd and Karl V. Kordesch

Thin Films-Interdiffusion and Reactions
Edited by J. M. Poate, M. N. Tu, and J. W. Mayer

Lithium Battery Technology
Edited by H. V. Venkatasetty

Quality and Reliability Methods for Primary Batteries
P. Bro and S. C. Levy

Techniques for Characterization of Electrodes and Electrochemical Processes
Edited by Ravi Varma and J. R. Selman

Electrochemical Oxygen Technology
Kim Kinoshita

Synthetic Diamond: Emerging CVD Science and Technology
Edited by Karl E. Spear and John P. Dismukes

Printed in Poland
by Amazon Fulfillment
Poland Sp. z o.o., Wrocław
28 October 2020

a8f93e98-689c-4e42-874c-3a2c450bf515R01